PHYSICS I

CAN

DUMMIES
A Wiley Brand

by STEVEN HOLZNER, PhD
with DANIEL FUNCH WOHNS, PhD

WILEY

U Can: Physics I For Dummies®

Published by:
John Wiley & Sons, Inc.
111 River Street
Hoboken, NJ 07030-5774
www.wiley.com

For general information on our other products and services, please contact our Customer Care Department within the U.S. at 877-762-2974, outside the U.S. at 317-572-3993, or fax 317-572-4002. For technical support, please visit www.wiley.com/techsupport.

Wiley publishes in a variety of print and electronic formats and by print-on-demand. Some material included with standard print versions of this book may not be included in e-books or in print-on-demand. If this book refers to media such as a CD or DVD that is not included in the version you purchased, you may download this material at http://booksupport.wiley.com. For more information about Wiley products, visit www.wiley.com.

Library of Congress Control Number: 2015937643

ISBN 978-1-119-09382-4 (pbk); ISBN 978-1-119-09372-5 (ebk); ISBN 978-1-119-09371-8 (ebk)

Manufactured in the United States of America

10 9 8 7 6 5 4 3 2 1

CONTENTS AT A GLANCE

TABLE OF CONTENTS

INTRODUCTION

Physics is what it's all about. What *what's* all about? Everything. Physics is present in every action around you. And because physics is everywhere, it gets into some tricky places, which means it can be hard to follow. Studying physics can be even worse when you're reading some dense textbook that's hard to follow.

For most people who come into contact with physics, textbooks that land with 1,200-page *whumps* on desks are their only exposure to this amazingly rich and rewarding field. And what follows are weary struggles as the readers try to scale the awesome bulwarks of the massive tomes. Has no brave soul ever wanted to write a book on physics from the *reader's* point of view? Yes, and here we come with such a book.

ABOUT THIS BOOK

This book is different. Instead of writing it from the physicist's or professor's point of view, we wrote it from the reader's point of view. After thousands of one-on-one tutoring sessions, we know where the usual book presentation of this stuff starts to confuse people, and we've taken great care to jettison the top-down kinds of explanations. You don't survive one-on-one tutoring sessions for long unless you get to know what really makes sense to people — what they want to see from *their* points of view. In other words, we designed this book to be crammed full of the good stuff — and *only* the good stuff. You also discover unique ways of looking at problems that professors and teachers use to make figuring out the problems simple.

This book is crammed with physics examples and physics problems. It's designed to show you how to tackle the kinds of problems you may encounter in physics classes.

In this book, you can find solutions to problems similar to the ones you're asked to solve elsewhere. And when you see how it's done, solving similar problems should be a breeze.

From time to time, we include sidebars to provide a little more insight into what's going on with a particular topic. They give you a little more of the story, such as how some famous physicist did what he did or an unexpected real-life application of the point under discussion. You can skip these sidebars, if you like, without missing any essential physics.

Some books have a dozen conventions that you need to know before you can start. Not this one. All you need to know is that variables and new terms appear in italics, like *this,* and that vectors — items that have both a magnitude and a direction — appear in **bold.** Web addresses appear in `monofont`.

FOOLISH ASSUMPTIONS

In writing this book, we made some assumptions about you:

- You have no or very little prior knowledge of physics.

- You have some math prowess. In particular, you know algebra and a little trig. You don't need to be an algebra pro, but you should know how to move items from one side of an equation to another and how to solve for values.

- You want physics concepts explained clearly and concisely, and you want examples that let you see those concepts in action.

- You want to work through some physics problems on your own. You'll find plenty of practice problems in this book along with step-by-step solutions in case you get stuck.

BEYOND THE BOOK

In addition to the material in the print or e-book you're reading right now, this product also comes with some access-anywhere goodies on the web. Check out these features:

- **Cheat Sheet (www.dummies.com/cheatsheet/ucanphysics1):** When you need a quick refresher, refer to the Cheat Sheet for some handy equations and the values of important constants.

- **Dummies.com articles (www.dummies.com/extras/ucanphysics1):** Each part in this book is supplemented by a relevant online article that provides additional tips and techniques related to the subject of that part. Read helpful articles that reveal how to draw a free-body diagram, how to use energy diagrams, the Carnot cycle, and ten physics heroes.

- **Online practice and study aids:** The online practice that comes free with this book offers 500 questions and answers that allow you to gain more practice with physics concepts. The beauty of the online questions is that you can customize your online practice to focus on the topics that give you the most trouble. So if you need help with projectile motion or angular momentum, just select those question types online and start practicing. Or if you're short on time but want to get a mixed bag of a limited number of questions, you can specify the number of questions you want to practice. Whether you practice a few hundred questions in one sitting or a couple dozen, and whether you focus on a few types of questions or practice every type, the online program keeps track of the questions you get right and wrong so you can monitor your progress and spend time studying exactly what you need.

To gain access to the online practice, all you have to do is register. Just follow these simple steps:

1. **Find your access code.**

    - **Print-book users:** If you purchased a hard copy of this book, turn to the inside of the front cover to find your access code.

    - **E-book users:** If you purchased this book as an e-book, you can get your access code by registering your e-book at www.dummies.com/go/getaccess. Simply select your book from the drop-down menu, fill in your personal information, and then answer the security question to verify your purchase. You'll then receive an email with your access code.

2. **Go to http://studyandprep.wiley.com.**

3. **Click on the product you want to access, and then click Login.**

4. **On the Register tab, enter your access code, and click Go.**

5. **Follow the instructions to create an account and set up your personal login.**

Now you're ready to go! You can come back to the online program as often as you want — simply log on with the username and password you created during your initial login. No need to enter the access code a second time.

Tip: If you have trouble with your access code or can't find it, contact Wiley Product Technical Support at 877-762-2974 or go to http://wiley.custhelp.com.

WHERE TO GO FROM HERE

You can leaf through this book; you don't have to read it from beginning to end. Like other *For Dummies* books, this one was designed to let you skip around as you like. This is your book, and physics is your oyster. You can jump into Chapter 1, which is where all the action starts; you can head to Chapter 2 for a discussion of the necessary algebra and trig you should know; or you can jump in anywhere you like if you know exactly what topic you want to study.

PART I

getting started with

PHYSICS

IN THIS PART . . .

- See what physics is all about.

- Brush up on basic algebra, trig, and physics measurements and units.

- Master the motion of displacement, velocity, and acceleration.

- Point yourself in the right direction with vectors.

1

USING PHYSICS TO UNDERSTAND YOUR WORLD

IN THIS CHAPTER

- Recognizing the physics in your world
- Understanding motion
- Handling the force and energy around you
- Getting hot under the collar with thermodynamics

Physics is the study of the world and universe around you. Luckily, the behavior of the matter and energy — the stuff of this universe — is not completely unruly. Instead, it strictly obeys laws, which physicists are gradually revealing through the careful application of the *scientific method,* which relies on experimental evidence and sound rigorous reasoning. In this way, physicists have been uncovering more and more of the beauty that lies at the heart of the workings of the universe, from the infinitely small to the mind-bogglingly large.

Physics is an all-encompassing science. You can study various aspects of the natural world (in fact, the word *physics* is derived from the Greek word *physika,* which means "natural things"), and accordingly, you can study different fields in physics: the physics of objects in motion, of energy, of forces, of gases, of heat and temperature, and so on. You enjoy the study of all these topics and many more in this book. In this chapter, we give an overview of physics — what it is, what it deals with, and why mathematical calculations are important to it — to get you started.

WHAT PHYSICS IS ALL ABOUT

Many people are a little on edge when they think about physics. For them, the subject seems like some highbrow topic that pulls numbers and rules out of thin air. But the truth is that physics exists to help you make sense of the world. Physics is a human adventure, undertaken on behalf of everyone, into the way the world works.

Remember: At its root, physics is all about becoming aware of your world and using mental and mathematical models to explain it. The gist of physics is this: You start by making an observation, you create a model to simulate that situation, and then you add some math to fill it out — and voilà! You have the power to predict what will happen in the real world. All this math exists to help you see what happens and why.

In this section, we explain how real-world observations fit in with the math. The later sections take you on a brief tour of the key topics that comprise basic physics.

Observing the world

You can observe plenty going on around you in your complex world. Leaves are waving, the sun is shining, light bulbs are glowing, cars are moving, computer printers are printing, people are walking and riding bikes, streams are flowing, and so on. When you stop to examine these actions, your natural curiosity gives rise to endless questions such as these:

- Why do I slip when I try to climb that snow bank?
- How distant are other stars, and how long would it take to get there?
- How does an airplane wing work?
- How can a thermos flask keep hot things warm *and* keep cold things cool?
- Why does an enormous cruise ship float when a paper clip sinks?
- Why does water roll around when it boils?

Any law of physics comes from very close observation of the world, and any theory that a physicist comes up with has to stand up to experimental measurements. Physics goes beyond qualitative statements about physical things — "If I push the child on the swing harder, then she swings higher," for example. With the laws of physics, you can predict precisely how high the child will swing.

Making predictions

Physics is simply about modeling the world (although an alternative viewpoint claims that physics actually uncovers the truth about the workings of the world; it doesn't just model it). You can use these mental models to describe how the world works: how blocks slide down ramps, how stars form and shine, how black holes trap light so it can't escape, what happens when cars collide, and so on.

When these models are first created, they sometimes have little to do with numbers; they just cover the gist of the situation. For example, a star is made up of this layer and then that layer, and as a result, this reaction takes place, followed by that one. And — pow! — you have a star. As time goes on, those models become more numeric, which is where physics students sometimes start having problems. Physics class would be a cinch if you could simply say, "That cart is going to roll down that hill, and as it gets toward the bottom, it's going to roll faster and faster." But the story is more involved than that — not only can you say that the cart is going to go faster, but in exerting your mastery over the physical world, you can also say how much faster it'll go.

There's a delicate interplay between theory, formulated with math, and experimental measurements. Often experimental measurements not only verify theories but also suggest ideas for new theories, which in turn suggest new experiments. Both feed off each other and lead to further discovery.

Many people approaching this subject may think of math as something tedious and overly abstract. However, in the context of physics, math comes to life. A quadratic equation may seem a little dry, but when you're using it to work out the correct angle to fire a rocket at for the perfect trajectory, you may find it more palatable! Chapter 2 explains all the math you need to know to perform basic physics calculations.

Reaping the rewards

So what are you going to get out of physics? If you want to pursue a career in physics or in an allied field such as engineering, the answer is clear: You'll need this knowledge on an everyday basis. But even if you're not planning to embark on a physics-related career, you can get a lot out of studying the subject. You can apply much of what you discover in an introductory physics course to real life:

- In a sense, all other sciences are based upon physics. For example, the structure and electrical properties of atoms determine chemical reactions; therefore, all of chemistry is governed by the laws of physics. In fact, you could argue that everything ultimately boils down to the laws of physics!

- Physics does deal with some pretty cool phenomena. Many videos of physical phenomena have gone viral on YouTube; take a look for yourself. Do a search for "non-Newtonian fluid," and you can watch the creeping, oozing dance of a cornstarch-water mixture on a speaker cone.

- More important than the applications of physics are the problem-solving skills it arms you with for approaching any kind of problem. Physics problems train you to stand back, consider your options for attacking the issue, select your method, and then solve the problem in the easiest way possible.

OBSERVING OBJECTS IN MOTION

Some of the most fundamental questions you may have about the world deal with objects in motion. Will that boulder rolling toward you slow down? How fast do you have to move to get out of its way? (Hang on just a moment while we get out our calculator. . . .) Motion was one of the earliest explorations of physics.

When you take a look around, you see that the motion of objects changes all the time. You see a motorcycle coming to a halt at a stop sign. You see a leaf falling and then stopping when it hits the ground, only to be picked up again by the wind. You see a pool ball hitting other balls in just the wrong way so that they all move without going where they should. Part I of this book handles objects in motion — from balls to railroad cars and most objects in between. In this section, we introduce motion in a straight line and rotational motion.

Measuring speed, direction, velocity, and acceleration

Speeds are big with physicists — how fast is an object going? Thirty-five miles per hour not enough? How about 3,500? No problem when you're dealing with physics. Besides speed, the direction an object is going is important if you want to describe its motion. If the home team is carrying a football down the field, you want to make sure they're going in the right direction.

When you put speed and direction together, you get a vector — the velocity vector. Vectors are a very useful kind of quantity. Anything that has both size and direction is best described with a *vector*. Vectors are often represented as arrows, where the length of the arrow tells you the magnitude (size), and the direction of the arrow tells you the direction. For a velocity vector, the length corresponds to the speed of the object, and the arrow points in the direction the object is moving. (To find out how to use vectors, head to Chapter 4.)

Everything has a velocity, so velocity is great for describing the world around you. Even if an object is at rest with respect to the ground, it's still on the Earth, which itself has a velocity. (And if everything has a velocity, it's no wonder physicists keep getting grant money — somebody has to measure all that motion.)

If you've ever ridden in a car, you know that velocity isn't the end of the story. Cars don't start off at 60 miles per hour; they have to accelerate until they get to that speed. Like velocity, acceleration has not only a magnitude but also a direction, so acceleration is a vector in physics as well. We cover speed, velocity, and acceleration in Chapter 3.

Round and round: Rotational motion

Plenty of things go round and round in the everyday world — CDs, DVDs, tires, pitchers' arms, clothes in a dryer, roller coasters doing the loop, or just little kids spinning from joy in their first snowstorm. That being the case, physicists want to get in on the action with measurements. Just as you can have a car moving and accelerating in a straight line, its tires can rotate and accelerate in a circle.

Going from the linear world to the rotational world turns out to be easy, because there's a handy physics *analog* (which is a fancy word for "equivalent") for everything linear in the rotational world. For example, distance traveled becomes angle turned. Speed in meters per second becomes angular speed in angle turned per second. Even linear acceleration becomes rotational acceleration.

So when you know linear motion, rotational motion just falls in your lap. You use the same equations for both linear and angular motion — just different symbols with slightly different meanings (angle replaces distance, for example). You'll be looping the loop in no time. Chapter 7 has the details.

WHEN PUSH COMES TO SHOVE: FORCES

Forces are a particular favorite in physics. You need forces to get motionless things moving — literally. Consider a stone on the ground. Many physicists (except, perhaps, geophysicists) would regard it suspiciously. It's just sitting there. What fun is that? What can you measure about that? After physicists had measured its size and mass, they'd lose interest.

But kick the stone — that is, apply a force — and watch the physicists come running over. Now something is happening — the stone started at rest, but now it's moving. You can find all kinds of numbers associated with this motion. For instance, you can connect the force you apply to something to its mass and get its acceleration. And physicists love numbers, because numbers help describe what's happening in the physical world.

Physicists are experts in applying forces to objects and predicting the results. Got a refrigerator to push up a ramp and want to know if it'll go? Ask a physicist. Have a rocket to launch? Same thing.

Springs and pendulums: Simple harmonic motion

Have you ever watched something bouncing up and down on a spring? That kind of motion puzzled physicists for a long time, but then they got down to work. They discovered that when you stretch a spring, the force isn't constant. The spring pulls back, and the more you pull the spring, the stronger it pulls back.

So how does the force compare to the distance you pull a spring? The force is directly proportional to the amount you stretch the spring: Double the amount you stretch the spring, and you double the amount of force with which the spring pulls back.

Physicists were overjoyed — this was the kind of math they understood. Force proportional to distance? Great — you can put that relationship into an equation, and you can use that equation to describe the motion of the object tied to the spring. Physicists got results telling them just how objects tied to springs would move — another triumph of physics.

This particular triumph is called *simple harmonic motion*. It's *simple* because force is directly proportional to distance, and so the result is simple. It's *harmonic* because it repeats over and over again as the object on the spring bounces up and down. Physicists were able to derive simple equations that could tell you exactly where the object would be at any given time.

But that's not all. Simple harmonic motion applies to many objects in the real world, not just things on springs. For example, pendulums also move in simple harmonic motion. Say you have a stone that's swinging back and forth on a string. As long as the arc it swings through isn't too high, the stone on a string is a pendulum; therefore, it follows simple harmonic motion. If you know how long the string is and how big of an angle the swing covers, you can predict where the stone will be at any time. We discuss simple harmonic motion in Chapter 13.

Absorbing the energy around you

You don't have to look far to find your next piece of physics. (You never do.) As you exit your house in the morning, for example, you may hear a crash up the street. Two cars have collided at a high speed, and locked together, they're sliding your way. Thanks to physics (and more specifically, Part III of this book), you can make the necessary measurements and predictions to know exactly how far you have to move to get out of the way.

Having mastered the ideas of energy and momentum helps at such a time. You use these ideas to describe the motion of objects with mass. The energy of motion is called *kinetic energy,* and when you accelerate a car from 0 to 60 miles per hour in 10 seconds, the car ends up with plenty of kinetic energy.

Where does the kinetic energy come from? It comes from *work,* which is what happens when a force moves an object through a distance. The energy can also come from *potential energy,* the energy stored in the object, which comes from the work done by a particular kind of force, such as gravity or electrical forces. Using gasoline, for example, an engine does work on the car to get it up to speed. But you need a force to accelerate something, and the way the engine does work on the car, surprisingly, is to use the force of friction with the road. Without friction, the wheels would simply spin, but because of a frictional force, the tires impart a force on the road. For every force between two objects, there is a reactive force of equal size but in the opposite direction. So the road also exerts a force on the car, which causes it to accelerate.

Or say that you're moving a piano up the stairs of your new place. After you move up the stairs, your piano has potential energy, simply because you put in a lot of work against gravity to get the piano up those six floors. Unfortunately, your roommate hates pianos and drops yours out the window. What happens next? The potential energy of the piano due to its height in a gravitational field is converted into kinetic energy, the energy of motion. You decide to calculate the final speed of the piano as it hits the street. (Next, you calculate the bill for the piano, hand it to your roommate, and go back downstairs to get your drum set.)

That's heavy: Pressures in fluids

Ever notice that when you're 5,000 feet down in the ocean, the pressure is different from at the surface? Never been 5,000 feet beneath the ocean waves? Then you may have noticed the difference in pressure when you dive into a swimming pool. The deeper you go, the higher the pressure is because of the weight of the water above you exerting a force downward. *Pressure* is just force per area.

Got a swimming pool? Any physicists worth their salt can tell you the approximate pressure at the bottom if you tell them how deep the pool is. When working with fluids, you have all kinds of other quantities to measure, such as the velocity of fluids through small holes, a fluid's density, and so on. Once again, physics responds with grace under pressure. You can read about forces in fluids in Chapter 8.

FEELING HOT BUT NOT BOTHERED: THERMODYNAMICS

Heat and cold are parts of your everyday life. Ever take a look at the beads of condensation on a cold glass of water in a warm room? Water vapor in the air is being cooled when it touches the glass, and it condenses into liquid water. The condensing water vapor passes thermal energy to the glass, which passes thermal energy to the cold drink, which ends up getting warmer as a result.

Thermodynamics can tell you how much heat you're radiating away on a cold day, how many bags of ice you need to cool a lava pit, and anything else that deals with heat energy. You can also take the study of thermodynamics beyond planet Earth. Why is space cold? In a normal environment, you radiate heat to everything around you, and everything around you radiates heat back to you. But in space, your heat just radiates away, so you can freeze.

Radiating heat is just one of the three ways heat can be transferred. You can discover plenty more about heat, whether created by a heat source like the sun or by friction, through the topics in Part IV.

2

REVIEWING PHYSICS MEASUREMENT
AND MATH FUNDAMENTALS

IN THIS CHAPTER

* Mastering measurements (and keeping them straight as you solve equations)
* Accounting for significant digits and possible error
* Brushing up on basic algebra and trig concepts

P hysics uses observations and measurements to make mental and mathematical models that explain how the world (and everything in it) works. This process is unfamiliar to most people, which is where this chapter comes in.

This chapter covers some basic skills you need for the coming chapters. We cover measurements and scientific notation, give you a refresher on basic algebra and trigonometry, and show you which digits in a number to pay attention to — and which ones to ignore. Continue on to build a physics foundation, solid and unshakable, that you can rely on throughout this book.

MEASURING THE WORLD AROUND YOU AND MAKING PREDICTIONS

Physics excels at measuring and predicting the physical world — after all, that's why physics exists. Measuring is the starting point — part of observing the world so you can then model and predict it. You have several different measuring sticks at your disposal: some for length, some for mass or weight, some for time, and so on. Mastering those measurements is part of mastering physics.

To keep like measurements together, physicists and mathematicians have grouped them into *measurement systems.* The most common measurement system you see in introductory physics is the meter-kilogram-second (MKS) system, referred to as SI (short for *Système International d'Unités,* the International System of Units), but you may also come across the foot-pound-second (FPS) system. Table 2-1 lists the primary units of measurement in the MKS system, along with their abbreviations.

These are the measuring sticks that will become familiar to you as you solve problems and triumph over the math in this book. Also for reference, Table 2-2 shows the primary units of measurement (and their abbreviations) in the CGS system. (Don't bother memorizing the ones you're not familiar with now; you can come back to them later as needed.)

Table 2-1 MKS Units of Measurement

Measurement	*Unit*	*Abbreviation*
Length	meter	m
Mass	kilogram	kg
Time	second	s
Force	newton	N
Energy	joule	J
Pressure	pascal	Pa
Electric current	ampere	A
Magnetic flux density	tesla	T
Electric charge	coulomb	C

Table 2-2 CGS Units of Measurement

Measurement	*Unit*	*Abbreviation*
Length	centimeter	cm
Mass	gram	g
Time	second	s
Force	dyne	dyn
Energy	erg	erg
Pressure	barye	Ba
Electric current	biot	Bi
Magnetic flux density	gauss	G
Electric charge	franklin	Fr

Warning: Because different measurement systems use different standard lengths, you can get several different numbers for one part of a problem, depending on the measurement you use. For example, if you're measuring the depth of the water in a swimming pool, you can use the MKS measurement system, which gives you an answer in meters, or the less common FPS system, in which case you determine the depth of the water in feet. The point? When working with equations, stick with the same measurement system all the way through the problem. If you don't, your answer will be a meaningless hodgepodge, because you're switching measuring sticks for multiple items as you try to arrive at a single answer. Mixing up the measurements causes problems — imagine baking a cake where the recipe calls for 2 cups of flour, but you use 2 liters instead.

Examples

Q. You're told to measure the length of a racecar track using the MKS system. What unit(s) will your measurement be in?

A. The correct answer is meters. The unit of length in the MKS system is the meter.

Q. You're told to measure the mass of a marble using the CGS system. What unit(s) will your measurement be in?

A. The correct answer is grams. The unit of mass in the CGS system is the gram.

Practice Questions

1. You're asked to measure the time it takes the moon to circle Earth using the MKS system. What will your measurement's units be?

2. You need to measure the force a tire exerts on the road as it's moving using the MKS system. What are the units of your answer?

3. You're asked to measure the amount of energy released by a firecracker when it explodes using the CGS system. What are the units of your answer?

Practice Answers

1. **seconds.** The unit of time in the MKS system is the second.

2. **newtons.** The unit of force in the MKS system is the newton.

3. **ergs.** The unit of energy in the CGS system is the erg.

ELIMINATING SOME ZEROS: USING SCIENTIFIC NOTATION

Physicists have a way of getting their minds into the darndest places, and those places often involve really big or really small numbers. Physics has a way of dealing with very large and very small numbers; to help reduce clutter and make them easier to digest, it uses *scientific notation*.

Remember: In scientific notation, you write a number as a decimal (with only one digit before the decimal point) multiplied by a power of ten. The power of ten (10 with an exponent) expresses the number of zeroes. To get the right power of ten for a vary large number, count all the places in front of the decimal point, from right to left, up to the place just to the right of the first digit (you don't include the first digit because you leave it in front of the decimal point in the result).

For example, say you're dealing with the average distance between the sun and Pluto, which is about 5,890,000,000,000 meters. You have a lot of meters on your hands, accompanied by a lot of zeroes. You can write the distance between the sun and Pluto as follows:

$$5,890,000,000,000 \text{ meters} = 5.89 \times 10^{12} \text{ meters}$$

The exponent is 12 because you count 12 places between the end of 5,890,000,000,000 (where a decimal would appear in the whole number) and the decimal's new place after the 5.

Scientific notation also works for very small numbers, such as the one that follows, where the power of ten is negative. You count the number of places, moving left to right, from the decimal point to just after the first nonzero digit (again leaving the result with just one digit in front of the decimal):

$$0.0000000000000000005339 \text{ meters} = 5.339 \times 10^{-19} \text{ meters}$$

Remember: If the number you're working with is larger than ten, you have a positive exponent in scientific notation; if it's smaller than one, you have a negative exponent. As you can see, handling super large or super small numbers with scientific notation is easier than writing them all out, which is why calculators come with this kind of functionality already built in.

Using unit prefixes

Scientists have come up with a handy notation that helps take care of variables that have very large or very small values in their standard units. Say you're measuring the thickness of a human hair and find it to be 0.00002 meters thick. You could use scientific notation to write this as 2×10^{-5} meters (20×10^{-6} meters), or you could use the unit prefix μ, which stands for *micro:* 20 μm. When you put μ in front of any unit, it represents 10^{-6} times that unit.

A more familiar unit prefix is k, as in *kilo,* which represents 10^3 times the unit. For example the kilometer, km, is 10^3 meters, which equals 1,000 meters. The following table shows other common unit prefixes that you may see.

Unit Prefix	Exponent
mega (M)	10^6
kilo (k)	10^3
centi (c)	10^{-2}
milli (m)	10^{-3}
micro (μ)	10^{-6}
nano (n)	10^{-9}
pico (p)	10^{-12}
femto (f)	10^{-15}

Examples

Q. How does the number 1,000 look in scientific notation?

A. The correct answer is 1.0×10^3. You have to move the decimal point three times to the left to get 1.0.

Q. What is 0.000037 in scientific notation?

A. The correct answer is 3.7×10^{-5}. You have to move the decimal point five times to the right to get 3.7.

Practice Questions

1. What is 0.0043 in scientific notation?

2. What is 430,000 in scientific notation?

3. What is 0.00000056 in scientific notation?

4. What is 6,700 in scientific notation?

Practice Answers

1. 4.3×10^{-3}. You have to move the decimal point three places to the right.

2. 4.3×10^5. You have to move the decimal point five places to the left.

3. 5.6×10^{-7}. You have to move the decimal point seven places to the right.

4. 6.7×10^3. You have to move the decimal point three places to the left.

FROM METERS TO INCHES AND BACK AGAIN: CONVERTING BETWEEN UNITS

Physicists use various measurement systems to record numbers from their observations. But what happens when you have to convert between those systems? Physics problems sometimes try to trip you up here, giving you the data you need in mixed units: centimeters for this measurement but meters for that measurement — and maybe even mixing in inches as well. Don't be fooled. You have to convert *everything* to the same measurement system before you can proceed. How do you convert in the easiest possible way? You use conversion factors, which we explain in this section.

Tip: To convert between measurements in different measuring systems, you can multiply by a conversion factor. A *conversion factor* is a ratio that, when you multiply it by the item you're converting, cancels out the units you don't want and leaves those that you do. The conversion factor must equal 1.

Here's how it works: For every relation between units — for example, 24 hours = 1 day — you can make a fraction that has the value of 1. If, for example, you divide both sides of the equation 24 hours = 1 day by 1 day, you get

$$\frac{24 \text{ hr}}{1 \text{ day}} = 1$$

Suppose you want to convert 3 days to hours. You can just multiply your time by the preceding fraction. Doing so doesn't change the value of the time because you're multiplying by 1. You can see that the unit of *days* cancels out, leaving you with a number of hours:

$$\frac{3 \text{ days}}{1} \times \frac{24 \text{ hr}}{1 \text{ day}} = \frac{3 \text{ days}}{1} \times \frac{24 \text{ hr}}{1 \text{ day}} = 72 \text{ hr}$$

Remember: Words such as *days, seconds,* and *meters* act like the variables x and y in that if they're present in both the numerator and the denominator, they cancel each other out.

To convert the other way — hours into days, in this example — you simply use the same original relation, 24 hours = 1 day, but this time divide both sides by 24 hours to get

$$1 = \frac{1 \text{ day}}{24 \text{ hr}}$$

Then multiply by this fraction to cancel the units from the bottom, which leaves you with the units on the top.

Consider the following problem. Passing the state line, you note that you've gone 4,680 miles in exactly three days. Very impressive. If you went at a constant speed, how fast were you going? Speed is just as you may expect — distance divided by time. So you calculate your speed as follows:

$$\frac{4,680 \text{ mi}}{3 \text{ days}} = 1,560 \text{ mi/day}$$

Your answer, however, isn't exactly in a standard unit of measure. You have a result in miles per day, which you write as miles/day. To calculate miles per hour, you need a conversion factor that knocks *days* out of the denominator and leaves *hours* in its place, so you multiply by *days/hour* and cancel out *days:*

$$\frac{\text{mi}}{\text{day}} \times \frac{\text{days}}{\text{hr}} = \frac{\text{mi}}{\text{hr}}$$

Your conversion factor is *days/hour.* When you multiply by the conversion factor, your work looks like this:

$$\frac{1,560 \text{ mi}}{1 \text{ day}} \times \frac{1 \text{ day}}{24 \text{ hr}}$$

Looking at the units when numbers make your head spin

Want an inside trick that teachers and instructors often use to solve physics problems? Pay attention to the units you're working with. We've had thousands of one-on-one problem-solving sessions with students in which we worked on homework problems, and we can tell you that this is a trick that instructors use all the time.

As a simple example, say you're given a distance and a time, and you have to find a speed. You can cut through the wording of the problem immediately because you know that distance (for example, meters) divided by time (for example, seconds) gives you speed (meters/second). Multiplication and division are reflected in the units. So, for example, because a rate like speed is given as a distance divided by a time, the units (in MKS) are meters/second. As another example, a quantity called *momentum* is given by velocity (meters/second) multiplied by mass (kilograms); it has units of kg · m/s.

As the problems get more complex, however, more items are involved — say, for example, a mass, a distance, a time, and so on. You find yourself glancing over the words of a problem to pick out the numeric values and their units. Have to find an amount of energy? Energy is mass times distance squared over time squared, so if you can identify these items in the question, you know how they're going to fit into the solution and you won't get lost in the numbers.

The upshot is that units are your friends. They give you an easy way to make sure you're headed toward the answer you want. So when you feel too wrapped up in the numbers, check the units to make sure you're on the right path. But remember: You still need to make sure you're using the right equations!

Note that because there are 24 hours in a day, the conversion factor equals exactly 1, as all conversion factors must. So when you multiply 1,560 miles/day by this conversion factor, you're not changing anything — all you're doing is multiplying by 1.

When you cancel out *days* and multiply across the fractions, you get the answer you've been searching for:

$$\frac{1,560 \ \text{mi}}{1 \ \text{day}} \times \frac{1 \ \text{day}}{24 \ \text{hr}} = 65 \ \text{mi/hr}$$

So your average speed is 65 miles per hour, which is pretty fast considering that this problem assumes you've been driving continuously for three days.

You don't *have* to use a conversion factor; if you instinctively know that you need to divide by 24 to convert from miles per day to miles per hour, so much the better. But if you're ever in doubt, use a conversion factor and write out the calculations, because taking the long road is far better than making a mistake. We've seen far too many people get everything in a problem right except for this kind of simple conversion.

Here are some handy conversions that you can come back to as needed:

- **Length:**
  - 1 m = 100 cm
  - 1 km = 1,000 m

- 1 in (inch) = 2.54 cm
- 1 m = 39.37 in
- 1 mile = 5,280 ft = 1.609 km
- 1 Å (angstrom) = 10^{-10} m
- **Mass:**
  - 1 kg = 1,000 g
  - 1 slug = 14.59 kg
  - 1 u (atomic mass unit) = 1.6605×10^{-27} kg
- **Force:**
  - 1 lb (pound) = 4.448 N
  - 1 N = 10^5 dynes
  - 1 N = 0.2248 lb
- **Energy:**
  - 1 J = 10^7 ergs
  - 1 J = 0.7376 ft-lb
  - 1 BTU (British thermal unit) = 1,055 J
  - 1 kWh (kilowatt hour) = 3.600×10^6 J
  - 1 eV (electron volt) = 1.602×10^{-19} J
- **Power:**
  - 1 hp (horsepower) = 550 ft-lb/s
  - 1 W (watt) = 0.7376 ft-lb/s

Examples

Q. A ball drops 5 meters. How many centimeters did it drop?

A. The correct answer is 500 centimeters. To perform the conversion, you do the following calculation:

$$5.0 \text{ m} \times \frac{100 \text{ cm}}{1 \text{ m}} = 500 \text{ cm}$$

Note that 100 centimeters divided by 1 meter equals 1 because there are 100 centimeters in a meter. In the calculation, the units you don't want — meters — cancel out.

Q. Convert 10 inches into meters.

A. The correct answer is 0.254 meters.

1. You know that 1 inch = 2.54 centimeters, so start with that conversion factor and convert 10 inches into centimeters:

$$10 \text{ in} \times \frac{2.54 \text{ cm}}{1 \text{ in}} = 25.4 \text{ cm}$$

2. Convert 25.4 centimeters into meters by using a second conversion factor:

$$2.54 \text{ cm} \times \frac{1 \text{ m}}{100 \text{ cm}} = 0.254 \text{ m}$$

Q. An SUV is traveling 2.78×10^{-2} kilometers per second. What's that in kilometers per hour?

A. The correct answer is 100 kilometers per hour.

1. You know that there are 60 minutes in an hour, so start by converting from kilometers per second to kilometers per minute:

$$2.78 \times 10^{-2} \frac{\text{km}}{\text{sec}} \times \frac{60 \text{ sec}}{1 \text{ min}} = 1.67 \text{ km/min}$$

2. Because there are 60 minutes in an hour, convert this to kilometers per hour by using a second conversion factor:

$$\frac{1.67 \text{ km}}{1 \text{ min}} \times \frac{60 \text{ min}}{1 \text{ hr}} = 100 \text{ km/hr}$$

Practice Questions

1. How many centimeters are in 2.35 meters?

2. How many inches are in 2.0 meters?

3. Given that there are 2.54 centimeters in 1 inch, how many centimeters are there in 1 yard?

4. How many inches are in an angstrom, given that 1 angstrom (Å) = 10^{-8} centimeters?

5. How many hours are in 1 year?

6. The three-toed sloth can move at a speed of about 10 feet per minute. How fast is that in kilometers per hour? (Round to the nearest hundredth of a kilometer per hour.)

Practice Answers

1. **235 cm.** Convert 2.35 meters into centimeters:

$$2.35 \, \cancel{m} \times \frac{100 \text{ cm}}{1 \, \cancel{m}} = 235 \text{ cm}$$

2. **79 in.** Convert 2.0 meters into inches:

$$2.0 \, \cancel{m} \times \frac{39.37 \text{ in}}{1 \, \cancel{m}} = 79 \text{ in}$$

3. **91.4 cm.** One yard is 3 feet, so convert that to inches:

$$3 \, \cancel{ft} \times \frac{12 \text{ in}}{1 \, \cancel{ft}} = 36 \text{ in}$$

Use a second conversion factor to convert that into centimeters:

$$36 \, \cancel{in} \times \frac{2.54 \text{ cm}}{1 \, \cancel{in}} = 91.4 \text{ cm}$$

4. **4.0×10^{-9} in.** Convert 1 angstrom to centimeters:

$$1 \, \cancel{Å} \times \frac{10^{-8} \text{ cm}}{1 \, \cancel{Å}} = 10^{-8} \text{ cm}$$

Use a second conversion factor to convert that into inches:

$$10^{-8} \, \cancel{cm} \times \frac{1 \text{ in}}{2.54 \, \cancel{cm}} = 4.0 \times 10^{-9} \text{ in}$$

5. **8,760 hr.** Convert 1 year into days:

$$1 \, \cancel{year} \times \frac{365 \text{ days}}{1 \, \cancel{year}} = 365 \text{ days}$$

Use a second conversion factor to convert that into hours:

$$365 \, \cancel{days} \times \frac{24 \text{ hr}}{1 \, \cancel{day}} = 8,760 \text{ hr}$$

6. **0.18 km/hr.** One foot is 12 inches, so convert that into centimeters:

$$12 \, \cancel{in} \times \frac{2.54 \text{ cm}}{1 \, \cancel{in}} = 30.48 \text{ cm}$$

Use a second conversion factor to convert centimeters into meters and then into kilometers:

$$30.48 \, \cancel{cm} \times \frac{1 \, \cancel{m}}{100 \, \cancel{cm}} \times \frac{1 \text{ km}}{1000 \, \cancel{m}} = 3.048 \times 10^{-4} \text{ km}$$

Convert 1 hour into minutes:

$$1 \text{ hr} \times \frac{60 \text{ min}}{1 \text{ hr}} = 60 \text{ min}$$

Use your results to convert 10 feet per minute into kilometers per hour:

$$10 \ \frac{\text{ft}}{\text{min}} \times \frac{60 \text{ min}}{1 \text{ hr}} \times \frac{3.048 \times 10^{-4} \text{ km}}{1 \text{ ft}} = 0.18 \text{ km/hr}$$

CHECKING THE ACCURACY AND PRECISION OF MEASUREMENTS

Accuracy and precision are important when making (and analyzing) measurements in physics. You can't imply that your measurement is more precise than you know it to be by adding too many significant digits, and you have to account for the possibility of error in your measurement system by adding a ± when necessary. This section delves deeper into the topics of significant digits, precision, and accuracy.

Knowing which digits are significant

This section is all about how to properly account for the known precision of the measurements and carry that through the calculations, how to represent numbers in a way that is consistent with their known precision, and what to do with calculations that involve measurements with different levels of precision.

Finding the number of significant digits

In a measurement, *significant digits* (or *significant figures*) are those that were actually measured. Say you measure a distance with your ruler, which has millimeter markings. You can get a measurement of 10.42 centimeters, which has four significant digits (you estimate the distance between markings to get the last digit). But if you have a very precise micrometer gauge, then you can measure the distance to within one-hundredth of that, so you may measure the same thing to be 10.4213 centimeters, which has six significant digits.

By convention, zeroes that simply fill out values down to (or up to) the decimal point aren't considered significant. When you see a number given as 3,600, you know the 3 and 6 are included because they're significant. However, knowing which, if any, of the zeros are significant can be tricky.

Tip: The best way to write a number so you leave no doubt about how many significant digits there are is to use scientific notation. For example, if you read of a measurement of 1,000 meters, you don't know if there are one, two, three, or four significant figures. But if it were written as 1.0×10^3 meters, you would know that there are two significant figures. If the measurement were written as 1.000×10^3 meters, then you would know that there are four significant figures.

Rounding answers to the correct number of digits

When you do calculations, you often need to round your answer to the correct number of significant digits. If you include any more digits, you claim a precision that you don't really have and haven't measured.

For example, if someone tells you that a rocket traveled 10.0 meters in 7.0 seconds, the person is telling you that the distance is known to three significant digits and the seconds are known to two significant digits (the number of digits in each of the measurements). If you want to find the rocket's speed, you can whip out a calculator and divide 10.0 meters by 7.0 seconds to come up with 1.428571429 meters per second, which looks like a very precise measurement indeed. But the result is too precise — if you know your measurements to only two or three significant digits, you can't say you know the answer to ten significant digits. Claiming as such would be like taking a meter stick, reading down to the nearest millimeter, and then writing down an answer to the nearest ten-millionth of a millimeter. You need to round your answer.

Remember: The rules for determining the correct number of significant digits after doing calculations are as follows:

- **When you multiply or divide numbers:** The result has the same number of significant digits as the original number that has the fewest significant digits. In the case of the rocket, where you need to divide, the result should have only two significant digits (the number of significant digits in 7.0). The best you can say is that the rocket is traveling at 1.4 meters per second, which is 1.428571429 rounded to one decimal place.

- **When you add or subtract numbers:** Line up the decimal points; the last significant digit in the result corresponds to the right-most column where all numbers still have significant digits. If you have to add 3.6, 14, and 6.33, you'd write the answer to the nearest whole number — the 14 has no significant digits after the decimal place, so the answer shouldn't, either. You can see what we mean by taking a look for yourself:

$$
\begin{array}{r}
3.6 \\
14 \\
+6.33 \\
\hline
23.93
\end{array}
$$

When you round the answer to the correct number of significant digits, your answer is 24.

Remember: When you round a number, look at the digit to the right of the place you're rounding to. If that right-hand digit is 5 or greater, round up. If it's 4 or less, round down. For example, you round 1.428 up to 1.43 and 1.42 down to 1.4.

Examples

Q. You're multiplying 12.01 centimeters by 9.7 centimeters. What should your answer be, keeping in mind that you should express it in significant digits?

A. The correct answer is 120 centimeters squared.

 1. The calculator says the product of 12.01 and 9.7 is 116.497.

 2. The number of significant digits in your result is the same as the smallest number of significant digits in any of the values being multiplied. That's two here (because of 9.7), so your answer rounds up to 120 centimeters squared.

Q. You're squaring 17.3 and then subtracting 79.9134. What is the result, with the correct number of significant digits?

A. The correct answer is 219.

 1. The calculator says the square of 17.3 is 299.29.

 2. The number of significant digits in a product is the same as the smallest number of significant digits in any of the values being multiplied (when you square 17.3, you're multiplying 17.3 by itself). There are three significant digits in 17.3, so you round your result to three digits, or 299.

 3. The calculator says 299 minus 79.9134 is 219.0866.

 4. The 299 has no significant digits after the decimal place, so the answer shouldn't either. You round your result to 219.

Practice Questions

 1. What is 19.3 multiplied by 26.12, taking into account significant digits?

 2. What is the sum of 7.9 grams, 19 grams, and 5.654 grams, taking into account significant digits?

 3. What do you get if you divide 1.93 meters by 0.069 seconds, keeping the correct number of significant digits?

 4. What do you get if you add 5.2 square meters to the result of 1.36 meters times 0.7130 meters, keeping the correct number of significant digits?

Practice Answers

1. **504.** The calculator says the product is 504.116. However, 19.3 has three significant digits, and 26.12 has four, so you use three significant digits in your answer. That makes the answer 504.

2. **33 g.** Here's how you do the sum of 7.9, 19, and 5.654:

$$
\begin{array}{r}
7.9 \\
19 \\
+\ 5.654 \\
\hline
32.554
\end{array}
$$

    The value 19 has no significant digits after the decimal place, so the answer shouldn't either, making it 33 grams (32.554 grams rounded up).

3. **28 m/s.** The calculator says the quotient of 1.93 and 0.069 is 27.9710144928, and your answer should have units of meters divided by seconds, or meters per second. Because 1.93 has three significant digits and 0.069 has two (the leading zero doesn't count), you use two significant digits in your answer. That makes the answer 28 meters per second.

4. **6.2 m².** The calculator says the product of 1.36 and 0.7130 is 0.96968, and if you add 5.2, you get 6.16968. Your answer has units of meters squared. Because 5.2 has only one significant digit after the decimal place, which is less than the three significant digits after the decimal place you keep after multiplying 1.36 and 0.7130, you use one significant digit after the decimal place in your answer. That makes the answer 6.2 meters squared.

Estimating accuracy

Physicists don't always rely on significant digits when recording measurements. Sometimes, you see measurements that use plus-or-minus signs to indicate possible error in measurement, as in the following:

$$5.36 \pm 0.05 \text{ meters}$$

The $\pm$ part (0.05 meters in the preceding example) is the physicist's estimate of the possible error in the measurement, so the physicist is saying that the actual value is between $5.36 + 0.05$ (that is, 5.41) meters and $5.36 - 0.05$ (that is, 5.31 meters), inclusive. Note that the possible error isn't the amount your measurement *differs* from the "right" answer; it's an indication of how precisely your apparatus can measure — in other words, how reliable your results are as a measurement.

ARMING YOURSELF WITH BASIC ALGEBRA

Physics deals with plenty of equations, and to be able to handle them, you should know how to move the variables in them around. Note that algebra doesn't just allow you to plug in numbers and find values of different variables; it also lets you rearrange equations so you can make substitutions in other equations, and these new equations show different physics concepts. If you can follow along with the derivation of a formula in a physics book, you can get a better understanding of why the world works the way it does. That's pretty important stuff! Time to travel back to basic algebra for a quick refresher.

You need to be able to isolate different variables. For instance, the following equation tells you the distance, s, that an object travels if it starts from rest and accelerates at rate of a for a time, t:

$$s = \frac{1}{2}at^2$$

Now suppose the problem actually tells you the time the object is in motion and the distance it travels and asks you to calculate the object's acceleration. By rearranging the equation algebraically, you can solve for the acceleration:

$$a = \frac{2s}{t^2}$$

In this case, you've multiplied both sides by 2 and divided both sides by t^2 to isolate the acceleration, a, on one side of the equation.

What if you have to solve for the time, t? By moving the number and variables around, you get the following equation:

$$t = \sqrt{\frac{2s}{a}}$$

Do you need to memorize all three of these variations on the same equation? Certainly not. You just memorize one equation that relates these three items — distance, acceleration, and time — and then rearrange the equation as needed. (If you need a review of algebra, get a copy of *Algebra I For Dummies*, by Mary Jane Sterling [Wiley].)

Example

Q. The equation for final speed, v_f — where the initial speed is v_i, the acceleration is a, and the time is t — is $v_f = v_i + at$. Solve for acceleration.

A. The correct answer is $a = (v_f - v_i)/t$. To solve for a, divide both sides of the equation by time, t.

Practice Questions

1. The equation for potential energy, *PE*, for a mass *m* at height *h*, where the acceleration due to gravity is *g*, is $PE = mgh$. Solve for *h*.

2. The equation relating final speed, v_f, to initial speed, v_i, in terms of acceleration *a* and distance *s* is $v_f^2 = v_i^2 + 2as$. Solve for *s*.

3. The equation relating distance *s* to acceleration *a*, time *t*, and speed *v* is $s = v_i t + \frac{1}{2}at^2$. Solve for v_i.

4. The equation for kinetic energy is $KE = \frac{1}{2}mv^2$. Solve for *v*.

Practice Answers

1. **$h = PE/mg$.** Divide both sides by *mg* to get your answer.

2. $s = \dfrac{v_f^2 - v_i^2}{2a}$. Divide both sides by 2*a* to get your answer.

3. $v_i = \dfrac{s}{t} - \dfrac{1}{2}at$. Subtract $\frac{1}{2}at^2$ from both sides:

$$s - \frac{1}{2}at^2 = v_i t$$

Divide both sides by *t* to get your answer.

4. $v = \sqrt{\dfrac{2KE}{m}}$. Multiply both sides by 2/*m*:

$$\frac{2}{m}KE = v^2$$

Take the square root to get your answer.

TACKLING A LITTLE TRIG

You need to know a little trigonometry, including the sine, cosine, and tangent functions, for physics problems. To find these values, start with a simple right triangle. Take a look at Figure 2-1, which displays a right triangle in all its glory, complete with labels I've provided for the sake of explanation. Note in particular the angle θ, which appears between one of the triangle's legs and the hypotenuse (the longest side, which is opposite the right angle). The side *y* is opposite θ, and the side *x* is adjacent to θ.

Figure 2-1:
A labeled triangle that you can use to find trig values.

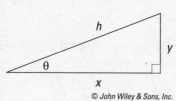

© John Wiley & Sons, Inc.

Remember: To find the trigonometric values of the triangle in Figure 2-1, you divide one side by another. Here are the definitions of sine, cosine, and tangent:

- $\sin\theta = \dfrac{y}{h}$

- $\cos\theta = \dfrac{x}{h}$

- $\tan\theta = \dfrac{y}{x}$

If you're given the measure of one angle and one side of the triangle, you can find all the other sides. Here are some other forms of the trig relationships — they'll probably become distressingly familiar before you finish any physics course, but you *don't* need to memorize them. If you know the preceding sine, cosine, and tangent equations, you can derive the following ones as needed:

- $x = h\cos\theta = \dfrac{y}{\tan\theta}$

- $y = h\sin\theta = x\tan\theta$

- $h = \dfrac{y}{\sin\theta} = \dfrac{x}{\cos\theta}$

To find the angle θ, you can go backward with the inverse sine, cosine, and tangent, which are written as $\sin^{-1}$, $\cos^{-1}$, and $\tan^{-1}$. Basically, if you input the sine of an angle into the $\sin^{-1}$ equation, you end up with the measure of the angle itself. Here are the inverses for the triangle in Figure 2-1:

- $\sin^{-1}\left(\dfrac{y}{h}\right) = \theta$

- $\cos^{-1}\left(\dfrac{x}{h}\right) = \theta$

- $\tan^{-1}\left(\dfrac{y}{x}\right) = \theta$

And here's one more equation, the Pythagorean theorem. It gives you the length of the hypotenuse when you plug in the other two sides:

$$h = \sqrt{x^2 + y^2}$$

If you need a more in-depth refresher, check out *Trigonometry For Dummies,* by Mary Jane Sterling (Wiley).

Example

Q. If the side labeled x in Figure 2-1 is 30 centimeters, and the angle θ is 30 degrees, what is the length of the hypotenuse, rounded to the nearest centimeter?

A. The correct answer is 35 centimeters.

1. Solve the equation $x = h\cos\theta$ for h to get

$$h = \frac{x}{\cos\theta}$$

2. Plug in the given values:

$$h = \frac{30 \text{ cm}}{\cos 30°}$$
$$= \frac{30 \text{ cm}}{\sqrt{3}/2}$$
$$\approx 35 \text{ cm}$$

Practice Questions

1. Given the hypotenuse h and the angle θ, what is the length x equal to?

2. If $x = 3$ m and $y = 4$ m, what is the length of the hypotenuse?

Practice Answers

1. $x = h\cos\theta$. Your answer comes from the definition of cosine.

2. **5 m.** Start with the Pythagorean theorem:

$$h = \sqrt{x^2 + y^2}$$

Plug in the numbers, and work out the answer:

$$h = \sqrt{(3 \text{ m})^2 + (4 \text{ m})^2}$$
$$= \sqrt{3^2 + 4^2} \text{ m}$$
$$= 5 \text{ m}$$

INTERPRETING EQUATIONS AS REAL-WORLD IDEAS

After teaching physics to college students for many years, we're very familiar with one of the biggest problems they face — getting lost in, and being intimidated by, the math.

Tip: Always keep in mind that the real world comes first and the math comes later. When you face a physics problem, make sure you don't get lost in the math; keep a global perspective about what's going on in the problem, because doing so helps you stay in control.

In physics, the ideas and observations of the physical world are the things that are important. Math operations are really only a simplified language for accurately describing what is going on. For example, here's a simple equation for speed:

$$v = \frac{s}{t}$$

In this equation, v is the speed, s is the distance, and t is the time. You can examine this equation's terms to see how this equation embodies simple common-sense notions of speed. Say that you travel a larger distance in the same amount of time. In that case, the right side of the equation must be larger, which means that your speed, on the left, is also greater. If you travel the same distance but it takes you more time, then the right side of this equation becomes smaller, which means that your speed is lower. The relationship between all the different components makes sense.

You can think of all the equations you come across in a similar way to make sure they make sense in the real world. If your equation behaves in a way that doesn't make physical sense, then you know that something must be wrong with the equation.

Bottom line: In physics, math is your friend. You don't need to get lost in it. Instead, you use it to formulate the problem and help guide you in its solution. Alone, each of these mathematical operations is very simple, but when you put them together, they're very powerful.

Be a genius: Don't focus on the math

Richard Feynman was a famous Nobel Prize winner in physics who had a reputation during the 1950s and '60s of being an amazing genius. He later explained his method: He attached the problem at hand to a real-life scenario, creating a mental image, while others got caught in the math. When someone would show him a long derivation that had gone wrong, for example, he'd think of some physical phenomenon that the derivation was supposed to explain. As he followed along, he'd get to the point where he suddenly realized the derivation no longer matched what happened in the real world, and he'd say, "No, that's the problem." He was always right, which mystified people who, awestruck, took him for a supergenius. Want to be a supergenius? Do the same thing: Don't let the math scare you.

3

EXPLORING THE NEED FOR SPEED

IN THIS CHAPTER

- Getting up to speed on displacement
- Dissecting different kinds of speed
- Going with acceleration
- Examining the link among acceleration, time, and displacement
- Connecting velocity, acceleration, and displacement

There you are in your Formula 1 racecar, speeding toward glory. You have the speed you need, and the pylons are whipping past on either side. You're confident that you can win, and coming into the final turn, you're far ahead. Or at least you think you are. Seems that another racer is also making a big effort, because you see a gleam of silver in your mirror. You get a better look and realize that you need to do something — last year's winner is gaining on you fast.

It's a good thing you know all about velocity and acceleration. With such knowledge, you know just what to do: You floor the gas pedal, accelerating out of trouble. Your knowledge of velocity lets you handle the final curve with ease. The checkered flag is a blur as you cross the finish line in record time. Not bad. You can thank your understanding of the issues in this chapter: displacement, velocity, and acceleration.

You already have an intuitive feeling for what we discuss in this chapter, or you wouldn't be able to drive or even ride a bike. Displacement is about where you are, speed is about how fast you're going, and anyone who's ever been in a car knows about acceleration. These characteristics of motion concern people every day, and physics has made an organized study of them. This knowledge has helped people to plan roads, build spacecraft, organize traffic patterns, fly, track the motion of planets, predict the weather, and even get mad in slow-moving traffic jams. Understanding movement is a vital part of understanding physics, and that's the topic of this chapter. Time to move on.

GOING THE DISTANCE WITH DISPLACEMENT

When something moves from Point A to Point B, displacement takes place in physics terms. In plain English, *displacement* is a distance in a particular direction.

Remember: Like any other measurement in physics (except for certain angles), displacement always has units — usually centimeters or meters. You may also use kilometers, inches, feet, miles, or even *light-years* (the distance light travels in one year, a whopper of a distance not fit for measuring with a meter stick: 5,865,696,000,000 miles, which is 9,460,800,000,000 kilometers or 9,460,800,000,000,000 meters).

In this section, we cover position and displacement in one to three dimensions.

Understanding displacement and position

You find displacement by finding the distance between an object's initial position and its final position. Say, for example, that you have a fine new golf ball that's prone to rolling around, shown in Figure 3-1. This particular golf ball likes to roll around on top of a large measuring stick. You place the golf ball at the 0 position on the measuring stick, as you see in Figure 3-1a.

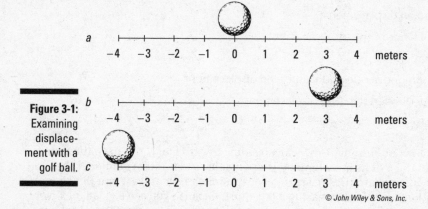

Figure 3-1: Examining displacement with a golf ball.

© John Wiley & Sons, Inc.

The golf ball rolls over to a new point, 3 meters to the right, as you see in Figure 3-1b. The golf ball has moved, so displacement has taken place. In this case, the displacement is just 3 meters to the right. Its initial position was 0 meters, and its final position is at +3 meters. The displacement is 3 meters.

Remember: In physics terms, you often see displacement referred to as the variable s (don't ask why).

Scientists, being who they are, like to go into even more detail. You often see the term s_i, which describes *initial position,* (the i stands for *initial*). And you may see the term s_f used to describe *final position.*

In these terms, moving from Figure 3-1a to 3-1b, s_i is at the 0-meter mark and s_f is at +3 meters. The displacement, s, equals the final position minus the initial position:

$$s = s_f - s_i$$
$$= 3 \text{ m} - 0 \text{ m} = 3 \text{ m}$$

Remember: Displacements don't have to be positive; they can be zero or negative as well. If the positive direction is to the right, then a negative displacement means that the object has moved to the left.

In Figure 3-1c, the restless golf ball has moved to a new location, which is measured as −4 meters on the measuring stick. The displacement is given by the difference between the initial and final position. If you want to know the displacement of the ball from its position in diagram B, take the initial position of the ball to be $s_i = 3$ meters; then the displacement is given by

$$s = s_f - s_i$$
$$= -4 \text{ m} - 3 \text{ m} = -7 \text{ m}$$

Tip: When working on physics problems, you can choose to place the origin of your position-measuring system wherever is convenient. The measurement of the position of an object depends on where you choose to place your origin; however, displacement from an initial position s_i to a final position s_f doesn't depend on the position of the origin because the displacement depends only on the *difference* between the positions, not the positions themselves.

Example

Q. You've taken the pioneers' advice to "Go West." You started in New York City and went west 10 miles the first day, 14 miles west the next day, and then back east 9 miles on the third day. What is your displacement from New York City after three days?

A. The correct answer is s = 15 miles west of New York City.

   1. You first went west 10 miles, so at the end of the first day, your displacement was 10 miles west.

   2. Next, you went west 14 miles, putting your displacement at 10 miles + 14 miles = 24 miles west of New York City.

   3. Finally, you traveled 9 miles east, leaving you at 24 miles – 9 miles = 15 miles west of New York City. So s = 15 miles west of New York City.

Practice Questions

   1. Suppose that the golf ball in Figure 3-1b now moves 1 more meter to the right. What is its new displacement from the origin, 0?

   2. Suppose that the golf ball in Figure 3-1b, which started 3 meters to the right of the origin, moves 5 meters to the left. What is its new position relative to the origin — in inches?

Practice Answers

   1. s = **+4 m.** The ball was originally at +3 meters and moved 1 meter to the right. So +3 + 1 = +4 meters.

   2. s = **–80 in.** The ball started at 3 meters and moved 5 meters to the left, putting it at +3 – 5 = –2 meters with respect to the origin. Then convert –2 meters into inches:

$$-2 \text{ m} \times \frac{39.37 \text{ in}}{1 \text{ m}} = -80 \text{ in}$$

Examining axes

Motion that takes place in the world isn't always in one dimension. Motion can take place in two or three dimensions. And if you want to examine motion in two dimensions, you need two intersecting meter sticks (or number lines), called *axes*. You have a horizontal axis — the *x*-axis — and a vertical axis — the *y*-axis. (For three-dimensional problems, watch for a third axis — the *z*-axis — sticking straight up out of the paper.)

Finding the distance

Take a look at Figure 3-2, where a golf ball moves around in two dimensions. The ball starts at the center of the graph and moves up to the right. In terms of the axes, the golf ball moves to +4 meters on the *x*-axis and +3 meters on the *y*-axis, which is represented as the point (4, 3); the *x* measurement comes first, followed by the *y* measurement: (*x*, *y*).

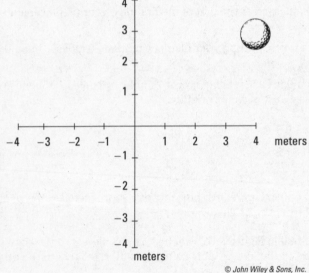

Figure 3-2:
A ball moving in two dimensions.

© John Wiley & Sons, Inc.

So what does this mean in terms of displacement? The change in the *x* position, Δx (Δ, the Greek letter delta, means "change in"), is equal to the final *x* position minus the initial *x* position. If the golf ball starts at the center of the graph — the origin of the graph, location (0, 0) — you have a change in the *x* location of

$$\Delta x = x_f - x_i$$
$$= 4 \text{ m} - 0 \text{ m} = 4 \text{ m}$$

The change in the *y* location is

$$\Delta y = y_f - y_i$$
$$= 3 \text{ m} - 0 \text{ m} = 3 \text{ m}$$

If you're more interested in figuring out the magnitude (size) of the displacement than in the changes in the *x* and *y* locations of the golf ball, that's a different story. The question now becomes: How far is the golf ball from its starting point at the center of the graph?

Remember: Using the *distance formula* — which is just the Pythagorean theorem solved for the hypotenuse — you can find the *magnitude of the displacement* of the golf ball, which is the distance it travels from start to finish. The Pythagorean theorem states that the sum of the squares of the legs of a right triangle $\left(a^2 + b^2\right)$ is equal to the square on the hypotenuse $\left(c^2\right)$. Here, the legs of the triangle are Δx and Δy, and the hypotenuse is s. Here's how to work the equation:

$$s = \sqrt{\Delta x^2 + \Delta y^2}$$
$$= \sqrt{\left(4 \text{ m}\right)^2 + \left(3 \text{ m}\right)^2}$$
$$= \sqrt{16 \text{ m}^2 + 9 \text{ m}^2}$$
$$= \sqrt{25 \text{ m}^2}$$
$$= 5 \text{ m}$$

So in this case, the magnitude of the ball's displacement is exactly 5 meters.

Determining direction

You can find the direction of an object's movement from the values of Δx and Δy. Because these are just the legs of a right triangle, you can use basic trigonometry to find the angle of the ball's displacement relative to the *x*-axis. The tangent of this angle is simply given by

$$\tan\theta = \frac{\Delta y}{\Delta x}$$

Therefore, the angle itself is just the inverse tangent of that:

$$\theta = \tan^{-1}\left(\frac{\Delta y}{\Delta x}\right)$$
$$= \tan^{-1}\left(\frac{3 \text{ m}}{4 \text{ m}}\right)$$
$$\approx 37°$$

The ball in Figure 3-2 has moved at an angle of 37° relative to the *x*-axis.

SPEED SPECIFICS: WHAT IS SPEED, ANYWAY?

There's more to the story of motion than just the actual movement. When displacement takes place, it happens in a certain amount of time. You may already know that speed is distance traveled per a certain amount of time:

$$\text{speed} = \frac{\text{distance}}{\text{time}}$$

For example, if you travel distance s in a time t, your speed, v, is

$$v = \frac{s}{t}$$

Remember: The variable v really stands for speed; true velocity also has a direction associated with it, whereas *speed* does not. For that reason, velocity is a vector (you usually see the velocity vector represented as **v** or $\vec{v}$). *Vectors* have both a magnitude (size) and a direction, so with velocity, you know not only how fast you're going but also in what direction. Speed is only a magnitude (if you have a certain velocity vector, in fact, the speed is the magnitude of that vector), so you see it represented by the term v (not in bold). You can read more about velocity and displacement as vectors in Chapter 4.

Remember: Just as you can measure displacement, you can measure the difference in time from the beginning to the end of the motion, and you usually see it written like this: $\Delta t = t_f - t_i$. Technically speaking (physicists love to speak technically), velocity is the change in position (displacement) divided by the change in time, so you can also represent it like this, if, say, you're moving along the x-axis:

$$v = \frac{\Delta x}{\Delta t} = \frac{x_f - x_i}{t_f - t_i}$$

Speed can take many forms, which you find out about in the following sections.

Reading the speedometer: Instantaneous speed

You already have an idea of what speed is; it's what you measure on your car's speedometer, right? When you're tooling along, all you have to do to see your speed is look down at the speedometer. There you have it: 75 miles per hour. Hmm, better slow it down a little — 65 miles per hour now. You're looking at your speed at this particular moment. In other words, you see your *instantaneous speed*.

Remember: Instantaneous speed is an important term in understanding the physics of speed, so keep it in mind. If you're going 65 miles per hour right now, that's your instantaneous speed. If you accelerate to 75 miles per hour, that becomes your instantaneous speed. Instantaneous speed is your speed at a particular instant of time. Two seconds from now, your instantaneous speed may be totally different.

Staying steady: Uniform speed

What if you keep driving 65 miles per hour forever? You achieve *uniform speed* in physics (also called *constant speed*). Uniform motion is the simplest speed variation to describe, because it never changes.

Uniform speed may be possible in the western portion of the United States, where the roads stay in straight lines for a long time and you don't have to change your speed. But uniform speed is also possible when you drive around a circle, too. Imagine driving around a racetrack; your velocity would change (because of the constantly changing direction), but your speed could remain constant as long as you keep your gas pedal pressed down the same amount. We discuss uniform circular motion in Chapter 7, but in this chapter, we stick to motion in straight lines.

Shifting speeds: Nonuniform motion

Nonuniform motion varies over time; it's the kind of speed you encounter more often in the real world. When you're driving, for example, you change speed often, and your changes in speed come to life in an equation like this, where v_f is your final speed and v_i is your initial speed:

$$\Delta v = v_f - v_i$$

The last part of this chapter is all about acceleration, which occurs in nonuniform motion. There, you see how changing speed is related to acceleration — and how you can accelerate even without changing speed!

Busting out the stopwatch: Average speed

Average speed is the total distance you travel divided by the total time it takes. Average speed is sometimes written as $\bar{v}$; a bar over a variable means *average* in physics terms.

Remember: Average speed differs from instantaneous speed, unless you're traveling in uniform motion (in which case your speed never varies). In fact, because average speed is the total distance divided by the total time, it may be very different from your instantaneous speed.

As an illustration of the difference between average speed and average velocity, consider the motion of the Earth around the sun. The Earth travels in its nearly circular orbit around the sun at an enormous average speed of something like 18 miles per second! However, if you consider one full revolution of the Earth, the Earth returns to its original position, relative to the sun, after one year. After one year, there's no displacement relative to the sun, so the Earth's average velocity over a year is zero, even though its average speed is enormous!

Remember: When considering motion, it's not only speed that counts but also direction. That's why velocity is important: It lets you record an object's speed and its direction. Pairing speed with direction enables you to handle cases like cross-country travel, where the direction can change.

You can relate total distance traveled, s, with average speed, $\bar{v}$, and time, t, like this:

$$s = \bar{v}t$$

Examples

Q. Suppose that you want to drive from New York City to Los Angeles to visit your uncle's family, a distance of about 2,781 miles. The trip takes you 4.0 days. What was your speed in miles per hour?

A. The correct answer is 29 miles per hour.

   1. Start by figuring out your speed (the distance traveled divided by the time taken to travel that distance):

$$\frac{2{,}781 \text{ mi}}{4 \text{ days}} = 695.25$$

   2. Okay, the speed is 695.25, but 695.25 *what?* This solution divides miles by days, so it's 695.25 miles per day — not exactly a standard unit of measurement. So what is that in miles per hour? To determine that, you cancel "days" out of this equation and put in "hours." Because 24 hours are in a day, you can multiply as follows (note that "days" cancel out, leaving miles over hours, or miles per hour):

$$\frac{695.25 \text{ mi}}{1 \text{ \cancel{day}}} \times \frac{1 \text{ \cancel{days}}}{24 \text{ hr}} = 28.97 \text{ mph}$$

So your speed was 28.97 miles per hour. Because 4.0 days only has two significant digits, you can round your answer to 29 miles per hour (see Chapter 2 for details about significant digits). That's your average speed, averaged over both day and night.

That answer seems pretty slow, because when you're driving, you're used to going 65 miles per hour. You've calculated an average speed over the whole trip, obtained by dividing the total distance by the total trip time, which includes non-driving time. You may have stopped at a hotel several nights, and while you slept, your instantaneous speed was 0 miles per hour; yet even at that moment, your overall average speed was still 29 miles per hour!

Q. Say that while you were driving in Ohio on your cross-country trip, you wanted to make a detour to visit your sister in Michigan after you dropped off a hitchhiker in Indiana. Your travel path looked like the straight lines in the following figure — first 80 miles to Indiana and then 30 miles to Michigan. If you drove at an average speed or a uniform speed of 55 miles per hour, what was the magnitude of your average velocity?

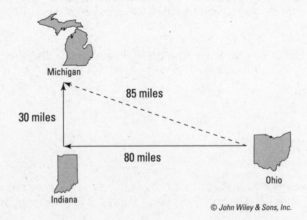

A. The correct answer is 43 miles per hour.

   1. You had to cover $80 + 30 = 110$ miles, so this trip took you 2.0 hours.

   2. The distance r between Ohio and Michigan as the crow flies is

$$r = \sqrt{(80 \text{ mi})^2 + (30 \text{ mi})^2}$$
$$\approx 85 \text{ mi}$$

   3. To find the magnitude of the average velocity, you divide the distance by the time and get

$$\frac{85 \text{ mi}}{2.0 \text{ h}} = 42.5 \text{ mph} \approx 43 \text{ mph}$$

Practice Questions

   1. Suppose that you used your new SpeedPass to get you through the tollbooths at both ends of your trip, which was 90.0 miles on the turnpike and took you 1 hour and 15 minutes. On your return home, you're surprised to find a traffic ticket for speeding in the mail. How fast did you go, on average, between the tollbooths? Was the turnpike authority justified in sending you a ticket, given that the speed limit was 65 miles per hour?

   2. Suppose that you and a friend are determined to find out who drives faster. You both start your trips in Chicago. Driving nonstop, you reach Los Angeles — a distance of 2,018 miles — in 1.290 days, and your friend, also driving nonstop, reaches Miami — a distance of 1,380.0 miles — in 0.8900 days. Who drove faster?

   3. Suppose that you're walking to school when you realize that you've left your lunch at home. You run 150 meters back home in 23 seconds. If it took you 61 seconds to walk the 150 meters, what was your average velocity for the 300 meters you traveled?

   4. Suppose that you're walking to school when you realize that you've left your lunch at home. You run 150 meters back home in 23 seconds. If it took you 61 seconds to walk the 150 meters, what was your average speed for the 300 meters you traveled?

Practice Answers

   1. v = **72 mph. The ticket was justified.** It took you 1 hour and 15 minutes, or 1.25 hours, to travel 90.0 miles. Divide 90.0 miles by 1.25 hours:

$$\frac{90.0 \text{ mi}}{1.25 \text{ hr}} = 72.0 \text{ mph}$$

This is faster than the speed limit of 65 miles per hour.

2. **Your speed was 1,564 miles per day; your friend's speed was 1,551 miles per day. You're faster.** Note that to simply compare speeds, you don't need to convert to miles per hour — miles per day will do fine. First, calculate your speed:

$$\frac{2{,}018 \text{ mi}}{1.29 \text{ days}} = 1{,}564 \text{ mi/day}$$

Next, calculate your friend's speed:

$$\frac{1{,}380 \text{ mi}}{0.89 \text{ days}} = 1{,}551 \text{ mi/day}$$

So you were faster than your friend.

3. **0 m/s.** Your displacement was 0 meters because you ended up where you started, so your average velocity was 0 meters per second.

4. **3.6 m/s.** The total time it took you to travel 300 meters was 23 seconds + 61 seconds = 84 seconds. Your average speed was the total distance traveled divided by the total time, or

$$\frac{300 \text{ m}}{84 \text{ s}} \approx 3.6 \text{ m/s}$$

SPEEDING UP (OR DOWN): ACCELERATION

Acceleration is a measure of how quickly your velocity changes. When you pass a parking lot's exit and hear squealing tires, you know what's coming next — someone is accelerating to cut you off. After he passes, he slows down right in front of you, forcing you to hit your brakes to slow down yourself. Good thing you know all about physics.

Remember: You may think that, with all this speeding up and slowing down, you'd use terms like *acceleration* and *deceleration*. Well, physics has no use for the term *deceleration*, because deceleration is just a particular kind of acceleration — one in which speed reduces.

Like speed, acceleration takes many forms that affect your calculations in various physics situations. In different physics problems, you have to take into account the direction of the acceleration (whether the acceleration is positive or negative in a particular direction), whether it's average or instantaneous, and whether it's uniform or nonuniform. This section tells you more about acceleration and explores its various forms.

Defining acceleration

In physics terms, *acceleration, a,* is the amount by which your velocity changes in a given amount of time, or

$$a = \frac{\Delta v}{\Delta t}$$

Given the initial and final velocities, v_i and v_f, and the initial and final times over which your speed changes, t_i and t_f, you can also write the equation like this:

$$a = \frac{\Delta v}{\Delta t} = \frac{v_f - v_i}{t_f - t_i}$$

Acceleration, like velocity, is actually a vector and is often written as **a,** in vector style (see Chapter 4). In other words, acceleration, like velocity but unlike speed, has a direction associated with it.

Determining the units of acceleration

You can calculate the units of acceleration easily enough by dividing velocity by time to get acceleration:

$$a = \frac{v_f - v_i}{t_f - t_i}$$

In terms of units, the equation looks like this:

$$a = \frac{v_f - v_i}{t_f - t_i} = \frac{\text{distance}/\text{time}}{\text{time}} = \text{distance}/\text{time}^2$$

Distance per time squared? Don't let that throw you. You end up with time squared in the denominator because you divide velocity by time. In other words, *acceleration* is the rate at which your velocity changes, because rates have time in the denominator. For acceleration, you see units of meters per second^2, centimeters per second^2, miles per second^2, feet per second^2, or even kilometers per hour^2.

Tip: It may be easier, for a given problem, to use units such as mph/s (miles per hour per second). This would be useful if the velocity in question had a magnitude of something like several miles per hour that changed typically over a number of seconds.

Looking at positive and negative acceleration

Just as for displacement and velocity, acceleration can be positive or negative. This section explains how positive and negative acceleration relate to changes in speed and direction.

Changing speed

The sign of the acceleration tells you whether you're speeding up or slowing down (depending on which direction you're traveling).

For example, say that you're driving at 75 miles per hour, and you see those flashing red lights in the rearview mirror. You pull over, taking 20 seconds to come to a stop. The officer appears by your window and says, "You were going 75 miles per hour in a 30-mile-per-hour zone." What can you say in reply?

You can calculate your rate of acceleration as you pulled over, which, no doubt, would impress the officer — look at you and your law-abiding tendencies! You whip out your calculator and begin entering your data. Remember that the acceleration is given in terms of the change in velocity divided by the change in time:

$$a = \frac{\Delta v}{\Delta t}$$

Plugging in the numbers, your calculations look like this:

$$a = \frac{\Delta v}{\Delta t}$$
$$= \frac{75 \text{ mph}}{20 \text{ s}}$$
$$\approx 3.8 \text{ mph/s}$$

Your acceleration was 3.8 miles per hour per second. But that can't be right! You may already see the problem here; take a look at the original definition of acceleration:

$$a = \frac{\Delta v}{\Delta t} = \frac{v_f - v_i}{t_f - t_i}$$

Your final speed was 0 miles per hour, and your original speed was 75 miles per hour, so plugging in the numbers here gives you this acceleration:

$$a = \frac{\Delta v}{\Delta t} = \frac{v_f - v_i}{t_f - t_i}$$
$$= \frac{0 - 75 \text{ mph}}{20 \text{ s}}$$
$$\approx -3.8 \text{ mph/s}$$

In other words, –3.8 mph/s, *not* +3.8 mph/s — a big difference in terms of solving physics problems (and in terms of law enforcement). If you accelerated at +3.8 miles per hour per second rather than –3.8 miles per hour per second, you'd end up going 150 miles per hour at the end of 20 seconds, not 0 miles per hour. And that probably wouldn't make the cop very happy.

Now you have your acceleration. You can turn off your calculator and smile, saying, "Maybe I was going a little fast, officer, but I'm very law abiding. Why, when I heard your siren, I accelerated at –3.8 miles per hour per second just to pull over promptly." The policeman pulls out his calculator and does some quick calculations. "Not bad," he says, impressed. And you know you're off the hook.

Accounting for direction

The sign of the acceleration depends on direction. If you slow down to a complete stop in a car, for example, and your original velocity was positive and your final velocity was 0, then your acceleration is negative because a positive velocity came down to 0. However, if you slow down to a complete stop in a car and your original velocity was negative and your final velocity was 0, then your acceleration would be positive because a negative velocity increased to 0.

Looking at positive and negative acceleration

When you hear that acceleration is going on in an everyday setting, you typically think that means the speed is increasing. However, in physics, that isn't always the case. An acceleration can cause speed to increase, decrease, and even stay the same!

Acceleration tells you the rate at which the velocity is changing. Because the velocity is a vector, you have to consider the changes to its magnitude and direction. The acceleration can change the magnitude and/or the direction of the velocity. Speed is only the magnitude of the velocity.

Here's a simple example that shows how a simple constant acceleration can cause the speed to increase and decrease in the course of an object's motion. Say you take a ball, throw it straight up in the air, and then catch it again. If you throw the ball upward with a speed of 9.8 meters per second, the velocity has a magnitude of 9.8 meters per second in the upward direction. Now the ball is under the influence of gravity, which, on the surface of the Earth, causes all free-falling objects to undergo a vertical acceleration of –9.8 meters per second^2. This acceleration is negative because its direction is vertically downward.

With this acceleration, what's the velocity of the ball after 1.0 second? Well, you know that

$$a = \frac{v_f - v_i}{t_f - t_i}$$

Rearrange this equation and plug in the numbers, and you find that the final velocity after 1.0 second is 0 meters per second:

$$v_f = v_i + a(t_f - t_i)$$
$$= 9.8 \text{ m/s} + (-9.8 \text{ m/s}^2)(1.0 \text{ s})$$
$$= 0 \text{ m/s}$$

After 1.0 second, the ball has zero velocity because it's reached the top of its trajectory, just at the point where it's about to fall back down again. So the acceleration has actually slowed down the ball because it was going in the direction opposite the velocity.

Now see what happens as the ball falls back down to Earth. The ball has zero velocity, but the acceleration due to gravity accelerates the ball downward at a rate of –9.8 meters per second^2. As the ball falls, it gathers speed before you catch it. What's its final velocity as you catch it, given that its initial velocity at the top of its trajectory is 0?

The time for the ball to fall back down to you is just the same as the time it took to reach the top of its trajectory, which is 1.0 second, so you can find the final velocity for this part of the ball's motion with this calculation:

$$v_f = v_i + a(t_f - t_i)$$
$$= 0 \text{ m/s} + (-9.8 \text{ m/s}^2)(1.0 \text{ s})$$
$$= -9.8 \text{ m/s}$$

So the final velocity is 9.8 meters per second directed straight downward. The magnitude of this velocity — that is, the speed of the ball — is 9.8 meters per second. The acceleration increases the speed of the ball as it falls because the acceleration is in the same direction as the velocity for this part of the ball's trajectory.

Remember: When you work with physics problems, bear in mind that acceleration can speed up or slow down an object, depending on the direction of the acceleration and the velocity of the object. Don't simply assume that just because something is accelerating its speed must be increasing. (By the way, if you want to see an example of how an acceleration can leave the speed of an object unchanged, take a look at the circular motion topic covered in Chapter 7.)

Examining average and instantaneous acceleration

Just as you can examine average and instantaneous speeds and velocities, you can also examine average and instantaneous acceleration. *Average acceleration* is the ratio of the change in velocity to the change in time. You calculate average acceleration, also written as $\bar{a}$, by taking the final velocity, subtracting the initial velocity, and dividing the result by the total time (final time minus the initial time):

$$\bar{a} = \frac{v_f - v_i}{t_f - t_i}$$

Remember: At any given point, the acceleration you measure is the instantaneous acceleration, and that number can be different from the average acceleration. For example, when you first see red flashing police lights behind you, you may jam on the brakes, which gives you a big acceleration, in the direction opposite to which you're moving (in everyday parlance, you'd say you just *decelerated,* but that term's a no-no in physics circles). But you lighten up a little and coast to a stop, so the acceleration is smaller. The average acceleration, however, is a single value, derived by dividing the overall change in velocity by the overall time.

Remember: Acceleration is the rate of change of velocity, not speed. If a velocity's direction changes without a change in speed, this is also a kind of acceleration.

Example

Q.   As they strap you into the jet on the aircraft carrier deck, the mechanic says you need to take off at a speed of at least 62.0 meters per second. You'll be catapulted at an acceleration of 31 meters per second². Is there going to be enough catapult to do the job? You ask how long the catapult is. "A hundred meters," says the mechanic, finishing strapping you in.

"Hmm," you think. "Will an acceleration of 31 meters per second² over a distance of 100 meters do the trick?"

A.   Yes — you will reach a speed of 62 meters per second after 62 meters.

1.   First, think of the distance that you need to be accelerated over as the size of the displacement from your initial position. To find this displacement, you can use the equation $s = \bar{v}t$, where s is the displacement, $\bar{v}$ is the average velocity, and t is the time — which means you have to find the time over which you're accelerated.

2. Use the equation that relates change in velocity, Δv, acceleration a, and change in time, Δt:

$$a = \frac{\Delta v}{\Delta t}$$

3. Solve for Δt:

$$\Delta t = \frac{\Delta v}{a}$$

4. Plug in the numbers:

$$\Delta t = \frac{\Delta v}{a}$$
$$= \frac{62 \text{ m/s}}{31 \text{ m/s}^2}$$
$$= 2.0 \text{ s}$$

5. Use $s = \bar{v}t$ to find the total distance you need to travel to get up to this speed, where $\bar{v} = (1/2)(v_i + v_f)t$, $v_i = 0$ m/s, and $v_f = 62$ m/s. So your equation is

$$s = \frac{1}{2}(v_i + v_f)t$$

6. Plug in the numbers to get

$$s = \frac{1}{2}(v_i + v_f)t$$
$$= \frac{1}{2}(0 \text{ m/s} + 62 \text{ m/s})(2.0 \text{ s})$$
$$= 62 \text{ m}$$

So it will take 62 meters of 31 meters per second² acceleration to get you to takeoff speed — and the catapult is 100 meters long. No problem.

Practice Questions

1. A rocket ship is going to land on the moon in exactly 2.0 hours. There's only one problem: It's going 17,000 miles an hour. What does its deceleration need to be, in miles per second², to land on the moon safely at 0.0 miles per hour?

2. You're stopped at a red light when you see a monster SUV careening toward you. In a lightning calculation, you determine you have 0.8 seconds before it hits you and you must be going at least 1.0 miles an hour forward at that time to avoid the SUV. What must your acceleration be, in miles per hour²? Can you avoid the SUV?

3. A bullet comes to rest in a block of wood in 1.0×10^{-2} seconds, with an acceleration of -8.0×10^4 meters per second². What was its original speed, in meters per second?

4. The light turns red, and you ease to a halt. Checking your stopwatch, you see that you stopped in 4.5 seconds. Your deceleration was 1.23×10^{-3} miles per second². What was your original speed in miles per hour?

Practice Answers

1. **6.6×10^{-4} mi/s^2.** Start by converting 17,000 miles per hour into miles per second:

$$\frac{17,000 \text{ mi}}{\text{hr}} \times \frac{1 \text{ hr}}{60 \text{ min}} \times \frac{1 \text{ min}}{60 \text{ s}} = 4.72 \text{ mi/s}$$

   To land on the moon, v_f must be 0 miles per second, and $t_f - t_i = 2.0$ hours, or $2.0 \times 3,600$ seconds = 7,200 seconds, so

$$a = \frac{\Delta v}{\Delta t} = \frac{v_f - v_i}{t_f - t_i} = \frac{0 - 4.72 \text{ mi/s}}{7,200 \text{ s}}$$

   Calculating this yields

$$\frac{-4.72 \text{ mi/s}}{7,200 \text{ s}} = 0.00066 \text{ mi/s}^2 = -6.6 \times 10^{-4} \text{ mi/s}^2$$

   So the rocket needs a constant acceleration of 6.6×10^{-4} miles per second^2 in a direction opposite to its velocity to land on the moon at a speed of 0 miles per second, touching down lightly.

2. **4.5×10^3 mi/hr^2. You will avoid the collision.** Start by converting 0.8 seconds into hours to get all the quantities in units you want — miles and hours:

$$0.8 \text{ s} \times \frac{1 \text{ min}}{60 \text{ s}} \times \frac{1 \text{ hr}}{60 \text{ min}} = 2.22 \times 10^{-4} \text{ hr}$$

   Calculate the acceleration needed to get you to 1.0 miles per hour:

$$a = \frac{\Delta v}{\Delta t} = \frac{v_f - v_i}{t_f - t_i} = \frac{1.0 \text{ mi/hr} - 0}{2.22 \times 10^{-4} \text{ hr} - 0} = 4.5 \times 10^3 \text{ mi/hr}^2$$

3. **800 m/s.** You can calculate the change in speed, because it's acceleration multiplied by time:

$$\Delta v = a\,(\Delta t) = (-8.0 \times 10^4 \text{ m/s}^2)(1.0 \times 10^{-2} \text{ s}) = -8.0 \times 10^2 \text{ m/s}$$

   So if the bullet lost 800 meters per second of speed to come to a rest ($v = 0$), it must have been going 800 meters per second originally.

4. **20 mph.** The change in speed is acceleration multiplied by time, so

$$\Delta v = a(\Delta t) = \left(1.23 \times 10^{-3} \text{ mi/s}^2\right)(4.5 \text{ s}) = 5.55 \times 10^{-3} \text{ mi/s}$$

   Convert this result to miles per hour:

$$\frac{5.55 \times 10^{-3} \text{ mi}}{\text{s}} \times \frac{60 \text{ s}}{1 \text{ min}} \times \frac{60 \text{ min}}{1 \text{ hr}} = 20 \text{ mph}$$

Understanding uniform and nonuniform acceleration

Acceleration can be uniform or nonuniform. Nonuniform acceleration requires a change in acceleration. For example, when you're driving, you encounter stop signs or stop lights often, and when you slow to a stop and then speed up again, you take part in nonuniform acceleration.

Other accelerations are very uniform (in other words, unchanging), such as the acceleration due to gravity near the surface of the Earth. This acceleration is 9.8 meters per second^2 downward, toward the center of the Earth, and it doesn't change (if it did, plenty of people would be pretty startled).

RELATING ACCELERATION, TIME, AND DISPLACEMENT

This section deals with four quantities of motion: acceleration, velocity, time, and displacement. You work the standard equation relating displacement and time to get velocity:

$$v = \frac{\Delta s}{\Delta t} = \frac{s_f - s_i}{t_f - t_i}$$

And you see the standard equation relating velocity and time to get acceleration:

$$a = \frac{\Delta v}{\Delta t} = \frac{v_f - v_i}{t_f - t_i}$$

But both of these equations only go one level deep, relating velocity to displacement and time and acceleration to velocity and time. What if you want to relate acceleration to displacement and time? This section shows you how you can cut velocity out of the equation.

Tip: When you're slinging around algebra, you may find it easier to write single quantities like v (to stand for Δv) rather than $v_f - v_i$. You can usually turn v into $v_f - v_i$ later if necessary.

Not-so-distant relations: Deriving the formula

You relate acceleration, displacement, and time by messing around with the equations until you get what you want. First, note that displacement equals average velocity multiplied by time:

$$s = \bar{v}t$$

You have a starting point. But what's the average velocity? If your acceleration is constant, your velocity increases in a straight line from 0 to its final value, as Figure 3-3 shows.

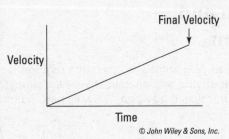

Figure 3-3: Increasing velocity under constant acceleration.

The average velocity is half the final velocity, and you know this because there's constant acceleration. Your final velocity is $v_f = at$, so your average velocity is half of this:

$$\bar{v} = \frac{1}{2}at$$

So far, so good. Now you can plug this average velocity into the $s = \bar{v}t$ equation and get

$$s = \bar{v}t = \left(\frac{1}{2}at\right)t$$

And this becomes

$$s = \frac{1}{2}at^2$$

You can also put in $t_f - t_i$ rather than just plain t:

$$s = \frac{1}{2}a(t_f - t_i)^2$$

Congrats! You've worked out one of the most important equations you need to know when you work with physics problems relating acceleration, displacement, time, and velocity.

Notice that when you derived this equation, you had an initial velocity of 0. What if you don't start off at 0 velocity, but you still want to relate acceleration, time, and displacement? What if you're initially going 100 miles per hour? That initial velocity would certainly add to the final distance you go. Because distance equals speed multiplied by time, the equation looks like this (don't forget that this assumes the acceleration is constant):

$$s = v_i(t_f - t_i) + \frac{1}{2}a(t_f - t_i)^2$$

Remember: You also see this written simply as the following (where t stands for Δt, the time over which the acceleration happened):

$$s = v_i t + \frac{1}{2}at^2$$

Calculating acceleration and distance

With the formula relating distance, acceleration, and time, you can find any of those values, given the other two. If you have an initial velocity, too, finding distance or acceleration isn't any harder. In this section, we work through some physics problems to show you how these formulas work, and then you can work through some practice questions to check your understanding.

Examples

Q. You climb into your drag racer, waving nonchalantly at the cheering crowd. You look down the quarter-mile track, and suddenly the flag goes down. You're off, getting a tremendous kick from behind as the car accelerates quickly. A brief 5.5 seconds later, you pass the end of the course and pop the chute.

You know the distance you went: 0.25 miles, or about 402 meters. And you know the time it took: 5.5 seconds. So just how hard was the kick you got — the acceleration — when you blasted down the track?

A. The correct answer is 27 meters per second^2.

1. You know

$$s = \frac{1}{2}at^2$$

You can rearrange this equation with a little algebra (just divide both sides by t^2 and multiply by 2) to get

$$a = \frac{2s}{t^2}$$

2. Plugging in the numbers, you get

$$a = \frac{2s}{t^2} = \frac{2(402 \text{ m})}{(5.5 \text{ s})^2} = 26.6 \text{ m/s}^2$$

What's 27 meters per second^2 in more understandable terms? The acceleration due to gravity, g, is 9.8 meters per second^2, so this is about $2.7g$. And that's quite a kick.

Q. There you are, the Tour de France hero, ready to give a demonstration of your bicycling skills. There will be a time trial of 8.0 seconds. Your initial speed is 6.0 meters per second, and when the whistle blows, you accelerate at 2.0 meters per second^2 for the 8.0 seconds allowed. At the end of the time trial, how far will you have traveled?

A. The correct answer is 110 meters.

1. Because you start off with an initial speed, you should use the following:

$$s = v_i t + \frac{1}{2}at^2$$

2. In this case, $a = 2.0$ m/s^2, $t = 8.0$ s, and $v_i = 6.0$ m/s, so when you plug in the numbers you get:

$$s = v_i t + \frac{1}{2} at^2$$

$$= (6.0 \text{ m/s})(8.0 \text{ s}) + \frac{1}{2}(2.0 \text{ m/s}^2)(8.0 \text{ s})^2$$

$$= 48.0 \text{ m} + \frac{1}{2}(128 \text{ m})$$

$$\approx 110 \text{ m}$$

3. You write the answer to two significant digits — 110 meters — because you know the time only to two significant digits (see Chapter 2 for info on rounding). In other words, you ride to victory in about 110 meters in 8.0 seconds. The crowd roars.

Practice Questions

1. The light turns green, and you accelerate at 10.0 meters per second^2. After 5.00 seconds, how far have you traveled?

2. A stone drops under the influence of gravity, 9.8 meters per second^2. How far does it drop in 1.2 seconds?

3. An eagle starts at a speed of 50.0 meters per second and, decelerating at 10.0 meters per second^2, comes to rest on a peak 5.00 seconds later. How far is the peak from the eagle's original position?

4. A trailer breaks loose from its truck on a steep incline. If the truck was moving uphill at 20.0 meters per second when the trailer broke loose and the trailer accelerates down the hill at 10.0 meters per second^2, how far downhill does the trailer go after 10.0 seconds?

5. A block of wood is shooting down a track at 10.0 meters per second and is slowing down because of friction. If it comes to rest in 20.0 seconds, what is its deceleration, in meters per second^2?

6. A minivan puts on the brakes and comes to a stop in 3.0 seconds. If the deceleration is 4.0 meters per second^2, how far does the minivan travel, in meters, before stopping?

Practice Answers

1. **125 m.** You want to relate distance to acceleration and time, so use this equation:

$$s = \frac{1}{2} at^2$$

Plug in the numbers:

$$s = \frac{1}{2} at^2 = \frac{1}{2}(10.0 \text{ m/s}^2)(5.00 \text{ s})^2 = 125 \text{ m}$$

2.   **−7.06 m (that's 7.06 meters downward).** To relate distance to acceleration and time, you use this equation:

$$s = \frac{1}{2}at^2$$

Substitute the numbers:

$$s = \frac{1}{2}at^2 = \frac{1}{2}\left(-9.8 \text{ m/s}^2\right)\left(1.2 \text{ s}\right)^2 = -7.06 \text{ m}$$

3.   **125 m.** To connect distance with speed, acceleration, and time, you use this equation:

$$s = v_i\left(t_f - t_i\right) + \frac{1}{2}a\left(t_f - t_i\right)^2$$

Plug in the numbers:

$$
\begin{aligned}
s &= v_i\left(t_f - t_i\right) + \frac{1}{2}a\left(t_f - t_i\right)^2 \\
&= \left(50.0 \text{ m/s}\right)\left(5.00 \text{ s}\right) - \frac{1}{2}\left(10.0 \text{ m/s}^2\right)\left(5.00 \text{ s}\right)^2 \\
&= 125 \text{ m}
\end{aligned}
$$

4.   **−300 m.** To relate acceleration to speed and time, use this equation:

$$s = v_i\left(t_f - t_i\right) + \frac{1}{2}a\left(t_f - t_i\right)^2$$

Plug in the numbers:

$$
\begin{aligned}
s &= v_i\left(t_f - t_i\right) + \frac{1}{2}a\left(t_f - t_i\right)^2 \\
&= \left(20.0 \text{ m/s}\right)\left(10.0 \text{ s}\right) - \frac{1}{2}\left(10.0 \text{ m/s}^2\right)\left(10.0 \text{ s}\right)^2 \\
&= -300 \text{ m}
\end{aligned}
$$

5.   **−5.00 × 10⁻¹ m/s².** To relate acceleration to speed and time, use this equation:

$$a = \frac{\Delta v}{\Delta t} = \frac{v_f - v_i}{t_f - t_i}$$

Plug in the numbers:

$$
\begin{aligned}
a &= \frac{\Delta v}{\Delta t} = \frac{v_f - v_i}{t_f - t_i} \\
&= \frac{0 - 10.0 \text{ m/s}}{20.0 \text{ s}} \\
&= -0.500 \text{ m/s}^2 = -5.00 \times 10^{-1} \text{ m/s}^2
\end{aligned}
$$

6.   **18 m.** To relate acceleration to velocity and time, use this equation:

$$a = \frac{\Delta v}{\Delta t} = \frac{v_f - v_i}{t_f - t_i}$$

Solve for v_i:

$$v_i = v_f - a(t_f - t_i) = 0 - (-4.0 \text{ m/s}^2)(3.0 \text{ s}) = 12 \text{ m/s}$$

Plug in the numbers:

$$s = v_i(t_f - t_i) + \frac{1}{2}a(t_f - t_i)^2$$

$$= (12 \text{ m/s})(3.0 \text{ s}) - \frac{1}{2}(4.0 \text{ m/s}^2)(3.0 \text{ s})^2$$

$$= 18 \text{ m}$$

LINKING VELOCITY, ACCELERATION, AND DISPLACEMENT

Say you want to relate displacement, acceleration, and velocity without having to know the time. Here's how it works. First, you solve the acceleration formula for the time:

$$t = \frac{v_f - v_i}{a}$$

Because displacement is $s = \bar{v}t$ and average velocity is $\bar{v} = (1/2)(v_i + v_f)$ when the acceleration is constant, you can get the following equation:

$$s = \frac{1}{2}(v_i + v_f)t$$

Substituting for the time, t, you get

$$s = \frac{1}{2}(v_i + v_f)\left(\frac{v_f - v_i}{a}\right)$$

After doing the algebra and simplifying, you get

$$s = \frac{v_f^2 - v_i^2}{2a}$$

Remember: Moving the $2a$ to the other side of the equation, you get an important equation of motion:

$$v_f^2 - v_i^2 = 2as$$

Whew. If you can memorize this one, you can relate velocity, acceleration, and displacement. Put this equation to work — you see it often in physics problems.

Examples

Q.   A drag racer's acceleration is 26.6 meters per second², and at the end of the race, his final speed is 146.3 meters per second. What is the total distance the drag racer traveled?

A.   The correct answer is 402 meters.

1.   To solve this problem, you need to relate speed, acceleration, and distance, so you start with this equation:

$$v_f^2 - v_i^2 = 2as = 2a(x_f - x_i)$$

2.   In this scenario, v_i is 0, which makes this equation simpler:

$$v_f^2 = 2as$$

3.   Solve for s:

$$s = \frac{1}{2a} v_f^2$$

4.   Plug in the numbers:

$$s = \frac{1}{2a} v_f^2 = \frac{1}{2(26.6 \text{ m/s}^2)}(146.3 \text{ m/s})^2 = 402 \text{ m}$$

So the answer is 402 meters, about a quarter of a mile — standard for a drag racing track.

Q.   Say you're in your rocket ship, happily speeding along at some 3.25 kilometers per second (about 7,280 miles per hour) when you see a sign: Speed Zone 215 km Ahead — New Speed Limit: 3.0 km/s.

You jam on the brakes (which are a retro rocket in the front of the rocket ship). The retro rocket is capable of accelerating your ship at –10.0 meters per second².

It's a tense moment. Will you get your speed down to below 3.0 kilometers per second in 215 kilometers of acceleration?

A.   You're safe – your final speed is 2.50 kilometer per second.

1.   Start with your old friend:

$$v_f^2 - v_i^2 = 2as$$

2.   Solve for the final speed:

$$v_f^2 - v_i^2 = 2as$$
$$\sqrt{v_f^2} = \sqrt{2as + v_i^2}$$
$$v_f = \sqrt{2as + v_i^2}$$

3. Plugging in the data gives you the following:

$$v_f = \sqrt{2as + v_i^2}$$
$$= \sqrt{2(-10.0 \text{ m/s}^2)(215,000 \text{ m}) + (3,250 \text{ m/s})^2}$$
$$= \sqrt{-4,300,000 \text{ m}^2/\text{s}^2 + 10,562,500 \text{ m}^2/\text{s}^2}$$
$$= \sqrt{6,262,500 \text{ m}^2/\text{s}^2}$$
$$\approx 2,500 \text{ m/s} = 2.50 \text{ km/s}$$

Whew, you think — 2.50 kilometers per second is well under the speed limit of 3.0 kilometers per second. You're safe.

Practice Questions

1. A bullet is accelerated over a meter-long rifle barrel at an acceleration of 400,000 meters per second². What is its final speed?

2. A dragster starts from rest and is accelerated at 5.0 meters per second². What is its speed 5.0×10^2 meters later?

3. A rocket is launched with an acceleration of 100.0 meters per second². After 100.0 kilometers, what is its speed in meters per second?

4. A motorcycle is going 40.0 meters per second and is accelerated at 6.00 meters per second². What is its speed after 2.00×10^2 meters?

5. A cheetah accelerates at 33 kilometers per hour per second. If it starts from rest, how far has it traveled when it reaches its top speed of 99 kilometers per hour?

6. You're traveling at 20 meters per second in your car when the traffic light turns yellow. If you're 30 meters from the intersection and you can brake at 8.0 meters per second², can you stop in time?

Practice Answers

1. v_f = **900 m/s.** Start with this equation:

$$v_f^2 - v_i^2 = 2as = 2a(x_f - x_i)$$

v_i is 0, so that makes things easier. Plug in the numbers:

$$v_f^2 = 2as$$
$$= 2(400,000 \text{ m/s}^2)1.0 \text{ m}$$
$$= 800,000 (\text{m/s})^2$$

Take the square root:

$$v_f = 894 \text{ m/s, which rounds to } 900 \text{ m/s with significant figures}$$

2. **$v_f = 71$ m/s.** You want to find speed in terms of distance and acceleration, so use this equation:

$$v_f^2 - v_i^2 = 2as = 2a\ (x_f - x_i)$$

Plug in the numbers:

$$v_f^2 = 2as$$
$$= 2\left(5.0 \text{ m/s}^2\right)500 \text{ m}$$
$$= 5{,}000\left(\text{m/s}\right)^2$$

Take the square root:

$$v_f = 70.7 \text{ m/s, which rounds to } 71 \text{ m/s with significant figures}$$

3. **$v_f = 4{,}472$ m/s.** You want to find the speed of the rocket, having been given distance and acceleration, so use this equation:

$$v_f^2 - v_i^2 = 2as = 2a(x_f - x_i)$$

One hundred kilometers is 100,000 meters, so plug in the numbers:

$$v_f^2 = 2as$$
$$= 2\left(100.0 \text{ m/s}^2\right)100{,}000 \text{ m}$$
$$= 2.000 \times 10^7 \left(\text{m/s}\right)^2$$

Take the square root to get the rocket's speed:

$$v_f = 4{,}472 \text{ m/s}$$

4. **$v_f = 63.2$ m/s.** To determine the motorcycle's final speed, use this equation:

$$v_f^2 - v_i^2 = 2as$$

$v_i = 40$ meters per second, so plug in the numbers:

$$v_f^2 - \left(40 \text{ m/s}\right)^2 = 2as$$
$$= 2\left(6.00 \text{ m/s}^2\right)200 \text{ m}$$
$$= 2{,}400\left(\text{m/s}\right)^2$$

That means that v_f^2 is

$$v_f^2 = 4{,}000\left(\text{m/s}\right)^2$$

Take the square root to get v_f:

$$v_f = 63.2 \text{ m/s}$$

5. $s = 41$ m. To determine the distance the cheetah has traveled when it reaches its top speed, use the following equation:

$$v_f^2 - v_i^2 = 2as$$

It starts at rest, so $v_i = 0$. The acceleration is 33 kilometers per hour per second, which is 118,800 kilometers per hour per hour, so plug in the numbers:

$$\left(99 \text{ km/hr}\right)^2 - 0^2 = 2\left(118{,}800 \text{ km/hr/hr}\right)s$$

$$s = \frac{\left(99 \text{ km/hr}\right)^2}{2\left(118{,}800 \text{ km/hr/hr}\right)}$$

$$s = 0.041 \text{ km}$$

That means that s is

$$s = 0.041 \text{ km} = 41 \text{ m}$$

6. **Yes, I can stop in 25 m.** To determine the distance it takes you to stop the car, use the following equation:

$$v_f^2 - v_i^2 = 2as$$

You want to stop, so $v_f = 0$. Plug in the numbers:

$$0^2 - \left(20 \text{ m/s}\right)^2 = 2\left(-8 \text{ m/s}^2\right)s$$

$$s = \frac{-\left(20 \text{ m/s}\right)^2}{2\left(-8 \text{ m/s}^2\right)}$$

$$= 25 \text{ m}$$

Twenty-five meters is shorter than the distance to the intersection, so you can stop in time.

4

FOLLOWING DIRECTIONS: MOTION IN TWO DIMENSIONS

IN THIS CHAPTER

- Mastering vector addition and subtraction
- Putting vectors into numerical coordinates
- Dividing vectors into components
- Identifying displacement, acceleration, and velocity as vectors
- Understanding projectile motion

Y ou aren't limited to moving left and right or forward and backward; you can move in more than one dimension. In the real world, you need to know which way you're going and how far to go. For example, when a person gives you directions, she may point and say something like, "The posse went 15 miles thataway!" When you're helping someone hang a door, the person may say, "Push hard to the left!" And when you swerve to avoid hitting someone in your car, you accelerate in another direction. All these statements involve vectors.

A *vector* is a quantity that has both a size (magnitude) and a direction. Because physics models everyday life, plenty of concepts in physics are vectors, too, including velocity, acceleration, and force. For that reason, you should snuggle up to vectors, because you see them in just about any physics course you take. Vectors are fundamental.

Many people who've had tussles with vectors decide they don't like them, which is a mistake — vectors are easy after you get a handle on them, and you get a handle on them in this chapter. We break down vectors from top to bottom and relate the characteristics of motion (displacement, velocity, and acceleration) to the concept of vectors. Here, balls fly through the air and roll off cliffs, baseball players race to make plays, and you find a great shortcut to the nearest park bench. Read on.

VISUALIZING VECTORS

In one dimension, displacement, velocity, and acceleration are either positive or negative (see Chapter 3). For example, they may be negative if they're to the left and positive to the right. The size of the displacement, velocity, or acceleration is given by the absolute size (regardless of sign) of the number representing it — this is the *magnitude*. The sign of the number indicates direction (left or right).

But what do you do if you have more than one dimension? If the object can move up and down as well as left and right, you can no longer use a single number to represent displacement, velocity, and acceleration. You need vectors. In this section, we represent vectors as arrows and show you what vector addition and subtraction look like.

Asking for directions: Vector basics

Remember: When you have a vector, you have to keep in mind two quantities: its direction and its magnitude. Quantities that have only a magnitude are called *scalars*. If you give a scalar magnitude a direction, you create a vector.

Visually, you see vectors drawn as arrows in physics, which is perfect because an arrow has both a clear direction and a clear magnitude (the length of the arrow). Take a look at Figure 4-1. The arrow represents a vector that starts at the arrow's foot (also called the *tail*) and ends at the head.

Figure 4-1:
A vector, represented by an arrow, has both a direction and a magnitude.

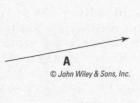

A

© John Wiley & Sons, Inc.

In physics, you use a letter in bold type to represent a vector. This is the notation we use in this book; in some books, however, you see a letter with an arrow on top like this: $\vec{A}$. The arrow means that this is not only a scalar value, which would be represented by A, but also something with direction.

Say that you tell some smarty-pants that you know all about vectors. When he asks you to give him a vector, **A,** you give him not only its magnitude but also its direction, because you need these two bits of info together to define this vector. That impresses him to no end! For example, you may say that **A** is a vector at 15° above the horizontal with a magnitude of 12 meters per second. Smarty-pants knows all he needs to know, including that **A** is a velocity vector.

Take a look at Figure 4-2, which features two vectors, **A** and **B.** They look pretty much the same — the same length and the same direction. In fact, these vectors are equal. Two vectors are *equal* if they have the same magnitude and direction, and you can write this as **A** = **B**.

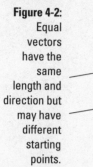
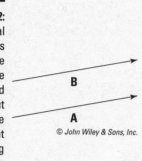

Figure 4-2:
Equal vectors have the same length and direction but may have different starting points.

© John Wiley & Sons, Inc.

Looking at vector addition from start to finish

Remember: Just as you can add two numbers to get a third number, you can add two vectors to get a *resultant* vector. To show that you're adding two vectors, put the arrows together so one arrow starts where the other arrow ends. The sum is a new arrow that starts at the base of the first arrow and ends at the head (pointy end) of the other.

Consider an example using displacement vectors. A *displacement vector* gives the change in position: the distance from the starting point to the ending point is the magnitude of the displacement vector, and the direction traveled is the direction of the displacement vector.

Assume, for example, that a passerby tells you that to get to your destination, you first have to follow vector **A** and then vector **B**. Just where is that destination? You work this problem just as you find the destination in everyday life. First, you drive to the end of vector **A,** and from that point, you drive to the end of vector **B,** just as you see in Figure 4-3.

When you get to the end of vector **B,** how far are you from your starting point? To find out, you draw a vector, **C,** from your starting point (foot, or tail, of the first vector) to your ending point (head of the second vector), as you see in Figure 4-4. This new vector represents your complete trip, from start to finish. In other words, **C = A + B.** The vector **C** is called the *sum,* the *result,* or the *resultant vector.*

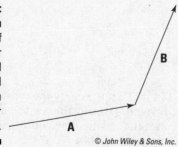

Figure 4-3:
Going from the tail of one vector to the head of a second gets you to your destination.

© John Wiley & Sons, Inc.

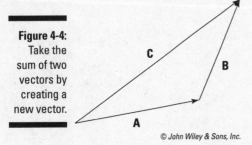

Figure 4-4:
Take the sum of two vectors by creating a new vector.

Going head-to-head with vector subtraction

You don't come across vector subtraction very often in physics problems, but it does pop up. To subtract two vectors, you put their feet (the non-pointy parts) together; then draw the resultant vector, which is the difference of the two vectors, from the head of the vector you're subtracting to the head of the vector you're subtracting it from.

To make heads or tails of this, check out Figure 4-5, where you subtract **A** from **C** (in other words, **C** − **A**). As you can see, the result is **B**, because **C** = **A** + **B**.

Figure 4-5:
Subtracting two vectors by putting their feet together and drawing the result.

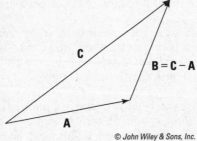

Tip: Another (and for some people, easier) way to do vector subtraction is to reverse the direction of the second vector (**A** in **C** − **A**) and use vector addition; that is, start with the first vector (**C**), put the reversed vector's foot (**A**) at the first vector's head, and draw the resultant vector.

PUTTING VECTORS ON THE GRID

Vectors may look good as arrows floating in space, but that's not exactly the most precise way of dealing with them. You can get numerical with vectors, taking them apart as you need them, by putting the arrows in a grid, on the coordinate plane. The coordinate plane allows you to work with vectors using (x, y) coordinates and algebra.

Adding vectors by adding coordinates

In this section, we explain how you can use the components of vectors to add vectors together. Doing so reduces the problem of adding vectors to a simple combination of adding numbers together, which is very useful when you solve problems.

Take a look at the vector addition problem **A** + **B** in Figure 4-6. Now that you have the vectors plotted on a graph, you can see how easy vector addition really is. If the measurements in Figure 4-6 are in meters, that means vector **A** is 5 meters to the right and 1 meter up, and vector **B** is 1 meter to the right and 4 meters up. To add them for the result, vector **C**, you add the horizontal parts together and the vertical parts together.

Figure 4-6:
Use vector
coordinates
to make
handling
vectors
easy.

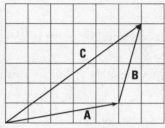

© John Wiley & Sons, Inc.

The resultant vector, **C**, ends up being 6 meters to the right and 5 meters up. You can see what that looks like in Figure 4-6: To get the horizontal part of the sum, you add the horizontal part of **A** (5 meters) to the horizontal part of **B** (1 meter). To get the vertical part of the sum, **C**, you just add the vertical part of **A** (1 meter) to the vertical part of **B** (4 meters).

Tip: If vector addition still seems cloudy, you can use a notation that was invented for vectors to help physicists and *For Dummies* readers keep it straight. Because **A** is 5 meters to the right (the positive *x*-axis direction) and 1 up (the positive *y*-axis direction), you can express it with (*x, y*) coordinates like this:

$$\mathbf{A} = (5, 1) \text{ m}$$

And because **B** is 1 meter to the right and 4 up, you can express it with (*x, y*) coordinates like this:

$$\mathbf{B} = (1, 4) \text{ m}$$

Having a notation is great, because it makes vector addition totally simple. To add two vectors together, you just add their *x* and *y* parts, respectively, to get the *x* and *y* parts of the result:

$$\mathbf{A} + \mathbf{B} = \mathbf{C}$$
$$(5, 1) \text{ m} + (1, 4) \text{ m} = (6, 5) \text{ m}$$

Remember: The whole secret of vector addition is breaking each vector up into its *x* and *y* parts and then adding those separately to get the resultant vector's *x* and *y* parts. Nothing to it. Now you can get as numerical as you like, because you're just adding or subtracting numbers. Getting those *x* and *y* parts can take a little work, but it's a necessary step. And when you have those parts, you're home free.

Example

Q. Assume you're looking for a hotel that's 20 miles due north and then 20 miles due east. What's the vector that points at the hotel from your starting location?

A. The correct answer is (20, 20) miles.

   1. Say that the east direction is along the positive *x*-axis and that north is along the positive *y*-axis.

   2. Write the problem in vector notation like this (east [positive *x*], north [positive *y*]):

   Step 1: (0, 20) mi

   Step 2: (20, 0) mi

   3. Add these two vectors together by adding the coordinates:

$$(0, 20) \text{ mi} + (20, 0) \text{ mi} = (20, 20) \text{ mi}$$

The resultant vector is (20, 20) miles. It points from your starting point directly to the hotel.

Practice Questions

   1. Suppose that you move to the right of the origin by 3.5 meters and then up 5.6 meters. What is your final vector from the origin, in vector component terms?

   2. A marble starts at the origin and moves to the right 5.0 centimeters. What is its new displacement in vector component terms?

Practice Answers

   1. **(3.5, 5.6) m.** You move to the right of the origin by 3.5 meters, leaving you at (3.5, 0) meters. Then you move up (in the positive *y* direction) by 5.6 meters, leaving you at (3.5, 5.6) meters.

   2. **(5.0, 0) cm.** The marble moves to the right — in the positive *x* direction — by 5.0 centimeters, so its final location is (5.0, 0) centimeters.

Changing the length: Multiplying and dividing a vector by a number

You can perform simple vector multiplication by a scalar (number). If you multiply a vector by a scalar, say 5, then the vector gets five times longer. Each of the components will also be multiplied by 5. Division works the same way — just divide each of the components. Here's an example and some practice problems to see how this works.

Example

Q. Say you're driving along at 150 miles per hour eastward on a racetrack and you see a competitor in your rearview mirror. If you double your velocity, what is your final velocity in vector component form?

A. The correct answer is (300, 0) miles per hour. Your original velocity is (150, 0) miles per hour. Multiply each component by 2 to get your final velocity:

$$2(150, 0) \text{ mph} = (300, 0) \text{ mph}$$

Practice Questions

1. What do you get if you divide the vector (3, –2) by 3?

2. What do you get if you multiply the vector (7, 3) by –1?

Practice Answers

1. **(1, –2/3).** Divide the components of the vector by 3:

$$\frac{(3, -2)}{3} = \left(\frac{3}{3}, \frac{-2}{3}\right) = \left(1, \frac{-2}{3}\right)$$

2. **(–7, –3).** Multiply the components of the vector by –1:

$$(7, 3) \times (-1) = (-7, -3)$$

A LITTLE TRIG: BREAKING UP VECTORS INTO COMPONENTS

Physics problems have a way of not telling you what you want to know directly. As the preceding section explains, a vector can be described by its components, which are enough to uniquely specify a vector. Because a vector, by definition, is a quantity that has both magnitude and direction, another way to specify a vector is to use its magnitude and direction directly. If you know one way of describing the vector, you can work out the other.

These are just two different ways of specifying the same thing, and each has its own use in physics problems. Here's why you may work with vector components:

- **When you have vectors in components, they're easy to add and subtract and manipulate generally.** When a problem gives you vectors in terms of their magnitude and direction (which is often the case), you typically need to calculate their components just so you can work through the problem.

- **Being able to treat the horizontal and vertical directions separately is useful because you can often split one difficult problem into two simple problems. Using components also helps when one direction is more important than the other.** For example, a problem may say that a ball is rolling on a table at angle of 15° with a speed of 7.0 meters per second and ask you how long the ball will take to roll off the table's edge if that edge is 1.0 meter away. In that case, you care only about how quickly the ball is moving horizontally, directly toward the table's edge — the speed in the vertical direction doesn't matter.

After you solve a problem, the answer usually needs to be in terms of the magnitude and direction. So after you find your answer in components, you often have to work out the magnitude and direction again.

This section shows you how you can take the magnitude and direction of a vector and work out its components as well as how you can take the components of a vector and work out its magnitude and direction.

Finding vector components

When you break a vector into its parts, those parts are called its *components*. For example, in the vector (4, 1), the x-axis (horizontal) component is 4, and the y-axis (vertical) component is 1. Typically, a physics problem gives you an angle and a magnitude to define a vector; you have to find the components yourself using a little trigonometry.

To see how to convert between the two ways of looking at vectors, take a look at vector **v** in Figure 4-7. The vector can be described as having a magnitude *v at an angle of θ*.

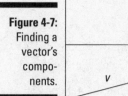

Figure 4-7:
Finding a
vector's
compo-
nents.

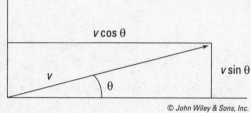

© John Wiley & Sons, Inc.

To convert this vector into the coordinate way of looking at vectors, you have to use the trigonometry shown in the figure. The x coordinate equals $v \cos \theta$, and the y coordinate equals $v \sin \theta$:

$$v_x = v \cos \theta$$
$$v_y = v \sin \theta$$

The two vector-component equations are worth knowing because you see them a lot in any beginning physics course. Make sure you know how they work, and always have them at your fingertips.

Of course, if you forget these equations, you can always retrieve them from basic trigonometry. You may remember that the sine and cosine of an angle in a right triangle are defined as the ratio of the opposite side and the adjacent side to the hypotenuse, like so: $\sin \theta = v_y / v$ and $\cos \theta = v_x / v$ (see Chapter 2). By multiplying both sides of these equations by v, you can express the x and y components of the vector as

$$v_x = v \cos \theta$$
$$v_y = v \sin \theta$$

Example

Q. Suppose that you've walked away from the origin so you're now 5.0 kilometers from the origin, at an angle of 45°. Resolve that into vector coordinates.

A. The correct answer is (3.5, 3.5) kilometers.

1. Apply the equation $v_x = v \cos \theta$ to find the x coordinate. That's 5.0 cos 45°, or 3.5.

2. Apply the equation $v_y = v \sin \theta$ to find the y coordinate. That's 5.0 sin 45°, or 3.5.

Practice Questions

1. Resolve a vector 9.0 meters long at 35° into its components.

2. Resolve a vector 4.0 meters long at 255° into its components.

Practice Answers

1. **(7.4, 5.2) m.** Apply the equation $v_x = v \cos \theta$ to find the x coordinate: $9.0 \cdot \cos 35°$, or 7.4. Then apply the equation $v_y = v \sin \theta$ to find the y coordinate: $9.0 \cdot \sin 35°$, or 5.2.

2. **(–1.0, –3.9) m.** Apply the equation $v_x = v \cos \theta$ to find the x coordinate: $4.0 \cdot \cos 255°$, or –1.0. Then apply the equation $v_y = v \sin \theta$ to find the y coordinate: $4.0 \cdot \sin 255°$, or –3.9.

Reassembling a vector from its components

Sometimes you have to find the angle and magnitude of a vector rather than the components. To find the magnitude, you use the Pythagorean theorem. And to find θ, you use the inverse tangent function (or inverse sine or cosine). This section shows you how these formulas work.

For example, assume you're looking for a hotel that's 20 miles due east and then 20 miles due north. From your present location, what is the angle (measured from east) of the direction to the hotel, and how far away is the hotel? You can write this problem in vector notation, like so (see the section "Putting Vectors on the Grid"):

Step 1: (20, 0) miles

Step 2: (0, 20) miles

When adding these vectors together, you get this result:

$$(20, 0)\,\text{mi} + (0, 20)\,\text{mi} = (20, 20)\,\text{mi}$$

The resultant vector is (20, 20) miles. That's one way of specifying a vector — use its components. But this problem isn't asking for the results in terms of components. The question wants to know the angle and distance to the hotel. In other words, looking at Figure 4-8, the problem asks, "What's h, and what's θ?"

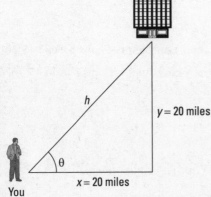

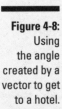

Figure 4-8:
Using the angle created by a vector to get to a hotel.

© John Wiley & Sons, Inc.

Finding the magnitude

If you know a vector's vertical and horizontal components, finding the vector's magnitude isn't so hard, because you just need to find the hypotenuse of a triangle. You can use the Pythagorean theorem $\left(x^2 + y^2 = h^2\right)$ to solve for h:

$$h = \sqrt{x^2 + y^2}$$

Plugging in the numbers gives you

$$\begin{aligned}
h &= \sqrt{x^2 + y^2} \\
&= \sqrt{\left(20 \text{ mi}\right)^2 + \left(20 \text{ mi}\right)^2} \\
&= \sqrt{400 \text{ mi}^2 + 400 \text{ mi}^2} \\
&= \sqrt{800 \text{ mi}^2} \\
&\approx 28 \text{ mi}
\end{aligned}$$

Finding and checking the angle

Remember: When you know the horizontal and vertical components of a vector, you can use the tangent to find the angle because $\tan\theta = y/x$. All you have to do is take the inverse tangent of y/x:

$$\theta = \tan^{-1}\left(\frac{y}{x}\right)$$

Suppose that you drive 20 miles east and 20 miles north. Here's how you find θ, the angle between your original position and your final one:

$$\begin{aligned}
\theta &= \tan^{-1}\left(\frac{y}{x}\right) \\
&= \tan^{-1}\left(\frac{20 \text{ mi}}{20 \text{ mi}}\right) \\
&= \tan^{-1}\left(1\right) \\
&= 45°
\end{aligned}$$

So the hotel is about 28 miles away (as you see from the earlier section "Finding the magnitude") at an angle of 45°.

Warning: Be careful when doing calculations with inverse tangents, because angles that differ by 180° have the same tangent. When you take the inverse tangent, you may need to add or subtract 180° to get the actual angle you want. The inverse tangent button on your calculator will always give you an angle between 90° and –90°. If your angle isn't in this range, you have to add or subtract 180°.

For this example, the answer of 45° must be correct. But consider a situation in which you'd need to add or subtract 180°: Suppose that you walk in completely the opposite direction to the hotel. You walk 20 miles west and 20 miles south ($x = -20$ miles, $y = -20$ miles), so if you use the same method to work out the angle, you get the following:

$$\theta = \tan^{-1}\left(\frac{y}{x}\right)$$

$$= \tan^{-1}\left(\frac{-20 \text{ mi}}{-20 \text{ mi}}\right)$$

$$= \tan^{-1}(1)$$

$$= 45°$$

You get the same answer for the angle even though you're walking in completely the opposite direction as before! That's because the tangents of angles that differ by 180° are equal. But if you look at the components of the vector ($x = -20$ miles, $y = -20$ miles), they're both negative, so the angle must be between −180° and 0°. If you subtract 180° from your answer of 45°, you get −135°, which is your actual angle.

An alternative method of working out the direction is to find the vector's magnitude (hypotenuse) and then use the components in terms of the sine and cosine of the angle:

- $x = h\cos\theta$
- $y = h\sin\theta$

Then you can write the cosine and sine of the angle as

$$\frac{x}{h} = \cos\theta$$

$$\frac{y}{h} = \sin\theta$$

Now all you have to do is take the inverse cosine or sine:

$$\theta = \cos^{-1}\left(\frac{x}{h}\right)$$

$$\theta = \sin^{-1}\left(\frac{y}{h}\right)$$

Here are some practice questions on finding a vector's magnitude and direction.

Example

Q.   Convert the vector given by the coordinates (1.0, 5.0) into magnitude/angle format.

A.   The correct answer is magnitude 5.1, angle 79°.

   1.   Apply the Pythagorean theorem to find the magnitude. Plug in the numbers to get 5.1.

   2.   Apply the equation $\theta = \tan^{-1}(y/x)$ to find the angle. Plug in the numbers to get $\tan^{-1}(5.0/1.0) = 79°$.

Practice Questions

   1.   Convert the vector (–1.0, 1.0) into magnitude/angle form.

   2.   Convert the vector (–5.0, –7.0) into magnitude/angle form.

Practice Answers

   1.   **Magnitude 1.4, angle 135°.** Apply the equation $v = \sqrt{x^2 + y^2}$ to find the magnitude, which is 1.4.

   Apply the equation $\theta = \tan^{-1}(y/x)$ to find the angle: $\tan^{-1}(1.0/-1.0) = -45°$. However, note that the angle must really be between 90° and 180° because the first vector component is negative and the second is positive. That means you should add 180° to –45°, giving you 135° (the tangent of 135° is also 1.0/–1.0 = –1.0).

   2.   **Magnitude 8.6, angle 234°.** Apply the equation $v = \sqrt{x^2 + y^2}$ to find the magnitude, which is 8.6.

   Apply the equation $\theta = \tan^{-1}(y/x)$ to find the angle: $\tan^{-1}(-7.0/-5.0) = 54°$. However, note that the angle must really be between 180° and 270° because both vector components are negative. That means you should add 180° to 54°, giving you 234° (the tangent of 234° is also –7.0/–5.0 = 7.0/5.0).

Adding vectors when you're given magnitude and angle

You're frequently asked to add vectors when solving physics problems. It's easiest to add vectors together in component form, but often you're given the magnitude and angle of the vectors instead. You just have to convert the vectors to component form, add the components, and convert back.

E

Example

Q. Add the two vectors in the following figure. One has a magnitude 5.0 and angle 45°, and the other has a magnitude 7.0 and angle 35°. State your answer in terms of magnitude and direction.

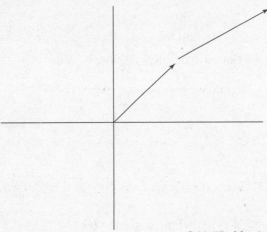

A. The correct answer is magnitude 12.0, angle 39°.

1. Resolve the two vectors into their components. For the first vector, apply the equation $v_x = v \cos \theta$ to find the x coordinate. That's 5.0 cos 45° = 3.5.

2. Apply the equation $v_y = v \sin \theta$ to find the y coordinate of the first vector. That's 5.0 sin 45°, or 3.5. So the first vector is (3.5, 3.5) in coordinate form.

3. For the second vector, apply the equation $v_x = v \cos \theta$ to find the x coordinate. That's 7.0 cos 35° = 5.7.

4. Apply the equation $v_y = v \sin \theta$ to find the y coordinate of the second vector. That's 7.0 sin 35° = 4.0. So the second vector is (5.7, 4.0) in coordinate form.

5. To add the two vectors, add them in coordinate form: (3.5, 3.5) + (5.7, 4.0) = (9.2, 7.5).

6. Convert (9.2, 7.5) into magnitude/angle form. Apply the equation $\theta = \tan^{-1}(y/x)$ to find the angle, which is $\tan^{-1}(7.5/9.2) = \tan^{-1}(0.82) = 39°$.

7. Apply the equation $v = \sqrt{x^2 + y^2}$ to find the magnitude, which is $v = \sqrt{9.2^2 + 7.5^2} = 11.9$. Converting to two significant digits gives you 12.

Q

Practice Questions

1. Add a vector whose magnitude is 13.0 and angle is 27° to one whose magnitude is 11.0 and angle is 45°. State your answer in terms of magnitude and direction.

2. Add a vector whose magnitude is 10.0 and angle is 257° to one whose magnitude is 11.0 and angle is 105°. State your answer in terms of magnitude and direction.

Practice Answers

1. **Magnitude 23.7, angle 35°.** For the first vector, use the equation $v_x = v \cos \theta$ to find the x coordinate: $13.0 \cdot \cos 27° = 11.6$. Use the equation $v_y = v \sin \theta$ to find the y coordinate of the first vector: $13.0 \cdot \sin 27°$, or 5.90. So the first vector is $(11.6, 5.90)$ in coordinate form.

    For the second vector, use the equation $v_x = v \cos \theta$ to find the x coordinate: $11.0 \cdot \cos 45° = 7.78$. Use the equation $v_y = v \sin \theta$ to find the y coordinate of the second vector: $11.0 \cdot \sin 45° = 7.78$. So the second vector is $(7.78, 7.78)$ in coordinate form.

    Add the two vectors in coordinate form: $(11.6, 5.90) + (7.78, 7.78) = (19.4, 13.7)$.

    Convert $(19.4, 13.7)$ into magnitude/angle form. Use the equation $\theta = \tan^{-1}(y/x)$ to find the angle: $\tan^{-1}(13.7/19.4) = \tan^{-1}(0.71) = 35°$.

    Apply the equation $v = \sqrt{x^2 + y^2}$ to find the magnitude, which is $v = \sqrt{19.4^2 + 13.7^2} = 23.7$.

2. **Magnitude 5.2, angle 170°.** For the first vector, use the equation $v_x = v \cos \theta$ to find the x coordinate: $10.0 \cdot \cos 257° = -2.25$. Use the equation $v_y = v \sin \theta$ to find the y coordinate of the first vector: $10.0 \cdot \sin 257°$, or -9.74. So the first vector is $(-2.25, -9.74)$ in coordinate form.

    For the second vector, use the equation $v_x = v \cos \theta$ to find the x coordinate: $11.0 \cdot \cos 105° = -2.85$. Use the equation $v_y = v \sin \theta$ to find the y coordinate of the second vector: $11.0 \cdot \sin 105° = 10.6$. So the second vector is $(-2.85, 10.6)$ in coordinate form.

    Add the two vectors in coordinate form: $(-2.25, -9.74) + (-2.85, 10.6) = (-5.10, 0.86)$.

    Convert the vector $(-5.10, 0.86)$ into magnitude/angle form. Use the equation $\theta = \tan^{-1}(y/x)$ to find the angle: $\tan^{-1}(0.86/-5.10) = \tan^{-1}(-0.17) = 170°$. Because x is negative and y is positive, this vector must be in the second quadrant.

    Apply the equation $v = \sqrt{x^2 + y^2}$ to find the magnitude, which is $v = \sqrt{5.10^2 + 0.86^2} = 5.2$.

FEATURING DISPLACEMENT, VELOCITY, AND ACCELERATION IN 2-D

When an object is moving in only one dimension (as in Chapter 3), you only have to deal with one component, which is just a single number — displacement is just a distance, velocity is just a speed, and acceleration is just speeding up or slowing down. So in one dimension, vectors just look like numbers: The magnitude of the vector is the size of the number, and the direction of the vector is just the sign of the number.

However, displacement, velocity, and acceleration are always vectors. In the real world, an object may be moving in two or more dimensions, so direction is important. In this section, we take another look at the equations for motion, except in more than one dimension so you can see more clearly how the equations are really vector equations.

Displacement: Going the distance in two dimensions

Displacement, which is the change in position (see Chapter 3), has a magnitude and a direction associated with it. When you have a change of position in a particular direction and of a particular distance, then these are given by the magnitude and direction of the displacement vector.

Instead of writing displacement as *s,* you should write it as **s,** a vector (if you're writing on paper, you can put an arrow over the *s* to signify its vector status). When you're talking about displacement in the real world, direction is as important as distance.

For example, say your dreams have come true: You're a big-time baseball or softball hero, slugging another line drive into the outfield. You take off for first base, which is 90 feet away. But 90 feet in which direction? Because you know how vital physics is, you happen to know that first base is 90 feet away at a 45° angle, as you can see in Figure 4-9.

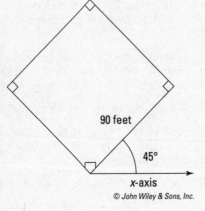

Figure 4-9:
A baseball diamond is a series of vectors created by the *x*-axis and *y*-axis.

© *John Wiley & Sons, Inc.*

Now you're set, all because you know that displacement is a vector. In this case, here's the displacement vector:

$$\mathbf{s} = 90 \text{ feet at } 45°$$

What's that in components?

$$\mathbf{s} = (s\cos\theta, s\sin\theta)$$
$$= (90\cos 45°, 90\sin 45°) \approx (64 \text{ feet}, 64 \text{ feet})$$

Sometimes, working with angles and magnitudes isn't as easy as working with *x* and *y* components. For example, say that you're at the park and ask directions to the nearest bench. The person you ask is very precise and deliberate and answers, "Go north 10.0 meters."

"North 10.0 meters," you say. "Thanks."

"Then east 20.0 meters. Then north another 50.0 meters."

"Hmm," you say. "North 10.0 meters, then 20.0 meters east, and then another 50.0 meters east . . . I mean north. Is that right?"

"Then 60.0 meters east."

You look at the person warily. "Is that it?"

"That's it," she says. "Nearest bench."

Okay, time for some physics. The first step is to translate all that north and east business into x and y coordinates like this: (x, y). So assuming that the positive x-axis points east and the positive y-axis points north (as on a map), the first step is 10.0 meters north, which becomes the following

$$(0, 10.0) \text{ m}$$

That is, the first step is 10.0 meters north, which translates into 10.0 meters in the positive y direction. Adding the second step, 20.0 meters east (the positive x direction), gives you

$$(0, 10.0) \text{ m}$$
$$+(20.0, 0) \text{ m}$$

The third step is 50.0 meters north, and adding that gives you

$$(0, 10.0) \text{ m}$$
$$+(20.0, 0) \text{ m}$$
$$+(0, 50.0) \text{ m}$$

And, finally, the fourth step is 60.0 meters east, which gives you

$$(0, 10.0) \text{ m}$$
$$+(20.0, 0) \text{ m}$$
$$+(0, 50.0) \text{ m}$$
$$+(60.0, 0) \text{ m}$$

Whew. Okay, what's the sum of all these vectors? You just add up the components:

$$(0, 10.0) \text{ m}$$
$$(20.0, 0) \text{ m}$$
$$(0, 50.0) \text{ m}$$
$$+(60.0, 0) \text{ m}$$
$$\overline{(80.0, 60.0) \text{ m}}$$

So the resulting vector is (80.0, 60.0) meters. Hmm, that seems a lot easier than the directions you got. Now you know what to do: Proceed 80.0 meters east and 60.0 meters north. See how easy adding vectors together is?

You can, if you like, go even further. You have the displacement to the nearest bench in terms of x and y components. But it looks like you'll have to walk 80.0 meters east and then 60.0 meters north to find the bench. Wouldn't it be easier if you just knew the direction to the bench and the total distance? Then you could cut the corner and just walk in a straight line directly to the bench.

This is an example where it's good to know how to convert from the (x, y) coordinate form of a vector into the magnitude/angle form. And you can do it with all the physics knowledge you have. Converting (80.0, 60.0) meters to the magnitude/angle form allows you to cut the corner when you walk to the bench, saving a few steps.

You know that the x and y components of a vector form a right triangle and that the total magnitude of the vector is equal to the hypotenuse of the right triangle, h. So the magnitude of h is

$$h = \sqrt{x^2 + y^2}$$

Plugging in the numbers gives you the following:

$$\begin{aligned} h &= \sqrt{(80.0 \text{ m})^2 + (60.0 \text{ m})^2} \\ &= \sqrt{6{,}400 \text{ m}^2 + 3{,}600 \text{ m}^2} \\ &= \sqrt{10{,}000 \text{ m}^2} \\ &= 100 \text{ m} \end{aligned}$$

Voilà! The bench is only 100 meters away. So instead of walking 80.0 meters east and then 60.0 meters north, a total distance of 140 meters, you need to walk only 100 meters. Your superior knowledge of vectors has saved you 40 meters.

But in what direction is the bench? You know it's 100 meters away — but 100 meters which way? You find the angle from the x-axis with this trig:

$$\tan\theta = \frac{y}{x}$$

$$\theta = \tan^{-1}\frac{y}{x}$$

So plugging in the numbers, you have

$$\theta = \tan^{-1}\left(\frac{60.0 \text{ m}}{80.0 \text{ m}}\right)$$

Therefore, the angle θ is the following (using the handy $\tan^{-1}$ button on your calculator):

$$\theta \approx 36.9°$$

And there you have it — the nearest bench is 100 meters away at 36.9° from the x-axis. You start off confidently in a straight line at 36.9° from the east, surprising the person who gave you directions, who was expecting you to take off in the goofy zigzag path she'd given you.

Velocity: Speeding in a new direction

Velocity, which is the rate of change of position (or speed in a particular direction), is a vector. Imagine that you just hit a ground ball on the baseball diamond and you're running along the first-base line, or the **s** vector, 90 feet at a 45° angle to the positive *x*-axis. But as you run, it occurs to you to ask, "Will my velocity enable me to evade the first baseman?" A good question, because the ball is on its way from the shortstop. Whipping out your calculator, you figure that you need 3.0 seconds to reach first base from home plate; so what's your velocity? To find your velocity, you quickly divide the **s** vector by the time it takes to reach first base:

$$\frac{\mathbf{s}}{3.0 \text{ s}}$$

This expression represents a displacement vector divided by a time, and time is just a scalar. The result must be a vector, too. And it is velocity, or **v**:

$$\frac{\mathbf{s}}{3.0 \text{ s}} = \frac{90 \text{ ft at } 45°}{3.0 \text{ s}} = 30 \text{ ft/s at } 45° = \mathbf{v}$$

Your velocity is 30 feet per second at 45°, and it's a vector, **v**.

Remember: Dividing a vector by a scalar gives you a vector with potentially different units and the same direction.

In this case, you see that dividing a displacement vector, **s**, by a time gives you a velocity vector, **v**. It has the same magnitude as when you divided a distance by a time, but now you see a direction associated with it as well, because the displacement, **s,** is a vector. So you end up with a vector result rather than the scalars you see in Chapter 3.

Acceleration: Getting a new angle on changes in velocity

What happens when you swerve, whether in a car or on a walk? You accelerate in a particular direction. And just like displacement and velocity, acceleration, **a**, is a vector.

Remember: When you take into account the vector nature of acceleration, velocity and displacement, the equations from Chapter 3 become

$$\mathbf{v}_f = \mathbf{v}_i + \mathbf{a}\,\Delta t$$

$$\mathbf{a} = \frac{\Delta \mathbf{v}}{\Delta t} = \frac{\mathbf{v}_f - \mathbf{v}_i}{t_f - t_i}$$

$$\mathbf{s} = \mathbf{v}_i(t_f - t_i) + \frac{1}{2}\mathbf{a}(t_f - t_i)^2$$

Example

Q. You're in a car traveling east at 88.0 meters per second; then you accelerate north at 5.00 meters per second^2 for 10.0 seconds. What is your final speed?

A. The correct answer is 101 meters per second.

1. Start with this vector equation:

$$\mathbf{v_f} = \mathbf{v_i} + \mathbf{a}\, \Delta t$$

2. This equation is simply vector addition, so treat the quantities involved as vectors. That is, $\mathbf{v_i} = (88, 0)$ meters per second and $\mathbf{a} = (0, 5)$ meters per second^2. Here's what the equation looks like when you plug in the numbers:

$$\mathbf{v_f} = (88.0, 0)\ \text{m/s} + (0, 5.00\ \text{m/s}^2)(10.0\ \text{s})$$

3. Do the math:

$$\mathbf{v_f} = (88.0, 0)\ \text{m/s} + (0, 5.00\ \text{m/s}^2)(10.0\ \text{s}) = (88.0, 50.0)\ \text{m/s}$$

4. You're asked to find the final speed, which is the magnitude of the velocity. Plug your numbers into the Pythagorean theorem.

$$v_f = \sqrt{88.0^2 + 50.0^2}\ \text{m/s}$$
$$\approx 101\ \text{m/s}$$

5. You can also find the final direction. Apply the equation $\theta = \tan^{-1}(y/x)$ to find the angle, which is $\tan^{-1}(50.0/88.0) = \tan^{-1}(0.57) = 29.6°$ in this case.

Practice Questions

1. You're going 40.0 meters per second east, and then you accelerate 10.0 meters per second^2 north for 10.0 seconds. What are the direction and magnitude of your final velocity?

2. A car is driving along an icy road at 10.0 meters per second east when it skids, accelerating at 15 meters per second^2 at 63° north of east for 1.0 second. What are the direction and magnitude of the car's final velocity?

Practice Answers

1. **Magnitude 108 m/s, angle 68° north of east.** Start with this equation: $\mathbf{v_f} = \mathbf{v_i} + \mathbf{a}\, \Delta t$.

   Plug in the numbers: $\mathbf{v_f} = (40.0, 0)\ \text{m/s} + (0, 10.0\ \text{m/s}^2)(10.0\ \text{s}) = (40.0, 100.0)\ \text{m/s}$.

Convert the vector (40.0, 100.0) meters per second into magnitude/angle form. Use the equation $\theta = \tan^{-1}(y/x)$ to find the angle: $\tan^{-1}(100.0/40.0) = \tan^{-1}(2.5) = 68°$.

Apply the equation $v = \sqrt{x^2 + y^2}$ to find the speed — the magnitude of the velocity, giving you 108 meters per second.

2.   **Magnitude 21.5 m/s, angle 39° north of east.** Start with this equation: $\mathbf{v_f} = \mathbf{v_i} + \mathbf{a}\,\Delta t$.

Convert the original velocity into vector component notation: (10.0, 0) meters per second.

Convert the acceleration into components. Use the equation $a_x = a \cos\theta$ to find the x coordinate of the acceleration: $(15 \text{ m/s}^2) \cdot \cos 63° = 6.8 \text{ m/s}^2$.

Use the equation $a_y = a \sin\theta$ to find the y coordinate of the acceleration: $(15 \text{ m/s}^2) \cdot \sin 63°$, or 13.4 m/s^2. So the acceleration is $(6.8, 13.4) \text{ m/s}^2$ in coordinate form.

Perform the vector addition: $(10.0, 0) \text{ m/s} + (6.8 \text{ m/s}^2, 13.4 \text{ m/s}^2)(1.0 \text{ s}) = (16.8 \text{ m/s}, 13.4 \text{ m/s})$.

Convert the vector (16.8, 13.4) meters per second into magnitude/angle form. Use the equation $\theta = \tan^{-1}(y/x)$ to find the angle: $\tan^{-1}(13.4/16.8) = \tan^{-1}(0.79) = 39°$.

Apply the equation $v = \sqrt{x^2 + y^2}$ to find the magnitude of the velocity, giving you 21.5 meters per second.

ACCELERATING DOWNWARD: MOTION UNDER THE INFLUENCE OF GRAVITY

Gravity problems present good examples of working with vectors in two dimensions. Because the acceleration due to gravity is only vertical, it's especially useful to treat the horizontal and vertical components separately. Because there's no acceleration in the horizontal direction, the horizontal component of motion is just uniform. The vertical component undergoes a constant acceleration of magnitude g, or 9.8 meters per second^2, directed straight down. You can use this idea to make the solutions to trajectory problems really easy.

In this section, you sling projectiles around and let gravity do its work on shaping their trajectories. You'll see that because the force of gravity only acts downward — that is, in the vertical direction — you can treat the vertical and horizontal components separately. We start with just vertical motion before going on to look at trajectories with both horizontal and vertical components to them. Armed with this information, you can calculate things like the time for a projectile to strike the ground or reach the top of its trajectory and the distance that a projectile will travel.

Shooting an object straight up

To start simply, figure out how far a projectile can travel straight up in the air. Say, for example, that on your birthday, your friends give you just what you've always wanted: a cannon.

It has a muzzle velocity of 860 meters per second, and it shoots 10-kilogram cannonballs. Anxious to show you how it works, your friends shoot it off. The only problem: The cannon is pointing straight up. How long do you have to get out of the way?

Example

Q. The cannon has a muzzle velocity of 860 meters per second; the cannonballs have a mass of 10 kilograms each. What is the maximum height the cannonball will reach if you fire the cannon straight up (ignoring air resistance)?

A. The correct answer is 38 kilometers.

1. Because the cannonball is moving straight up and down, use this equation:

$$v_f^{\,2} - v_i^{\,2} = 2as$$

2. At the cannonball's maximum height, its vertical velocity will be 0. The acceleration is 9.8 meters per second^2 downward.

3. Solve for s:

$$s = \frac{v_f^{\,2} - v_i^{\,2}}{2a}$$

4. Plug in what you know — v_f is 0 meters per second, v_i is 860 meters per second, and the acceleration is g downward (g being 9.8 meters per second^2, the acceleration due to gravity on the surface of the Earth), or $-g$:

$$s = \frac{v_f^{\,2} - v_i^{\,2}}{2a} = \frac{0^2 - \left(860 \text{ m/s}\right)^2}{2\left(-9.8 \text{ m/s}^2\right)} \approx 3.8 \times 10^4 \text{ m}$$

The correct answer is about 38 kilometers. Notice that the mass of the cannonball does not affect the answer to this question.

Practice Questions

1. How long would it take a cannonball shot 28 kilometers straight up to reach its maximum height?

2. A firework explodes at the top of its trajectory at a height of 300 meters. If the firework was launched vertically, what was its initial velocity?

Practice Answers

1. **88 s.** Use the following equation to find the time the cannonball will take to reach its maximum height:

$$v_f = v_i + at$$

The cannonball has velocity $v_f = 0$ meters per second at the top and acceleration $a = -g = -9.8$ meters per second^2. Solve for time:

$$t = \frac{v_i}{g}$$

Enter the numbers into your calculator:

$$t = \frac{v_i}{g} = \frac{860 \text{ m/s}}{9.8 \text{ m/s}^2} \approx 88 \text{ s}$$

2. **80 m/s upward.** The firework is moving straight up and down, so you use this equation:

$$v_f{}^2 - v_i{}^2 = 2as$$

At the firework's maximum height, its vertical velocity will be 0. The acceleration is $-g$. Solve for v_i:

$$v_f{}^2 - 2as = v_i{}^2$$

Plug in what you know — v_f is 0 meters per second, s is 300 meters, and the acceleration is g downward (g being 9.8 meters per second^2, the acceleration due to gravity on the surface of the Earth), or $-g$:

$$\sqrt{0^2 - 2(-9.8 \text{ m/s}^2)(300 \text{ m})} = v_i$$
$$\sqrt{5880 \text{ m}^2/\text{s}^2} = v_i$$
$$76.68 \text{ m/s} \approx v_i$$

Accounting for significant digits, the initial velocity is 80 meters per second upward.

Projectile motion: Firing an object at an angle

Firing projectiles at an angle introduces a horizontal component to the motion. However, the force of gravity acts only in the vertical direction, so the horizontal component of the trajectory is uniform. You can tackle this kind of problem by separating out the horizontal and vertical components of the motion.

Example

Q. One of your devious friends decides to fire a cannonball at an angle across a horizontal field, as shown in the following figure. Where will the cannonball land if the initial speed of the cannonball is 860 meters per second and it's shot at an angle of 30° with respect to the horizontal?

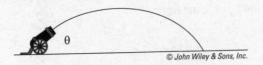

© John Wiley & Sons, Inc.

A. The correct answer is 65 kilometers.

1. Break the initial velocity into x and y components:

$$v_x = v_i \cos\theta$$
$$v_y = v_i \sin\theta$$

2. The acceleration is $-g$ in the y direction.

3. Find the x and y components of the displacement:

$$x = v_x t = (v_i \cos\theta)t$$

$$y = v_y t - \frac{1}{2}gt^2$$

4. The cannonball hits the ground when $y = 0$. Solve the second equation to find the time: $t = 2v/g$.

5. Use the time to find the range of the cannon in the x direction:

$$s = v_x t = \frac{2v_x v_y}{g} = \frac{2v_i^2 \sin\theta \cos\theta}{g}$$

6. Plug in numbers to find the range:

$$s = \frac{2(860 \text{ m/s})^2 \sin 30° \cos 30°}{9.8 \text{ m/s}^2}$$

$$\approx 6.5 \times 10^4 \text{ m}$$

The correct answer is about 65 kilometers, accounting for significant digits.

Practice Questions

1.   A pebble rolls horizontally off of a cliff with a speed of 4.3 meters per second. If the pebble falls 11 meters before hitting the ground, where does the pebble land?

2.   You want to launch a water balloon at your friend who is 7.0 meters away. If you want the water balloon to start and end 1.0 meter above the ground and you throw the water balloon at an angle of 45°, how fast should you throw it to hit your friend?

Practice Answers

1.   **6.4 m from the cliff.** The initial velocity is in the x direction:

$$v_x = v_i$$
$$v_y = 0$$

The acceleration is $-g$ in the y direction.

Find the x and y components of the displacement:

$$x = v_x t = v_i t$$

$$y = -\frac{1}{2} g t^2$$

Solve the second equation for time:

$$t = \sqrt{-\frac{2y}{g}}$$

Plug into the x component of the displacement:

$$x = v_i \sqrt{-\frac{2y}{g}}$$

Plug into your calculator:

$$x = (4.3 \text{ m/s}) \sqrt{-\frac{2(-11 \text{ m})}{9.8 \text{ m/s}^2}}$$

$$x \approx 6.4 \text{ m}$$

2. **8.4 m/s.** Break the initial velocity into x and y components:

$$v_{ix} = v_i \cos\theta$$
$$v_{iy} = v_i \sin\theta$$

The acceleration is $-g$ in the y direction.

Find the x and y components of the displacement:

$$x = v_{ix}t = \left(v_i \cos\theta\right)t$$

$$y = y_i + v_{iy}t - \frac{1}{2}gt^2 = y_i + \left(v_i \cos\theta\right)t - \frac{1}{2}gt^2$$

The water balloon explodes when $y = y_i = 1$ m. Solve the second equation to find the time: $t = 2v_{iy}/g$.

Use the time to find the range of the cannon in the x direction:

$$s = v_{ix}t = \frac{2v_{ix}v_{iy}}{g} = \frac{2v_i^2 \sin\theta \cos\theta}{g}$$

Solve for the initial speed:

$$\sqrt{\frac{sg}{2\sin\theta \cos\theta}} = v_i$$

Plug in numbers to find the initial speed:

$$\sqrt{\frac{(7\text{ m})(9.8\text{ m/s}^2)}{2\sin 45^\circ \cos 45^\circ}} = v_i$$

$$8.4\text{ m/s} = v_i$$

PART II

may the forces be
with
YOU

WEB EXTRA *Free-body diagrams are your friends when solving force problems. Head to www.dummies.com/ extras/ucanphysics1 for an article on how to draw free-body diagrams.*

IN THIS PART . . .

- Nail down Newton's laws of motion to understand how force affects motion.

- See how gravitational force and friction affect objects moving down an inclined plane.

- Wrap your mind around what happens when objects move in a circle.

- Figure out fluid flow and pressure in fluids.

5

WHEN PUSH COMES TO SHOVE: FORCE

You can't get away from forces in your everyday world; you use force to open doors, type at a keyboard, steer a car, drive a bulldozer through a wall, climb the stairs of the Statue of Liberty (not everyone, necessarily), take your wallet out of your pocket — even to breathe or talk. You unknowingly take force into account when you cross bridges, walk on ice, lift a hot dog to your mouth, unscrew a jar's cap, or flutter your eyelashes at your sweetie. Force is integrally connected to making objects move, and physics takes a big interest in understanding how it works.

Force is fun stuff. Like other physics topics, you may assume it's difficult, but that's before you get into it. Like your old buddies displacement, velocity, and acceleration (see Chapters 3 and 4), force is a vector, meaning it has a magnitude and a direction (unlike, say, speed, which just has a magnitude).

This chapter is where you find Newton's famous three laws of motion. You've heard these laws before in various forms, such as "For every action, there's an equal and opposite reaction." That's not quite right; it's more like "For every force, there's an equal and opposite force," and this chapter is here to set the record straight. In this chapter, we use Newton's laws as a vehicle to focus on force and how it affects the world.

Newton, Einstein, and the laws of physics

In the 17th century, Sir Isaac Newton was the first to put the relationship among force, mass, and acceleration into equation form. (He's also famous for watching apples drop off trees and developing the consequent mathematical expression of gravity.)

As with other advances in physics, Newton made observations first, modeled them mentally, and then expressed those models in mathematical terms. Newton expressed his model by using three assertions, which have come to be known as Newton's laws. But don't forget that physics just models the world, and as such, it's all subject to later revision.

Newton's laws have been heavily revised by the likes of Albert Einstein and his theory of relativity. Newton's laws are based on ideas of space and time and mass that make sense to most people in everyday terms: Everyone agrees when two events are simultaneous, mass is a constant that doesn't depend on speed, and so on. But Einstein's theory of relativity takes the speed of light as a constant for all observers however they're moving, and this leads to some very different ideas of space and time, which in turn brings about very different laws of motion. However, Einstein's theory becomes important only for motion close to the speed of light. At speeds that you see around you every day, Newton's laws of motion are extremely accurate and, therefore, are still very important to understand.

NEWTON'S FIRST LAW: RESISTING WITH INERTIA

Newton's laws explain what happens with forces and motion, and his first law states, "An object continues in a state of rest, or in a state of motion at a constant velocity along a straight line, unless compelled to change that state by a net force." Translation? If you don't apply a *net,* or unbalanced, force to an object at rest or in motion, it will stay at rest or in that same motion along a straight line. Forever.

For example, when scoring a hockey goal, the hockey puck slides toward the open goal in a straight line because the ice it slides on is nearly frictionless. If you're lucky, the puck won't come into contact with the opposing goalie's stick, which would cause it to change its motion.

Newton's first law may not seem very intuitive because most things don't seem to continue moving in straight lines forever. Left to themselves, most moving things come to a halt. The idea that the natural tendency of an object in motion is to come to a halt was Aristotle's, and it was accepted wisdom for 2,000 years. It took the tremendous insight of Newton to see that the natural state of motion is actually to continue in a straight line at constant velocity. Only when acted on by a force does the motion change.

Remember: In everyday life, objects don't coast around in straight lines at constant velocity. This is because most objects around you are subject to friction forces. So, for example, when you slide a coffee mug across your desk, it slows and comes to a stop (or spills over). That's not to say Newton's first law is invalid, just that friction provides a force to change the mug's motion to stop it.

Saying that if you don't apply a force to an object in motion, it will stay in motion at constant velocity forever sounds an awful lot like a *perpetual-motion machine,* a theoretical machine that would run indefinitely without the input of any energy. Interestingly, such a machine is perfectly possible according to Newton's laws. In practice, you just can't get away from forces that will ultimately affect an object in motion. Even in the farthest reaches of space, the rest of the mass in the universe pulls at you, if only very slightly. And that means your motion is affected. So much for perpetual motion!

What Newton's first law really says is that the only way to get something to change its motion is to use force. It also says that an object in motion tends to stay in motion, which introduces the idea of inertia.

Resisting change: Inertia and mass

Inertia is the natural tendency of an object to resist any change in its motion, which means that it tends to stay at rest or in constant motion along a straight line. Inertia is a quality of mass, and the mass of an object is really just a measurement of its inertia. To get a stationary object to move — that is, to change its current state of motion — you have to apply a force to overcome its inertia.

Remember: Be careful to distinguish between mass and weight. The *weight* of an object is the force of gravity on it, so weight depends on where the mass is. For example, a 1-kilogram object would have a different weight on the moon than it does on Earth, but the mass would be the same. Even in deep space, with no significant gravitational field and therefore no weight, the mass would still be 1 kilogram. If you tried to push this object in space, you'd feel a resistance to the acceleration, which is inertia. The larger the mass of the object, the more resistance you would feel.

Say, for example, you're at your summer vacation house, taking a look at the two boats at your dock: a dinghy and an oil tanker. If you apply the same net force to each with your foot, the boats respond in different ways. The dinghy scoots away and glides across the water. The oil tanker moves away more slowly (what a strong leg you have!). That's because they both have different masses and, therefore, different amounts of inertia. When responding to the same net force, an object with little mass — and a small amount of inertia — will have greater acceleration than an object with large mass, which has a large amount of inertia.

Remember: Because objects have inertia, they resist changing their motion, which is why you have to start applying forces to get changes in velocity and therefore acceleration. Mass ties force and acceleration together.

Measuring mass

The units of mass (and, therefore, inertia) depend on your measuring system. In the meter-kilogram-second (MKS) system, or the International System of Units (SI), mass is measured in kilograms (under the influence of the Earth's gravity, 1 kilogram of mass weighs about 2.205 pounds). What's the unit of mass in the foot-pound-second system? Brace yourself: It's the *slug*. Under the influence of the Earth's gravity, a slug has a weight of about 32 pounds. If you need to convert between the slug and the kilogram, then you'll be pleased to know that 1 slug is equal to about 14.59 kilograms.

Warning: Mass isn't the same as weight. Mass is a measure of inertia; when you put that mass into a gravitational field, you get weight. So, for example, a slug is a certain amount of mass. When you subject that slug to the gravitational pull on the surface of the Earth, it has weight. And that weight is about 32 pounds. If you took that same slug of mass to the moon, which doesn't have as much gravitational pull as Earth, the slug would weigh only around 5.3 pounds, which is about 1/6 of its weight on Earth.

NEWTON'S SECOND LAW: RELATING FORCE, MASS, AND ACCELERATION

Newton's first law says that an object remains in uniform motion unless acted on by a net force. When a net force is applied, the object accelerates. Newton's second law details the relationship among net force, the mass, and the acceleration:

- **The acceleration of an object is in the direction of the net force.** If you push or pull an object in a particular direction, it accelerates in that direction.

- **The acceleration has a magnitude proportional to the magnitude of the net force.** If you push twice as hard (and no other forces are present), the acceleration is twice as big.

- **The magnitude of the acceleration is inversely proportional to the mass of the object.** That is, the larger the mass, the smaller the acceleration for a given force (which is just as you'd expect from inertia).

Remember: All these features of the relation among net force (ΣF), acceleration (a), and mass (m) are contained in the following equation:

$$\sum F = ma$$

Note that you use the term ΣF to describe the net force because the Greek letter sigma, Σ, stands for "sum"; therefore, ΣF means the sum of all the separate forces acting on the object. If this is not zero, then there's a net force.

Naming units of force

So what are the units of force? Well, $\Sigma F = ma$, so in the MKS or SI system, force must have these units:

$$\text{kilogram-meters/second}^2$$

This is a *derived* unit because you reach it by using a formula. Because most people think this unit line looks a little awkward, the MKS units are given a special name: *newtons* (named after guess who). Newtons are often abbreviated as simply N. Table 5-1 shows unit names for force in the MKS and foot-pound-second systems of measurement.

Table 5-1 Units of Force

System of Measurement	Derived Unit	Special Unit Name
Meter-kilogram-second (MKS) or SI	kilogram-meters/second^2 (kg·m/s^2)	newton (N)
Foot-pound-second	slug-feet/second^2 (slug·ft/s^2)	pound (lb)

So how do these units relate to each other? Well, 1.0 pound is about 4.448 newtons.

Relating the formula to the real world

You can see that the equation $\Sigma F = ma$ is consistent with Newton's first law of motion (which deals with inertia), because if there's no net force (ΣF) acting on a mass m, then the left-hand side of this equation is 0; therefore, the acceleration must also be 0 — just as you'd expect from the first law.

If you rearrange the net-force equation to solve for acceleration, you can see that if the size of the net force doubles, then so does the size of the acceleration (if you push twice as hard, the object accelerates twice as much), and if the mass doubles, then the acceleration halves (if the mass is twice as big, it accelerates half as much — inertia):

$$a = \frac{\sum F}{m}$$

Take a look at the hockey puck in Figure 5-1 and imagine it's sitting there all lonely in front of a net. These two should meet.

In a totally hip move, you decide to apply your knowledge of physics to this one. You figure that if you apply the force of your stick to the puck for a tenth of a second, you can accelerate it in the appropriate direction. You try the experiment, and sure enough, the puck flies into the net. Score! Figure 5-1 shows how you made the goal. You applied a net force to the puck, which has a certain mass, and off it went — accelerating in the direction you pushed it.

What's its acceleration? That depends on the force you apply (along with any other forces that may be acting on the puck), because $\Sigma F = ma$.

Figure 5-1:
Acceler-
ating a
hockey
puck.

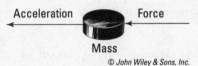

Acceleration   Force

Mass
© John Wiley & Sons, Inc.

Examples

Q. You're at rest on an ice rink when you get hit from behind with a force of 50 newtons as someone bumps you. If your mass is 70 kilograms, what is your acceleration?

A. The correct answer is 0.714 meters per second^2.

   1. Use the equation $F = ma$: Solving for a gives you $a = F/m$.

   2. Plug in the numbers: $a = F/m = (50.0 \text{ N})/(70.0 \text{ kg}) = 0.714 \text{ m/s}^2$.

Q. You push a rowboat on a calm lake (assuming no friction) with a force of 40.0 newtons. If the rowboat has a mass of 80.0 kilograms, how far has it gone in 10.0 seconds?

A. The correct answer is 25 meters.

   1. Use the equation $F = ma$, and solve for the acceleration, giving you $a = F/m$.

   2. Use the equation $s = \frac{1}{2}at^2$ and substitute F/m for a:

   $$s = \frac{Ft^2}{2m}$$

   3. Plug in the numbers: $s = \frac{Ft^2}{2m} = \frac{40.0\left(10.0^2\right)}{2\left(80.0\right)} = 25 \text{ m}$.

Practice Questions

1. You come home to find a delivered package with a mass of 100 kilograms blocking the door. If you push it with a force of 100 newtons, what will its acceleration be if no friction is involved?

2. You're gliding across a frictionless lake in a sailboat. If your mass is 70 kilograms and the boat's mass is 200 kilograms, with what force does the wind need to blow you to give you an acceleration of 0.30 meters per second^2?

3. You have control of a space station, which has a mass of 400,000 kilograms. To give it an acceleration of 2.0 meters per second^2, what force do you need to apply with the rockets?

4. You find a stone in the forest and give it a push of 50.0 newtons. It accelerates at 2.0 meters per second^2. What is its mass?

5. You're applying a force of 17 newtons to a hockey puck with a mass of 0.17 kilograms. Starting from rest, how far has the puck gone in 0.10 seconds?

6. A space station with a mass of 1.0×10^5 kilograms is moving toward a satellite at 5.0 meters per second. If you want to avoid crashing them together and have only 1.0×10^3 seconds in which to act, what force do you need to apply to stop the space station from colliding with the satellite?

7. Your 1,000-kilogram car needs a push. Starting at rest, how hard do you have to push to get it up to 5.0 meters per second in 1.0×10^2 seconds?

Practice Answers

1. **1 m/s^2.** Solving $F = ma$ for a gives you $a = F/m$.

   Plug in the numbers: $a = F/m = 100$ N/100 kg $= 1$ m/s^2.

2. **81 N.** Use the equation $F = ma$.

   Plug in the numbers: $F = ma = (70$ kg $+ 200$ kg$)(0.30$ m/s$^2) = 81$ N.

3. **800,000 N.** Use the equation $F = ma$.

   Plug in the numbers: $F = ma = (400,000$ kg$)(2.0$ m/s$^2) = 800,000$ N.

4. **25 kg.** Use the equation $F = ma$, and solve for the mass, giving you $m = F/a$.

   Plug in the numbers: $m = F/a = 50$ N/2.0 m/s$^2 = 25$ kg.

5. **0.50 m.** Use the equation $F = ma$, and solve for the acceleration, giving you $a = F/m$.

    Use the equation $s = \frac{1}{2}at^2$, and substitute F/m for a:

    $$s = \frac{Ft^2}{2m}$$

    Plug in the numbers: $s = \frac{Ft^2}{2m} = \frac{(17\ \text{N})(0.10\ \text{s})^2}{2(0.17\ \text{kg})} = 0.50\ \text{m}$

6. **–5.0 × 10² N.** Use the equation $F = ma$, and solve for the acceleration, giving you $a = F/m$.

    Use the equation $v_f = v_i + at$. In this question, $v_f = 0$, so $at = -v_i$.

    This becomes $at = \frac{Ft}{m} = -v_i$; solving for F gives you $F = \frac{-mv_i}{t}$.

    Plug in the numbers, and you get $F = -500\ \text{N} = -5.0 \times 10^2\ \text{N}$ (it's negative because it's opposite to the direction of travel).

7. **50 N.** Use the equation $F = ma$, and solve for the acceleration, giving you $a = F/m$.

    Use the equation $v_f = v_i + at$. In this question, $v_f = 5$ m/s and $v_i = 0$, so $at = v_f$.

    This becomes $at = \frac{Ft}{m} = v_f$; solving for F gives you $F = \frac{mv_f}{t}$.

    Plug in the numbers, and you get $F = 50\ \text{N}$.

Vector addition: Gathering net forces

Remember: Most books shorten $\Sigma F = ma$ to simply $F = ma$, which is what we do, too, but we must note that F stands for *net force*. An object you apply force to responds to the net force — that is, the vector sum of all the forces acting on it.

Take a look, for example, at all the forces (represented by arrows) acting on the ball in Figure 5-2. Which way will the golf ball be accelerated?

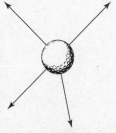

Figure 5-2: A ball may face many forces that act on it.

© John Wiley & Sons, Inc.

Because Newton's second law talks about net force, the problem becomes easier. All you have to do is add the various forces together as vectors to get the resultant, or net, force vector, ΣF, as Figure 5-3 shows. When you want to know how the ball will accelerate, you can apply the equation $\Sigma F = ma$.

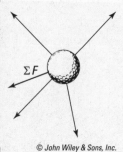

Figure 5-3:
The net force vector factors in all forces to determine the ball's acceleration.

Remember: A *free-body diagram,* like Figure 5-2, shows all the forces acting on an object, making it easier to determine their components and find the net force. If you draw a free-body diagram before you start writing down equations, you can avoid making mistakes.

Examples

Q. Suppose that you have two forces as shown in the following figure: **A** = 5.0 newtons at 40°, and **B** = 7.0 newtons at 125°. What is the net force, Σ**F**?

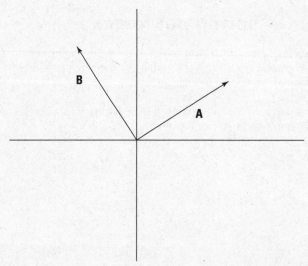

A.   The correct answer is magnitude 8.9 newtons, angle 91°.

   1.   Convert force **A** into vector component notation. Use the equation $A_x = A \cos \theta$ to find the x coordinate of the force: (5.0 N) cos 40° = 3.8 N.

   2.   Use the equation $A_y = A \sin \theta$ to find the y coordinate of the force: (5.0 N) sin 40°, or 3.2 N. That makes the vector **A** = (3.8, 3.2) N in coordinate form.

   3.   Convert the vector **B** into components. Use the equation $B_x = B \cos \theta$ to find the x coordinate of the acceleration: (7.0 N) cos 125° = –4.0 N.

   4.   Use the equation $B_y = B \sin \theta$ to find the y coordinate of the second force: (7.0 N) sin 125°, or 5.7 N. That makes the force **B** = (–4.0, 5.7) N in coordinate form.

   5.   Perform the vector addition to find the net force: (3.8, 3.2) N + (–4.0, 5.7) N = (–0.2, 8.9) N.

   6.   Convert the vector (–0.2, 8.9) into magnitude/angle form. Use the equation $\theta = \tan^{-1}(y/x)$ to find the angle: $\tan^{-1}(-44.5) = 91°$.

   7.   Apply the equation $h = \sqrt{x^2 + y^2}$ to find the magnitude of the net force, giving you 8.9 N.

Q.   Suppose that the forces acting on the hockey puck in the following figure are **A** = 9.0 newtons at 0°, and **B** = 14.0 newtons at 45°. What is the acceleration of the puck, given that its mass is 1.00×10^{-1} kilograms?

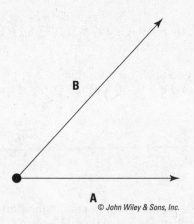

B

A

A.   The correct answer is magnitude 213 meters per second², angle 28°.

   1.   Convert force **A** into vector component notation. Use the equation $A_x = A \cos \theta$ to find the x coordinate of the force: (9.0 N) cos 0° = 9.0 N.

   2.   Use the equation $A_y = A \sin \theta$ to find the y coordinate of the force: (9.0 N) sin 0°, or 0.0. That makes the vector **A** = (9.0, 0.0) N in coordinate form.

   3.   Convert the vector **B** into components. Use the equation $B_x = B \cos \theta$ to find the x coordinate of the acceleration: (14.0 N) cos 45° = 9.9 N.

   4.   Use the equation $B_y = B \sin \theta$ to find the y coordinate of the second force: (14.0 N) sin 45°, or 9.9 N. That makes the force **B** = (9.9, 9.9) N in coordinate form.

5. Perform the vector addition to find the net force: (9.0, 0.0) N + (9.9, 9.9) N = (18.9, 9.9) N.

6. Convert the vector (18.9, 9.9) N into magnitude/angle form. Use the equation $\theta = \tan^{-1}(y/x)$ to find the angle of the net force: $\tan^{-1}(0.52) = 28°$.

7. Apply the equation $h = \sqrt{x^2 + y^2}$ to find the magnitude of the net force, giving you 21.3 N.

8. Calculate the magnitude of the acceleration from the magnitude of the net force, 21.3 N: $a = F/m = (21.3\text{ N})/(0.100\text{ kg}) = 213\text{ m/s}^2$.

Practice Questions

1. Add two forces: **A** is 8.0 N at 53°, and **B** is 9.0 N at 19°.

2. Add two forces: **A** is 12.0 N at 129°, and **B** is 3.0 N at 225°.

3. Assume that the two forces acting on a 0.10-kilogram hockey puck are as follows: **A** is 16.0 N at 53°, and **B** is 21.0 N at 19°. What is the acceleration of the hockey puck?

4. Two forces act on a 1.0×10^3-kilogram car. **A** is 220 N at 64°, and **B** is 90.0 N at 80°. What is the car's acceleration?

5. Suppose that two forces act on a 100-kilogram boat. **A** is 100 N at 10°, and **B** is 190 N at 210°. What is the boat's acceleration?

6. A marble with a mass of 1.0 gram is hit by two other marbles that each apply a force for 0.3 seconds. If force **A** is 0.010 N at 63° and **B** is 0.050 N at 135°, what is the acceleration of the original marble?

Practice Answers

1. **Magnitude: 16 N; Angle: 35°.** Convert force **A** into vector component notation. Use the equation $A_x = A \cos \theta$ to find the x coordinate of force **A**: (8.0 N) cos 53° = 4.8 N.

   Use the equation $A_y = A \sin \theta$ to find the y coordinate of force **A**: (8.0 N) sin 53° = 6.4 N. That makes force **A** = (4.8, 6.4) N in coordinate form.

   Convert the vector **B** into components. Use the equation $B_x = B \cos \theta$ to find the x coordinate of force **B**: (9.0 N) cos 19° = 8.5 N.

   Use the equation $B_y = B \sin \theta$ to find the y coordinate of the second force: (9.0 N) sin 19° = 2.9 N. That makes force **B** = (8.5, 2.9) N in coordinate form.

   Perform vector addition to find the net force: (4.8, 6.4) N + (8.5, 2.9) N = (13.3, 9.3) N.

   Convert the force vector (13.3, 9.3) N into magnitude/angle form. Use the equation $\theta = \tan^{-1}(y/x)$ to find the angle: $\tan^{-1}(0.70) = 35°$.

   Apply the equation $h = \sqrt{x^2 + y^2}$ to find the magnitude of the net force, giving you 16 N.

2. **Magnitude: 12 N; Angle: 143°.** Convert force **A** into vector component notation. Use the equation $A_x = A \cos \theta$ to find the x coordinate of force **A**: 12.0 cos 129° = –7.6.

   Use the equation $A_y = A \sin \theta$ to find the y coordinate of force **A**: 12.0 sin 129° = 9.3 N. That makes force **A** = (–7.6, 9.3) N in coordinate form.

   Convert force **B** into components. Use the equation $B_x = B \cos \theta$ to find the x coordinate of force **B**: 3.0 cos 225° = –2.1 N.

   Use the equation $B_y = B \sin \theta$ to find the y coordinate of force **B**: (3.0 N) sin 225° = –2.1 N. That makes force **B** = (–2.1, –2.1) N in coordinate form.

   Perform vector addition to find the net force: (–7.6, 9.3) N + (–2.1, –2.1) N = (–9.7, 7.2) N.

   Convert the force vector (–9.7, 7.2) N into magnitude/angle form. Use the equation $\theta = \tan^{-1}(y/x)$ to find the angle: $\tan^{-1}(-0.74)$ = 143°.

   Apply the equation $h = \sqrt{x^2 + y^2}$ to find the magnitude of the net force, giving you 12 N.

3. **Magnitude: 354 m/s²; Angle: 34°.** Convert force **A** into its components. Use the equation $A_x = A \cos \theta$ to find the x coordinate of force **A**: (16.0 N) cos 53° = 9.6 N.

   Use the equation $A_y = A \sin \theta$ to find the y coordinate of force **A**: (16.0 N) sin 53° = 12.8 N. That makes force **A** = (9.6, 12.8) N in coordinate form.

   Convert force **B** into its components. Use the equation $B_x = B \cos \theta$ to find the x coordinate of force **B**: (21.0 N) cos 19° = 19.9 N.

   Use the equation $B_y = B \sin \theta$ to find the y coordinate of force **B**: (21.0 N) sin 19° = 6.8 N. That makes force **B** = (19.9, 6.8) N in coordinate form.

   Perform vector addition to find the net force: (9.6, 12.8) N + (19.9, 6.8) N = (29.5, 19.6) N.

   Convert the vector (29.5, 19.6) N into magnitude/angle form. Use the equation $\theta = \tan^{-1}(y/x)$ to find the angle of the net force: $\tan^{-1}(0.66)$ = 34°. The direction of the acceleration is the same as the direction of the net force: 34°.

   Apply the equation $h = \sqrt{x^2 + y^2}$ to find the magnitude of the net force, giving you 35.4 N.

   Use the magnitude of the force and the mass to find the magnitude of the acceleration: $a = F/m$ = (35.4 N)/(0.10 kg) = 354 m/s².

4. **Magnitude: 0.31 m/s²; Angle: 69°.** Convert force **A** into its components. Use the equation $A_x = A \cos \theta$ to find the x coordinate of force **A**: 220 cos 64° = 96 N.

   Use the equation $A_y = A \sin \theta$ to find the y coordinate of the force: (220 N) sin 64° = 198 N. That makes force **A** = (96, 198) N in coordinate form.

   Convert force **B** into its components. Use the equation $B_x = B \cos \theta$ to find the x coordinate of force **B**: (90 N) cos 80° = 16 N.

   Use the equation $B_y = B \sin \theta$ to find the y coordinate of force **B**: (90 N) sin 80° = 89 N. That makes force **B** = (16, 89) N in coordinate form.

   Perform vector addition to find the net force: (96, 198) N + (16, 89) N = (112, 287) N.

   Convert the vector (112, 287) N into magnitude/angle form. Use the equation $\theta = \tan^{-1}(y/x)$ to find the angle of the net force: $\tan^{-1}(2.56)$ = 69°. The direction of the acceleration is the same as the direction of the net force: 69°.

Apply the equation $h = \sqrt{x^2 + y^2}$ to find the magnitude of the net force, giving you 308 N.

Use the magnitude of the force and the mass to find the magnitude of the acceleration: $a = F/m = (308 \text{ N})/(1{,}000 \text{ kg}) = 0.31 \text{ m/s}^2$.

5. **Magnitude: 1 m/s²; Angle: 229°.** Convert force **A** into its components. Use the equation $A_x = A \cos\theta$ to find the x coordinate of force **A**: $(100 \text{ N}) \cos 10° = 98 \text{ N}$.

Use the equation $A_y = A \sin\theta$ to find the y coordinate of force **A**: $(100 \text{ N}) \sin 10° = 17 \text{ N}$. That makes force **A** = (98, 17) N in coordinate form.

Convert force **B** into its components. Use the equation $B_x = B \cos\theta$ to find the x coordinate of force **B**: $(190 \text{ N}) \cos 210° = -165 \text{ N}$.

Use the equation $B_y = B \sin\theta$ to find the y coordinate of force **B**: $(190 \text{ N}) \sin 210° = -95$. That makes force **B** = (–165, –95) N in coordinate form.

Perform vector addition to find the net force: (98, 17) N + (–165, –95) N = (–67, –78) N.

Convert the vector (–67, –78) N into magnitude/angle form. Use the equation $\theta = \tan^{-1}(y/x)$ to find the angle of the net force: $\tan^{-1}(1.2) = 49°$. But that answer's not right because both components are negative, which means that the angle is actually between 180° and 270°. Add 180° to 49° to get 229°. The acceleration is in the same direction as the net force.

Apply the equation $h = \sqrt{x^2 + y^2}$ to find the magnitude of the net force, giving you 102 N.

Use the magnitude of the force and the mass to find the magnitude of the acceleration: $a = F/m = (102 \text{ N})/(100 \text{ kg}) = 1.0 \text{ m/s}^2$.

6. **Magnitude: 54 m/s²; Angle: 125°.** Convert force **A** into its components. Use the equation $A_x = A \cos\theta$ to find the x coordinate of force **A**: $(0.01 \text{ N}) \cos 63° = 4.5 \times 10^{-3} \text{ N}$.

Use the equation $A_y = A \sin\theta$ to find the y coordinate of force **A**: $(0.01 \text{ N}) \sin 63° = 8.9 \times 10^{-3} \text{ N}$. That makes force **A** = $(4.5 \times 10^{-3}, 8.9 \times 10^{-3})$ N in coordinate form.

Convert force **B** into its components. Use the $B_x = B \cos\theta$ to find the x coordinate of force **B**: $(0.05 \text{ N}) \cos 135° = -3.5 \times 10^{-2} \text{ N}$.

Use the equation $B_y = B \sin\theta$ to find the y coordinate of force **B**: $(0.05 \text{ N}) \sin 135° = 3.5 \times 10^{-2} \text{ N}$. That makes force **B** = $(-3.5 \times 10^{-2}, 3.5 \times 10^{-2})$ N in coordinate form.

Perform vector addition to find the net force: $(4.5 \times 10^{-3}, 8.9 \times 10^{-3})$ N + $(-3.5 \times 10^{-2}, 3.5 \times 10^{-2})$ N = $(-3.1 \times 10^{-2}, 4.4 \times 10^{-2})$ N.

Convert the vector $(-3.1 \times 10^{-2}, 4.4 \times 10^{-2})$ N into magnitude/angle form. Use the equation $\theta = \tan^{-1}(y/x)$ to find the angle of the net force: $\tan^{-1}(1.42) = 125°$. The acceleration is in the same direction as the net force.

Apply the equation $h = \sqrt{x^2 + y^2}$ to find the magnitude of the net force, giving you $5.4 \times 10^{-2} \text{ N}$.

Use the magnitude of the force and the mass to find the magnitude of the acceleration: $a = F/m = (5.4 \times 10^{-2} \text{ N})/(0.001 \text{ kg}) = 54 \text{ m/s}^2$.

Pulleys: Supporting double the force

A rope and pulley can act together to change the direction of the force you apply, but not for free. To change the direction of your force from $-F$ (that is, downward) to $+F$ (upward on the mass), the pulley's support has to respond with a force of $2F$.

Here's how this works: When you pull a rope in a pulley system to lift a stationary object, you lift the mass if you exert enough force to overcome its weight, mg, where g is the acceleration due to gravity at the surface of the Earth, 9.8 meters per second^2. Take a look at Figure 5-4, in which a rope goes over a pulley and down to a mass m.

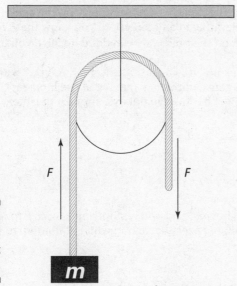

Figure 5-4: Using a pulley to exert force.

The rope and pulley together function not only to transmit the force, F, you exert but also to change the direction of that force, as you see in the figure. The force you exert downward is exerted upward on the mass, because the rope, going over the pulley, changes the force's direction. In this case, if F is greater than mg, you can lift the mass. If you apply no force on the object, then the only force acting on it is gravity, $F_{gravity}$, and so the object accelerates at a rate of $-mg$ (the negative sign indicates that the acceleration is downward) because

$$F_{gravity} = -mg$$

If you apply a force on the rope of magnitude F, then it is transmitted by the rope and pulley to the object as an upwardly directed force of the same magnitude. Therefore, the total force on the object is given by the sum of these two forces, $F_{gravity} + F$. The force F, acting alone without gravity, would accelerate the object upward at a rate that you can call a:

$$F = ma$$

When the two forces act together, you have the following sum:

$$F_{gravity} + F = -mg + ma$$
$$= m(a - g)$$

So you can see that if F is greater than mg, then a is greater than g and the object accelerates upward.

Assume that you lift the mass and it hangs in the air. In this case, F must equal mg to hold the mass stationary. The direction of your force is being changed from downward to upward. How does that happen?

To figure this out, consider the force that the pulley's support exerts on the ceiling. What's that force? Because the pulley isn't accelerating in any direction, you know that $\Sigma F = 0$ on the pulley. That means that all the forces on the pulley, when added up, give you 0.

From the pulley's point of view, two forces pull downward: the force F you pull with and the force mg that the mass exerts on you (because nothing is moving at the moment). That's $2F$ downward. To balance all the forces and get 0 total, the pulley's support must exert a force of $2F$ upward.

Example

Q. In the following figure, the mass m isn't moving, and you're applying a force F to hold it stationary. What force is the pulley's support exerting, and in which direction, to keep the pulley where it is?

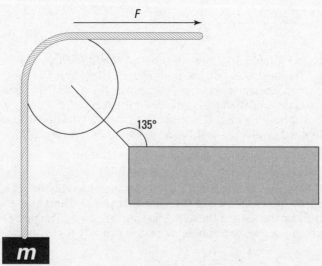

A. The correct answer is magnitude $mg\sqrt{2}$, angle 135°.

1. Because the pulley isn't moving, $\Sigma F = 0$ on the pulley.

2. The force due to the mass's weight is

$$F_{mass} = \left(0, -mg\right)$$

3. The force of the rope on the pulley must be of magnitude mg and directed to the right because you're holding the mass stationary and the rope transmits the force you're applying. That force looks like this:

$$F_{rope} = \left(mg, 0\right)$$

4. The force exerted on the pulley by the rope and the mass by adding the vectors F_{mass} and F_{rope} is

$$F_{mass+rope} = F_{mass} + F_{rope}$$
$$= \left(0, -mg\right) + \left(mg, 0\right)$$
$$= \left(mg, -mg\right)$$

5. Because the sum of all the forces on the pulley is 0, the force of the support is

$$F_{support} = -F_{mass+rope}$$

6. Plug in your previous result to find the force of the support:

$$F_{support} = -F_{mass+rope}$$
$$= -\left(mg, -mg\right)$$
$$= \left(-mg, mg\right)$$

7. The magnitude is equal to

$$F_{support} = \sqrt{\left(-mg\right)^2 + \left(mg\right)^2} \approx mg\sqrt{2}$$

8. Find the angle:

$$\theta = \tan^{-1}\left(\frac{F_{support,y}}{F_{support,x}}\right)$$
$$= \tan^{-1}\left(\frac{mg}{-mg}\right)$$
$$= -45°$$

This answer doesn't make sense, because a vector that points at −45° points down and to the right.

9. Angles that differ by a multiple of 180° give the same tangent, so you can add –45° to 180° to get 135°, which is up and to the left, and agrees with the figure. The force points at 135°.

Tip: If you get confused about the signs when doing this kind of work, check your answers against the directions you know the force vectors actually go in. A picture's worth a thousand words, even in physics!

Practice Questions

Use the following figure to answer the practice questions.

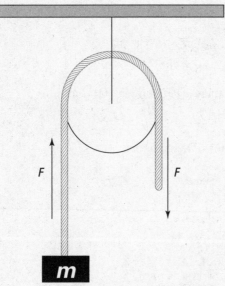

© John Wiley & Sons, Inc.

1. In the figure, the force *F* is 120 newtons and the mass *m* of the object is 5.0 kilograms. What is the acceleration of the mass?

2. In the figure, the force *F* is 45 newtons and the mass *m* of the object is 12 kilograms. What is the acceleration of the mass?

Practice Answers

1. **14 m/s^2 upward.** The pulley changes the direction of the force but doesn't change its magnitude, so the mass has a force of 120 newtons pulling it up, and a force of mg pulling it down. Use Newton's second law:

    $$120\,\text{N} - mg = ma$$

    Solve for the acceleration:

    $$\frac{120\,\text{N}}{m} - g = a$$

    Plug in the numbers:

    $$\frac{120\,\text{N}}{5.0\,\text{kg}} - 9.8\,\text{m/s}^2 = a$$

    $$14\,\text{m/s}^2 \approx a$$

2. **6.1 m/s^2 downward.** The pulley changes the direction of the force but doesn't change its magnitude, so the mass has a force of 45 newtons pulling it up, and a force of mg pulling it down. Use Newton's second law:

    $$45\,\text{N} - mg = ma$$

    Solve for the acceleration:

    $$\frac{45\,\text{N}}{m} - g = a$$

    Plug in the numbers:

    $$\frac{45\,\text{N}}{12\,\text{kg}} - 9.8\,\text{m/s}^2 = a$$

    $$-6.1\,\text{m/s}^2 \approx a$$

Finding equilibrium

In physics, an object is in *equilibrium* when it has zero acceleration — when the net force acting on it is zero. The object doesn't actually have to be at rest — it can be going 1,000 miles per hour as long as the net force on it is zero and it isn't accelerating. Forces may be acting on the object, but they all add up, as vectors, to zero.

Example

Q. Look at the following figure, where you've started your own grocery store and bought a wire rated at 45 newtons to hang the sign with. The wire provides force F_1, and a brace provides force F_2. If the sign weighs 24.0 newtons, can the wire support the sign?

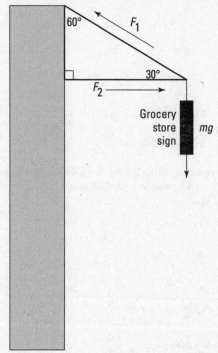

© John Wiley & Sons, Inc.

A. The correct answer is no.

1. The sign is in equilibrium, which means that the net force on it is zero.

2. The only forces with components in the vertical direction are the force of the wire and the weight of the sign, so the y component of F_1 must equal the weight of the sign:

$$F_{1y} = F_1 \sin 30° = mg$$

3. Solve for the tension in the wire:

$$F_1 = \frac{mg}{\sin 30°}$$

4. Plug in the numbers:

$$F_1 = \frac{24.0 \text{ N}}{\sin 30°} = 48.0 \text{ N}$$

The tension in the wire is greater than 45 newtons, so the wire isn't strong enough.

Practice Questions

Use the following figure to answer the practice questions.

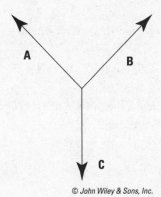

© John Wiley & Sons, Inc.

1. The figure shows three connected ropes. If rope A has a tension of 10 newtons at 135° and rope B has a tension of 10 newtons at 45°, what must the tension in rope C be to keep things in equilibrium?

2. You have three ropes tied together in equilibrium as shown in the figure. The tension in rope A is 15 newtons at 135°, and the tension in rope B is 15 newtons at 45°. What must the tension in rope C be?

Practice Answers

1. **14 N downward.** Convert tension **A** into vector component notation. Use the equation $A_x = A \cos \theta$ to find the x coordinate of the tension from rope A: (10.0 N) cos 135° = –7.07 N.

Use the equation $A_y = A \sin \theta$ to find the y coordinate of the tension from rope A: (10.0 N) sin 135°, or 7.07 N. That makes the tension **A** = (–7.07, 7.07) N in coordinate form.

Convert the tension **B** into components. Use the equation $B_x = B \cos \theta$ to find the x coordinate of the tension from rope B: (10.0 N) cos 45° = 7.07 N.

Use the equation $B_y = B \sin \theta$ to find the y coordinate of the second tension: (10.0 N) sin 45°, or 7.07 N. That makes the tension **B** = (7.07, 7.07) N in coordinate form.

Perform vector addition to find the net tension: (–7.07, 7.07) N + (7.07, 7.07) N = (0, 14.1) N.

To counteract the total tension from ropes A and B, the tension in rope C must be 14.1 N downward (that is, –14.1 N). With significant figures, this answer rounds to 14 N downward.

2. **21 N downward.** Convert tension **A** into vector component notation. Use the equation $A_x = A \cos\theta$ to find the x coordinate of tension **A**: 15.0 cos 135° = –10.6 N.

Use the equation $A_y = A \sin\theta$ to find the y coordinate of tension **A**: (15.0 N) sin 135° = 10.6. That makes tension **A** = (–10.6, 10.6) N in coordinate form.

Convert tension **B** into components. Use the equation $B_x = B \cos\theta$ to find the x coordinate of tension **B**: (15.0 N) cos 45° = 10.6 N.

Use the equation $B_y = B \sin\theta$ to find the y coordinate of tension **B**: (15.0 N) sin 45° = 10.6 N. That makes tension **B** = (10.6, 10.6) N in coordinate form.

Perform vector addition to find the net tension: (–10.6, 10.6) N + (10.6, 10.6) N = (0, 21.2) N.

To counteract this tension, the tension in rope C must have a magnitude of 21.2 N and point downward. With significant figures, the magnitude is rounded to 21 N.

NEWTON'S THIRD LAW: LOOKING AT EQUAL AND OPPOSITE FORCES

Newton's third law of motion is famous, especially in wrestling and drivers' ed circles, but you may not recognize it in all its physics glory: "Whenever one body exerts a force on a second body, the second body exerts an oppositely directed force of equal magnitude on the first body."

The more popular version of this, which we're sure you've heard many times, is "For every action, there's an equal and opposite reaction." But for physics, it's better to express the originally intended version, in terms of forces, not actions (which, from what we've seen, can apparently mean everything from voting trends to temperature forecasts!).

Seeing Newton's third law in action

Here's a real-world example to show you how Newton's third law of motion works. Say that you're in your car, speeding up with constant acceleration. To do this, your car has to exert a force against the road; otherwise, the car wouldn't be accelerating. And the road has to exert the same force on your car. You can see what this looks like, tire-wise, in Figure 5-5.

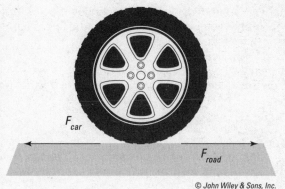

Figure 5-5: Equal forces acting on a car tire and the road during acceleration.

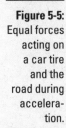

F_{car}

F_{road}

The two forces in the Figure 5-5 are equal in magnitude but opposite in direction. However, they don't cancel out because the two forces are acting on different bodies — one on the car and the other on the road. The force that the car exerts on the road is equal and opposite to the force the road exerts on the car. The force on the car accelerates it.

So why doesn't the road accelerate? The car accelerates, so shouldn't the road accelerate in the opposite direction? Believe it or not, it does; Newton's law is in full effect. Your car pushes the Earth, affecting the motion of the Earth in just the tiniest amount. Given the fact that the Earth is about 6,000,000,000,000,000,000,000 times as massive as your car, however, any effects aren't too noticeable!

Similarly, when a hockey player slaps a puck, the puck accelerates away from the spot of contact, and so does the hockey player. If hockey pucks weighed 1,000 pounds — with a mass of about 31 slugs, or 450 kilograms — you'd notice this effect much more; in fact, the puck wouldn't move much at all, but the player would hurtle off in the opposite direction after striking it. (More on what happens in this case in Part III of this book.)

Pulling hard enough to overcome friction

Because of Newton's third law, whenever you apply a force to an object, say, by pulling it, then the object applies an equal and opposite force on you. Here's an example that lets you work out how much force you're subject to when you drag something along. For fantasy physics purposes, say that a hockey game ends, and you get the job of dragging a 31-slug hockey puck off the rink. You use a rope to do the trick, as shown in Figure 5-6. The force you exert on the rope is equal to the force the rope exerts on you by Newton's third law.

Figure 5-6: Pulling a heavy puck with a rope to exert equal force on both ends.

Hockey puck

$F_{friction}$

F_{rope}

Ice rink

Remember: Physics problems are very fond of using ropes, including ropes with pulleys, because with ropes, the force you apply at one end is the same as the force that the rope exerts on what you tie it to at the other end.

In this case, the massive hockey puck will have some friction that resists you — not a terrific amount, given that it slides on top of ice, but still, some. Therefore, the net force on the puck is

$$\sum F = F_{rope} - F_{friction}$$

Because F_{rope} is greater than $F_{friction}$, the puck will accelerate and start to move. In fact, if you pull on the rope with a constant force, the puck will accelerate at a constant rate, which obeys the equation

$$\sum F = F_{rope} - F_{friction} = ma$$

Because some of the force you exert on the puck goes into accelerating it and some goes into overcoming the force of friction, the force you exert on the puck is the same as the force it exerts on you (but in the opposite direction), as Newton's third law predicts:

$$F_{rope} = F_{friction} + ma$$

6

GETTING DOWN WITH GRAVITY, INCLINED PLANES, AND FRICTION

Many physics problems involve *inclined planes* — those ramps that you're always seeing balls and carts roll down in physics classroom labs. Gravitational force is what makes carts roll down ramps, of course, but there's more to it than that. In the classic beginning physics problem, you have to resolve the gravitational force along and perpendicular to the ramp to find the acceleration of the cart along the ramp.

In the real world, you also have friction. For example, if you're unloading a refrigerator from a truck using a ramp and the refrigerator slides down the ramp by itself, friction happens, and you have to take that into account.

In this chapter, we guide you through the wonderful world of inclined planes, providing you with plenty of practice questions to ensure that you come out a pro at handling this type of physics problem.

ACCELERATION DUE TO GRAVITY: ONE OF LIFE'S LITTLE CONSTANTS

When you're on or near the surface of Earth, the pull of gravity is constant. It's a constant force directed straight down with magnitude equal to mg, where m is the mass of the object being pulled by gravity and g is the magnitude of the acceleration due to gravity:

$$g = 9.8 \text{ meters/second}^2 = 32.2 \text{ feet/second}^2$$

Acceleration is a vector, meaning it has a direction and a magnitude (see Chapter 4), so this equation really boils down to **g**, an acceleration straight down toward the center of the Earth. The fact that $\mathbf{F}_{gravity} = m\mathbf{g}$ is important because it says that the acceleration of a falling body doesn't depend on its mass:

$$\mathbf{F}_{gravity} = m\mathbf{a} = m\mathbf{g}$$

In other words, $m\mathbf{a} = m\mathbf{g}.$

Remember: Because **a** = **g** for a falling object, a heavier object doesn't fall faster than a lighter one. Gravity gives any freely falling body the same acceleration downward (*g* near the surface of Earth), assuming that no other forces, such as air resistance, are present.

FINDING A NEW ANGLE ON GRAVITY WITH INCLINED PLANES

Plenty of gravity-oriented problems in introductory physics involve inclined planes, or ramps. Gravity accelerates objects down ramps — but not the full force of gravity; only the component of gravity acting along the ramp accelerates the object. That's why an object rolling down a steep ramp rolls quickly: The ramp slopes sharply downward, close to the direction of gravity, so most of the force of gravity can act along the ramp.

Finding the force of gravity along a ramp

The first step in working with ramps of any kind is to resolve the forces that you're dealing with, and that means using vectors. (For more on vectors, check out Chapter 4.) For example, take a look at the cart in Figure 6-1; it's on an inclined plane, ready to roll.

The force on the cart is the force due to gravity, $F_g = mg$. So how fast will the cart accelerate along the ramp? To get the answer, you should resolve the gravitational force — not in the horizontal and vertical directions, however, but along the ramp's inclined plane and perpendicular to that plane.

Remember: The reason you resolve the gravitational force in these directions is because the force along the plane provides the cart's acceleration while the force perpendicular to the ramp, called the *normal force,* does not. (When you start introducing friction into the picture, you'll see that the force of friction is proportional to the normal force — that is, it's proportional to the force with which the object going down the ramp presses against the ramp.)

Try out some questions involving force and ramps.

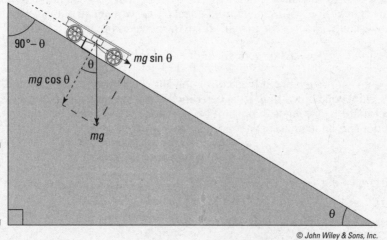

Figure 6-1:
Racing a cart down a ramp.

Examples

Q. In Figure 6-1, what are the components of the gravitational forces along the ramp and normal to the ramp?

A. The correct answer is $F_g(\sin \theta)$ along the ramp, $F_g(\cos \theta)$ normal (perpendicular to the ramp).

1. To figure out the angle between $\mathbf{F}_g$ and the ramp, look at the following figure. The angle of the ramp is given by the angle ABC. The angle at the top of the ramp is the complement of this because the angles of a triangle add up to $180°$, so the angle $BDE = 90° - \theta$. The angle BCA must be equal to the angle BDE because the triangles EBD and ABC are similar, so you can say that the angle $BCA = 90° - \theta$.

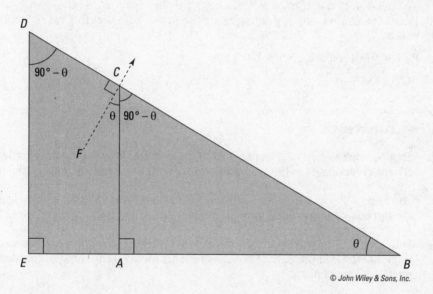

© John Wiley & Sons, Inc.

2. The angle between $\mathbf{F}_g$ and the ramp is $90° - \theta$. So what's the component of $\mathbf{F}_g$ along the ramp? Knowing the angle between $\mathbf{F}_g$ and the ramp, you can figure out the component of $\mathbf{F}_g$ along the ramp as usual:

$$\text{component of } \mathbf{F}_g \text{ along the ramp} = F_g \cos (90° - \theta)$$

3. Apply the following equation:

$$\sin \theta = \cos (90° - \theta)$$

$$\text{component of } \mathbf{F}_g \text{ along the ramp} = F_g(\cos [90° - \theta]) = F_g(\sin \theta) \text{ along the ramp}$$

Notice that this makes sense because as θ goes to $0°$, the force along the ramp also goes to 0, and as θ goes to $90°$, the force along the ramp goes to $\mathbf{F}_g$.

4. Solve for the normal force, F_N, perpendicular to the ramp: $F_g \cos\theta$. The normal force, $\mathbf{F_N}$, must exactly balance the component of the force of gravity perpendicular to the inclined plane. Apply the following equation:

$$\cos\theta = \sin(90° - \theta)$$

$$F_N = \text{force perpendicular to the ramp} = F_g\big(\sin[90° - \theta]\big) = F_g(\cos\theta)$$

Q. Suppose that you have a block of ice on a ramp at 40°, and it slides down. What is its acceleration?

A. The correct answer is 6.3 meters per second².

1. What's important here is the force along the ramp: $F_g \sin\theta = mg(\sin\theta)$.

2. The acceleration of the ice is $F/m = mg(\sin\theta)/m = g(\sin\theta)$. In other words, the acceleration is the component of **g** acting along the ramp. Note that this result is independent of mass.

3. Plug in the numbers: $g(\sin\theta) = 6.3$ m/s².

Practice Questions

1. Suppose that the cart in Figure 6-1 has a mass of 3.0 kilograms and the angle $\theta = 45°$. What are the components of the gravitational force on the cart along and normal to the ramp?

2. You have a block of ice with a mass of 10.0 kilograms on a ramp at an angle of 23°. What are the components of the gravitational force on the ice along and normal to the ramp?

3. You have a refrigerator with a mass of 1.00×10^2 kilograms on a ramp at an angle of 19°. What are the components of the gravitational force on the refrigerator along and normal to the ramp?

4. You're unloading a couch on a cart from a moving van. The couch gets away from you on the 27° ramp. Neglecting friction, what is its acceleration?

5. You have a block of ice with a mass of 10.0 kilograms on a ramp with an angle of 23° when it slips away from you. What is its acceleration down the ramp?

6. You're sliding down a toboggan run at 35°. What is your acceleration?

Practice Answers

1. **21 N along the ramp, 21 N normal to the ramp.** The forces on the cart are $F_g(\sin\theta)$ along the ramp and $F_g(\cos\theta)$ normal to the ramp.

   Plug the numbers into $F_g = mg$: $3.0(9.8) = 29$ N.

   The force along the ramp is $F_g(\sin\theta) = 29(\sin 45°) = 21$ N.

   The force normal to the ramp is $F_g(\cos\theta) = 29(\cos 45°) = 21$ N.

2. **38 N along the ramp, 90 N normal to the ramp.** The forces on the ice are $F_g(\sin \theta)$ along the ramp and $F_g(\cos \theta)$ normal to the ramp.

    Plug the numbers into $F_g = mg$: $10.0(9.8) = 98$ N.

    The force along the ramp is $F_g(\sin \theta) = 98(\sin 23°) = 38$ N.

    The force normal to the ramp is $F_g(\cos \theta) = 98(\cos 23°) = 90$ N.

3. **320 N along the ramp, 930 N normal to the ramp.** The forces on the ice are $F_g(\sin \theta)$ along the ramp and $F_g(\cos \theta)$ normal to the ramp.

    Plug the numbers into $F_g = mg$: $(1.00 \times 10^2)(9.8) = 980$ N.

    The force along the ramp is $F_g(\sin \theta) = 980(\sin 19°) = 320$ N.

    The force normal to the ramp is $F_g(\cos \theta) = 980(\cos 19°) = 930$ N.

4. **4.4 m/s².** The force along the ramp is $F_g(\sin \theta) = mg(\sin \theta)$.

    The acceleration of the cart is $F/m = mg(\sin \theta)/m = g(\sin \theta)$.

    Plugging in the numbers gives you $g(\sin \theta) = 4.4$ m/s².

5. **3.8 m/s².** The force along the ramp is $F_g(\sin \theta) = mg(\sin \theta)$.

    The acceleration of the ice is $F/m = mg(\sin \theta)/m = g(\sin \theta)$.

    Plugging in the numbers gives you $g(\sin \theta) = 3.8$ m/s².

6. **5.6 m/s².** The force along the ramp is $F_g(\sin \theta) = mg(\sin \theta)$.

    Your acceleration is $F/m = mg(\sin \theta)/m = g(\sin \theta)$.

    Plugging in the numbers gives you $g(\sin \theta) = 5.6$ m/s².

Figuring the speed along a ramp

When objects slide (frictionlessly) down a ramp, they're acted on by a force, which means that they're accelerated and therefore their speed changes. The equation to use in physics problems like these is

$$v_f^2 - v_i^2 = 2as$$

Remember: Finding the object's final speed under these circumstances is easy when you remember that $a = g(\sin \theta)$, s is the length of the ramp, and v_i is usually 0.

Example

Q. Say you have a block of ice on a ramp at 20°, and it slides down a ramp of 5.0 meters. What is its final speed at the bottom of the ramp?

A. The correct answer is 5.8 meters per second.

1. The force along the ramp is $F_g(\sin\theta) = mg(\sin\theta)$.

2. The acceleration of the ice is $F/m = mg(\sin\theta)/m = g(\sin\theta)$.

3. Use the equation $v_f^2 = 2as = 2gs(\sin\theta)$. Plug in the numbers: $v_f^2 = 34 \ \text{m}^2/\text{s}^2$, which means $v_f = 5.8 \ \text{m/s}$.

Practice Questions

1. Starting from rest, you go down a 100-meter ski jump of 60°. What is your speed at takeoff?

2. You're heading down a toboggan run of 1 kilometer at an angle of 18°. What is your final speed?

3. You have a block of ice on a ramp with an angle of 23° when it slips away from you. What is its speed at the bottom of the 6.0-meter ramp?

4. A cart starts at the top of a 50-meter slope at an angle of 38°. What is the cart's speed at the bottom?

Practice Answers

1. **40 m/s.** The force along the ramp is $F_g(\sin\theta) = mg(\sin\theta)$.

   Your acceleration is $F/m = mg(\sin\theta)/m = g(\sin\theta)$.

   Use the equation $v_f^2 = 2as = 2gs(\sin\theta)$.

   Solving for $v_f = 41 \ \text{m/s}$. Round to 40 m/s to account for significant figures.

2. **80 m/s.** The force along the ramp is $F_g(\sin\theta) = mg(\sin\theta)$.

   Your acceleration is $F/m = mg(\sin\theta)/m = g(\sin\theta)$.

   Use the equation $v_f^2 = 2as = 2gs(\sin\theta)$.

   Solving for $v_f = 78 \ \text{m/s}$. Round to 80 m/s to account for significant figures.

3. **6.8 m/s.** The force along the ramp is $F_g(\sin\theta) = mg(\sin\theta)$.

   The block of ice's acceleration is $F/m = mg(\sin\theta)/m = g(\sin\theta)$.

   Use the equation $v_f^2 = 2as = 2gs(\sin\theta)$.

   Solving for $v_f = 6.8$ m/s.

4. **25 m/s.** The force along the ramp is $F_g(\sin\theta) = mg(\sin\theta)$.

   The cart's acceleration is $F/m = mg(\sin\theta)/m = g(\sin\theta)$.

   Use the equation $v_f^2 = 2as = 2gs(\sin\theta)$.

   Solving for $v_f = 25$ m/s.

GETTING STICKY WITH FRICTION

You know all about friction. It's the force that holds objects in motion back — or so it may seem. Actually, friction is essential for everyday living. Imagine a world without friction: no way to drive a car on the road, no way to walk on pavement, no way to pick up that tasty sandwich. Friction may seem like an enemy to the hearty physics follower, but it's also your friend.

Friction comes from the interaction of surface irregularities. If you introduce two surfaces that have plenty of microscopic pits and projections, you produce friction. And the harder you press those two surfaces together, the more friction you create as the irregularities interlock more and more.

Physics has plenty to say about how friction works. For example, imagine that you decide to put all your wealth into a huge gold ingot (a bar of gold), only to have someone steal your fortune. The thief applies a force to the ingot to accelerate it away as the police start after him. Thankfully, the force of friction comes to your rescue, because the thief can't accelerate away nearly as fast as he thought — all that gold drags heavily along the ground. See Figure 6-2, which shows the forces on the gold ingot.

Figure 6-2: The forces acting on a bar of gold.

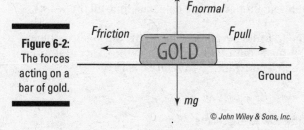

© John Wiley & Sons, Inc.

So if you want to get quantitative here, what would you do? You'd say that the pulling force, F_{pull}, minus the force due to friction, $F_{friction}$, is equal to the net force in the *x*-axis direction, which gives you the acceleration in that direction:

$$F_{pull} - F_{friction} = ma$$

That looks straightforward enough. But how do you calculate $F_{friction}$? You start by calculating the normal force.

Calculating friction and the normal force

Remember: The force of friction, $F_{friction}$, always acts to oppose the force you apply when you try to move an object. Friction is proportional to the force with which an object pushes against the surface you're trying to slide it along.

As you can see in Figure 6-2, the force with which the gold ingot presses against the ground in this situation is just its weight, or *mg.* The ground presses back with the same force in accordance with Newton's third law. The force that pushes up against the ingot, perpendicular to the surface, is called the *normal force,* and its symbol is $\mathbf{F_N}$. The normal force isn't necessarily equal to the force due to gravity; it's the force perpendicular to the surface an object is sliding on. In other words, the normal force is the force pushing the two surfaces together, and the stronger the normal force, the stronger the force due to friction.

In the case of Figure 6-2, because the ingot slides along the horizontal ground, the normal force has the same magnitude as the weight of the ingot, so $F_{normal} = mg$. You have the normal force, which is the force pressing the ingot and the ground together. But where do you go from there? You find the force of friction.

Conquering the coefficient of friction

The force of friction comes from the surface characteristics of the materials that come into contact. How can physics predict those characteristics theoretically? It doesn't. Detailed knowledge of the surfaces that come into contact is something people have to measure themselves (or they can check a table of information after someone else has done all the work).

What you measure is how the normal force (a force perpendicular to the surface an object is sliding on) relates to the friction force. It turns out that to a good degree of accuracy, the two forces are proportional, and you can use a constant, μ, to relate the two:

$$F_{friction} = \mu F_{normal}$$

Usually, you see this equation written in the following terms:

$$F_F = \mu F_N$$

This equation tells you that when you have the normal force, F_N, all you have to do is multiply it by a constant to get the friction force, F_F. This constant, μ, is called the *coefficient of friction,* and it's something you measure for contact between two particular surfaces. (*Note:* Coefficients are simply numbers; they don't have units.)

Here are a couple of things to remember:

- **The equation $F_F = \mu F_N$ relates the magnitude of the force of friction to the magnitude of the normal force.** The normal force is always directed perpendicular to the surface, and the friction force is always directed parallel to the surface. $\mathbf{F_F}$ and $\mathbf{F_N}$ are perpendicular to each other.

- **The force due to friction is generally independent of the contact area between the two surfaces.** This means that even if you have an ingot that's twice as long and half as high, you still get the same frictional force when dragging it over the ground. This makes sense, because if the area of contact doubles, you may think that you should get twice as much friction. But because you've spread out the gold into a longer ingot, you halve the force on each square centimeter, because less weight is above it to push down.

Example

Q. You're pushing a refrigerator along your kitchen floor and need to apply 100 newtons to get it moving. If your refrigerator has a mass of 100 kilograms, what is the coefficient of friction?

A. The correct answer is 0.1.

1. The force due to friction is $F_F = \mu F_N$, so $\mu = F_F/F_N$.

2. The force due to friction, F_F, is 100 newtons, and the normal force, F_N, is (100 kg) g = 980 N.

3. Use the equation $\mu = F_F/F_N$ to get the coefficient of friction (note that μ has no units): 100/980 = 0.1.

Practice Questions

1. You're pushing a 15-kilogram box of books across the carpet and need to apply 100 newtons. What is the coefficient of friction?

2. You're pushing a 70-kilogram easy chair from one room to the next. If you need to apply 200 newtons, what is the coefficient of friction?

Practice Answers

1. **0.7.** Solve for μ. The force due to friction is $F_F = \mu F_N$, so $\mu = F_F/F_N$.

   The force due to friction, F_F, is 100 N, and the normal force, F_N, is (15 kg)g = 148 N.

   Use the equation $\mu = F_F/F_N$ to get the coefficient of friction (note that μ has no units): 100/148 = 0.68. Round to 0.7 to account for significant figures.

2.   **0.3.** Solve for μ. The force due to friction is $F_F = \mu F_N$, so $\mu = F_F/F_N$.

The force due to friction, F_F, is 200 newtons, and the normal force, F_N, is (70 kg)g = 690 N.

Use the equation $\mu = F_F/F_N$ to get the coefficient of friction: 200/690 = 0.29. Round to 0.3 to account for significant figures.

ON THE MOVE: UNDERSTANDING STATIC AND KINETIC FRICTION

Okay, are you ready to get out your lab coat and start calculating the forces due to friction? Not so fast — you need to know whether the objects in contact with each other are moving. You have two different coefficients of friction for each pair of surfaces because two different physical processes are involved:

- **Static:** When two surfaces aren't moving but are pressing together, they have the chance to interlock on the microscopic level. That's static friction. The coefficient of static friction is μ_s.

- **Kinetic:** When the surfaces are sliding, the microscopic irregularities don't have the same chance to connect, and you get kinetic friction. Kinetic friction is weaker than static friction; however, for most hard, smooth surfaces, these two coefficients are quite similar. The coefficient of kinetic friction is μ_k.

Therefore, you must account for two different coefficients of friction for each pair of surfaces: a static coefficient of friction, μ_s, and a kinetic coefficient of friction, μ_k.

You can notice yourself that static friction is stronger than kinetic friction. Imagine that a box you're unloading onto a ramp starts to slide. To make it stop, you can put your foot in its way, and after you stop it, the box is more likely to stay put and not start sliding again. That's because static friction, which happens when the box is at rest, is greater than kinetic friction, which happens when the box is sliding.

Starting motion with static friction

You experience static friction when you push something that starts at rest. This is the friction that you have to overcome to get something to slide.

Example

Q.   Say that the static coefficient of friction between the ingot from Figure 6-2 and the ground is 0.30, and the ingot has a mass of 1,000 kilograms (quite a fortune in gold). What's the horizontal force that a thief has to exert to get the ingot moving?

A.   The correct answer is 2,900 newtons.

    1.   The magnitude of the force of friction is related to the magnitude of the normal force by

$$F_F = \mu_s F_N$$

    2.   Because the surface is flat, the normal force — the force that presses the two surfaces together — is in the opposite direction of the ingot's weight and has the same magnitude. That means that

$$F_F = \mu_s mg$$

where m is the mass of the ingot and g is the acceleration due to gravity near the surface of the Earth.

    3.   Plug in the numbers:

$$F_F = \mu_s mg$$
$$= (0.30)(1,000 \text{ kg})(9.8 \text{ m/s}^2)$$
$$\approx 2,900 \text{ N}$$

The thief needs about 2,900 newtons of force just to get the ingot started. There are 4.448 newtons to a pound, so that translates to about 650 pounds of force. Pretty respectable force for any thief.

Practice Questions

    1.   You're standing at the top of a ski slope and need 15 newtons of force to get yourself moving. If your mass is 60 kilograms, what is the coefficient of static friction, μ_s?

    2.   You've started to pull a garbage can out to the curb. If the can has a mass of 20 kilograms and you need to apply 70 newtons to get the can moving, what is the coefficient of static friction, μ_s?

Practice Answers

    1.   **0.03.** Solve for μ_s. The force due to friction is $F_F = \mu_s F_N$, so $\mu_s = F_F/F_N$.

        The force due to friction, F_F, is 15 newtons, and the normal force, F_N, is (60 kg)g = 590 N.

        Use the equation $\mu_s = F_F/F_N$ to get the coefficient of static friction: 15/590 = 0.03.

    2.   **0.4.** Solve for μ_s. The force due to friction is $F_F = \mu_s F_N$, so $\mu_s = F_F/F_N$.

        The force due to friction, F_F, is 70 newtons, and the normal force, F_N, is (20 kg)g = 196 N.

        Use the equation $\mu_s = F_F/F_N$ to get the coefficient of static friction: 70/196 = 0.36. From significant figures, this answer rounds to 0.4.

Sustaining motion with kinetic friction

The force due to kinetic friction, which occurs when two surfaces are already sliding, isn't quite as strong as static friction, but that doesn't mean you can predict what the coefficient of kinetic friction is going to be, even if you know the coefficient of static friction — someone has to measure both forces.

Example

Q. Say that your gold ingot from Figure 6-2, which has a mass of 1,000 kilograms, has a coefficient of kinetic friction, μ_k, of 0.18. How much force does the thief need to pull the ingot along at a constant speed during his robbery?

A. The correct answer is 1,800 newtons.

1. The magnitude of the kinetic coefficient of friction is related to the magnitude of the normal force by

$$F_F = \mu_k F_N = \mu_k mg$$

2. Put in the numbers:

$$F_F = \mu_k mg$$
$$= (0.18)(1{,}000 \text{ kg})(9.8 \text{ m/s}^2)$$
$$\approx 1{,}800 \text{ N}$$

The thief needs approximately 1,800 newtons of force to keep your gold ingot sliding while evading the police. That converts to about 400 pounds of force (4.448 newtons to a pound) — not exactly the kind of force you can keep going while trying to run at top speed, unless you have some friends helping you. Lucky you! Physics states that the police are able to recover your gold ingot. The cops know all about friction — taking one look at the prize, they say, "We got it back. You drag it home."

Practice Questions

1. You're dragging your pet dog toward the bathtub. If he has a mass of 15.0 kilograms and the coefficient of kinetic friction, μ_k, is 0.10, how much force will you need to apply to keep him going?

2. You're pushing a box of books with a mass of 25 kilograms across a level surface. If the coefficient of kinetic friction, μ_k, is 0.27, how much force will you need to apply to keep the box moving?

Practice Answers

1. **15 N.** You need to overcome the force due to kinetic friction: $F = \mu_k F_N$. You need to find the normal force.

   The equation for normal force is $F_N = mg$. Use the normal force to calculate the force due to kinetic friction: $\mu_k mg$.

   Plug in the numbers: $F = \mu_k mg = (0.10)(15.0 \text{ kg})(9.8 \text{ m/s}^2) = 15$ N.

2. **66 N.** You need to overcome the force due to kinetic friction: $F = \mu_k F_N$. You need to find the normal force.

   The equation for normal force is $F_N = mg$. Use the normal force to calculate the force due to kinetic friction: $\mu_k mg$.

   Plug in the numbers: $F = \mu_k mg(\cos \theta) = (0.27)(25 \text{ kg})(9.8 \text{ m/s}^2) = 66$ N.

A not-so-slippery slope: Handling uphill and downhill friction

Frictional forces depend on the normal force acting. However, when frictional forces are acting on a ramp, the angle of the ramp tilts the normal force at an angle. When you work out the frictional forces, you need to take this into account. Here are a few examples and practice questions so you can see how this works.

Examples

Q. Suppose that you're pushing a 40.0-kilogram plastic crate up a 19.0° ramp. The coefficient of kinetic friction, μ_k, is 0.100, and you need to apply a force to keep the crate moving. What is the force you will need to apply?

A. The correct answer is 165 newtons.

1. The force due to kinetic friction along the ramp is $F_k = \mu_k F_N$.
2. The normal force, F_N, is $F_N = mg(\cos \theta)$. That makes the force due to static friction $\mu_k mg(\cos \theta)$.
3. The force due to gravity along the ramp is $F_g(\sin \theta)$.

4. Both of these forces point down the ramp and need to be overcome by the force pushing up the ramp. So in other words, F_{push} is

$$F_{push} = F_g(\sin\theta) + F_k$$

5. Plug in your earlier results: $F = F_g(\sin\theta) + F_k = mg(\sin\theta) + \mu_k mg(\cos\theta)$.

6. Putting in the numbers gives you F = 128 N + 37 N = 165 N.

Q. The following figure shows a box on a ramp. Suppose that the box contains a new flat-screen TV that you're pushing up the ramp and into your house.

Suppose that the flat-screen TV's box has a mass of 1.00×10^2 kilograms, and the ramp has an angle of 23°. What is the force needed to get the box moving up the ramp if the coefficient of static friction is 0.20?

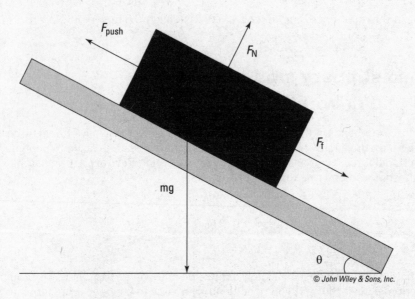

© John Wiley & Sons, Inc.

A. The correct answer is 563 newtons.

1. The force due to static friction along the ramp is $F_s = \mu_s F_N$.

2. The normal force, F_N, is $F_N = mg(\cos\theta)$. That makes the force due to static friction $\mu_s mg(\cos\theta)$.

3. The force due to gravity along the ramp is $F_g(\sin\theta)$.

4. Both of these forces point down the ramp and need to be overcome by the force pushing up the ramp. So in other words, F_{push} is

$$F_{push} = F_g(\sin\theta) + F_s$$

5. Plug in these forces:

$$F_{push} = F_g(\sin\theta) + F_s = mg(\sin\theta) + \mu_s mg(\cos\theta)$$

6. Plug in the numbers: F_{push} = 383 N + 180 N = 563 N.

Q. A plastic crate slips down a 19° ramp with a coefficient of kinetic friction, μ_k, of 0.10. What is its acceleration as it slides?

A. The correct answer is 2.3 meters per second^2.

1. The object is sliding down the ramp — you're not pushing it — which means the force of kinetic friction is opposing (not adding to) the component of gravity along the ramp. So the net force on the crate is

$$F = F_g(\sin\theta) - F_k = mg(\sin\theta) - \mu_k mg(\cos\theta)$$

2. The force due to gravity along the ramp is $F_g(\sin\theta)$.

3. The force due to kinetic friction along the ramp is $F_k = \mu_k F_N$. This force points up the ramp, because friction opposes the motion of the crate.

4. The net force on the crate is

$$F = F_g(\sin\theta) - F_k = mg(\sin\theta) - \mu_k mg(\cos\theta)$$

5. Because $a = F/m$, the acceleration of the crate is

$$a = g(\sin\theta) - \mu_k g(\cos\theta)$$

6. Plug in the numbers: a = 3.2 m/s^2 – 0.93 m/s^2 = 2.3 m/s^2.

Practice Questions

1. You're dragging your little brother up the 25° wheelchair ramp at the doctor's office. If he has a mass of 40.0 kilograms and the coefficient of static friction, μ_s, is 0.15, how much force will you need to apply to get him moving?

2. Suppose that you're struggling to get a 20.0-kilogram block of ice moving up a 40.0° ramp. If the coefficient of static friction, μ_s, is a low 0.050, how much force will you need to apply to overcome the weight pulling the block down the ramp and static friction?

3. You're pushing a box of books with a mass of 25 kilograms up a 40° ramp. If the coefficient of kinetic friction, μ_k, is 0.27, how much force will you need to apply to keep the box moving up the ramp?

4. You want to keep a 120-kilogram refrigerator moving up a 23° ramp. If the coefficient of kinetic friction, μ_k, is 0.20, how much force will you need to keep it moving?

5.   You drop a 1.0-kilogram book on a 15° ramp, and the coefficient of kinetic friction is 0.30. Will the book slide down the ramp?

6.   A refrigerator breaks away from the movers and slides down a 23° ramp that has a coefficient of kinetic friction of 0.25. What is its acceleration?

Practice Answers

1.   **220 N.** Calculate the forces you need to overcome: The force due to gravity is $mg(\sin\theta)$; and the force due to friction, F_F, is $F = \mu_s F_N$. You need to find the normal force.

The equation for normal force is $F_N = mg(\cos\theta)$. Use the normal force to calculate the force due to friction: $\mu_s mg(\cos\theta)$.

The total force you have to overcome is $F = mg(\sin\theta) + \mu_s mg(\cos\theta)$.

Plug in the numbers: $F = mg(\sin\theta) + \mu_s mg(\cos\theta) = 166\ N + 53\ N = 219\ N$. The final answer should be rounded to 220 N. (The value for g is only known to two significant figures [9.8] because it varies in the third digit depending on location on Earth.)

2.   **130 N.** Calculate the forces you need to overcome: The force due to gravity is $mg(\sin\theta)$; and the force due to friction, F_F, is $F = \mu_s F_N$. You need to find the normal force.

The equation for normal force is $F_N = mg(\cos\theta)$. Use the normal force to calculate the force due to friction: $\mu_s mg(\cos\theta)$.

The total force you have to overcome is $F = mg(\sin\theta) + \mu_s mg(\cos\theta)$.

Plug in the numbers: $F = mg(\sin\theta) + \mu_s mg(\cos\theta) = 126\ N + 7.5\ N = 133\ N$. Note that most of this force is due to the component of the weight along the ramp. The final answer should be rounded to 130 N.

3.   **210 N.** Calculate the forces you need to overcome: The force due to gravity is $mg(\sin\theta)$; and the force due to kinetic friction is $F = \mu_k F_N$. You need to find the normal force.

The equation for normal force is $F_N = mg(\cos\theta)$. Use the normal force to calculate the force due to kinetic friction: $\mu_k mg(\cos\theta)$.

The force you have to overcome to keep the box moving is $F = mg(\sin\theta) + \mu_k mg(\cos\theta)$.

Plug in the numbers: $F = mg(\sin\theta) + \mu_k mg(\cos\theta) = 160\ N + 50\ N = 210\ N$.

4.   **680 N.** Calculate the forces you need to overcome: The force due to gravity is $mg(\sin\theta)$; and the force due to kinetic friction is $F = \mu_k F_N$. You need to find the normal force.

The equation for normal force is $F_N = mg(\cos\theta)$. Use the normal force to calculate the force due to kinetic friction: $\mu_k mg(\cos\theta)$.

The force you have to overcome to keep the refrigerator moving is $F = mg(\sin\theta) + \mu_k mg(\cos\theta)$.

Plug in the numbers: $F = mg(\sin\theta) + \mu_k mg(\cos\theta) = 460\ N + 220\ N = 680\ N$.

5. **No, the book will not slide.** Calculate the forces on the book, and if the force down the ramp is larger than the force up the ramp, the book will slide. The force down the ramp is $mg(\sin \theta)$.

    The force due to kinetic friction pointing up the ramp is $F = \mu_k F_N$, which means you need to find the normal force.

    The equation for the normal force is $F_N = mg(\cos \theta)$. Use the normal force to calculate the force due to kinetic friction: $\mu_k mg(\cos \theta)$.

    The net force on the book down the ramp is $F = mg(\sin \theta) - \mu_k mg(\cos \theta)$.

    Plug in the numbers: $F = mg(\sin \theta) - \mu_k mg(\cos \theta) = 2.5 \, N - 2.8 \, N = -0.3 \, N$ along the ramp, so the book will not slide. (Remember that the coefficient of static friction is always greater than the coefficient of kinetic friction.)

6. **1.5 m/s² down the ramp.** Calculate the forces on the refrigerator: The force down the ramp is $mg(\sin \theta)$; and the force due to kinetic friction pointing up the ramp is $F = \mu_k F_N$. You need to find the normal force.

    The equation for the normal force is $F_N = mg(\cos \theta)$. Use the normal force to calculate the force due to kinetic friction: $\mu_k mg(\cos \theta)$.

    The net force on the refrigerator down the ramp is $F = mg(\sin \theta) - \mu_k mg(\cos \theta)$.

    Divide by m to get the acceleration: $a = g \sin \theta - \mu_k g \cos \theta$.

    Plug in the numbers: $a = g(\sin \theta) - \mu_k g(\cos \theta) = 3.8 \, m/s^2 - 2.3 \, m/s^2 = 1.5 \, m/s^2$.

7

CIRCLING AROUND ROTATIONAL MOTION AND ORBITS

~~~~~~

IN THIS CHAPTER

- Working with centripetal acceleration
- Feeling the pull of centripetal force
- Incorporating angular displacement, velocity, and acceleration
- Orbiting with Newton's laws and gravity

~~~~~~

Circular motion can include rockets' moving around planets, race cars' whizzing around a track, or bees' buzzing around a hive. In this chapter, you look at the velocity and acceleration of objects that are moving in circles. This discussion leads to more general forms of rotational motion, where it's useful to talk about motion in angular terms.

Angular equivalents exist for displacement, velocity, and acceleration. Instead of dealing with linear displacement as a distance, you deal with angular displacement as an angle. Angular velocity indicates what angle you sweep through in so many seconds, and angular acceleration gives you the rate of change in the angular velocity. All you have to do is take linear equations and substitute the angular equivalents: angular displacement for displacement, angular velocity for velocity, and angular acceleration for acceleration.

CENTRIPETAL ACCELERATION: CHANGING DIRECTION TO MOVE IN A CIRCLE

In order to keep an object moving in circular motion, its velocity constantly changes direction. Because velocity changes, you have acceleration. Specifically, you have *centripetal acceleration* — the acceleration needed to keep an object moving in circular motion. At any point, the velocity of the object is perpendicular to the radius of the circle.

If the string holding the ball in Figure 7-1 breaks at the top, bottom, left, or right moment you see in the illustration, which way would the ball go? If the velocity points to the left, the ball would fly off to the left. If the velocity points to the right, the ball would fly off to the right. And so on. That's not intuitive for many people, but it's the kind of physics question that may come up in introductory courses.

Remember: The velocity of an object in circular motion is always at right angles to the radius of the object's path. At any one moment, the velocity points along the tiny section of the circle's circumference where the object is, so the velocity is tangential to the circle.

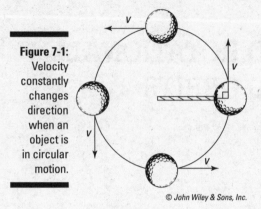

Figure 7-1:
Velocity
constantly
changes
direction
when an
object is
in circular
motion.

© John Wiley & Sons, Inc.

Keeping a constant speed with uniform circular motion

An object with *uniform circular motion* travels in a circle with a constant speed. Practical examples may be hard to come by, unless you see a race car driver on a perfectly circular track with his accelerator stuck, a clock with a seconds hand that's in constant motion, or the moon orbiting the Earth.

Take a look at Figure 7-2, where a golf ball tied to a string is whipping around in circles. The golf ball is traveling at a uniform speed as it moves around in a circle, so you can say it's traveling in uniform circular motion.

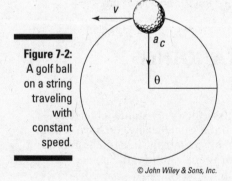

Figure 7-2:
A golf ball
on a string
traveling
with
constant
speed.

© John Wiley & Sons, Inc.

Remember: An object in uniform circular motion doesn't travel with a uniform velocity because its direction changes all the time.

Any object that travels in uniform circular motion always takes the same amount of time to move completely around the circle. That time is called its *period,* designated by T.

If you're swinging a golf ball around on a string at a constant speed, you can easily relate the ball's speed to its period. You know that the distance the ball must travel each time around the circle equals the circumference of the circle, which is $2\pi r$ (where r is the radius of the circle), so you can get the equation for finding an object's period by first finding its speed:

$$v = \frac{\text{circumference}}{\text{period}} = \frac{2\pi r}{T}$$

Remember: If you solve for T, you get the equation for the period:

$$T = \frac{2\pi r}{v}$$

Another time measurement you'll see in physics problems is *frequency*. Whereas the period is the time an object takes to go around in a circle, the frequency is the number of circles the object makes per second. The frequency, f, is connected to the period like this:

$$f = \frac{1}{T}$$

Example

Q. The moon's orbital radius is 3.85×10^8 meters, and its period is about 27.3 days. What is its speed as it goes around Earth?

A. The correct answer is 1,020 meters per second, when rounded for significant figures.

1. Convert 27.3 days to seconds:

$$27.3 \text{ days} \times \frac{24 \text{ hr}}{\text{day}} \times \frac{60 \text{ min}}{\text{hr}} \times \frac{60 \text{ s}}{\text{min}} = 2.36 \times 10^6 \text{ s}$$

2. Use the equation for the period to solve for speed:

$$v = \frac{2\pi r}{T}$$

3. Plug in the numbers:

$$v = \frac{2\pi r}{T} = \frac{2\pi \left(3.85 \times 10^8 \text{ m}\right)}{2.36 \times 10^6 \text{ s}} = 1,024 \text{ m/s}$$

Practice Questions

1. You have a ball on a string, and you're whipping it around in a circle. If the radius of its circle is 1.0 meters and its period is 1.0 seconds, what is its speed?

2. You have a toy plane on a wire, and it's traveling around in a circle. If the radius of its circle is 10.0 meters and its period is 0.75 seconds, what is its speed?

Practice Answers

1. **6.3 m/s.** Use the equation for the period to solve for speed:

$$v = \frac{2\pi r}{T}$$

Plug in the numbers:

$$v = \frac{2\pi r}{T} = \frac{2\pi (1.0 \text{ m})}{1.0 \text{ s}} = 6.3 \text{ m/s}$$

2. **84 m/s.** Use the equation for the period to solve for speed:

$$v = \frac{2\pi r}{T}$$

Plug in the numbers:

$$v = \frac{2\pi r}{T} = \frac{2\pi (10.0 \text{ m})}{0.75 \text{ s}} = 84 \text{ m/s}$$

Finding the magnitude of the centripetal acceleration

When an object travels in uniform circular motion, its speed is constant, which means that the magnitude of the object's velocity doesn't change. Therefore, acceleration can have no component in the same direction as the velocity; if it did, the velocity's magnitude would change.

However, the velocity's direction is constantly changing — it always bends so that the object maintains movement in a constant circle. To make that happen, the object's centripetal acceleration is always concentrated toward the center of the circle, perpendicular to the object's velocity at any one time. The acceleration changes the direction of the object's velocity while keeping the magnitude of the velocity constant.

In the ball's case (refer to Figures 7-1 and 7-2), the string exerts a force on the ball to keep it going in a circle — a force that provides the ball's centripetal acceleration. To provide that force, you have to constantly pull on the ball toward the center of the circle. (Picture what it feels like, force-wise, to whip an object around on a string.) You can see the centripetal acceleration vector, $\mathbf{a}_c$, in Figure 7-2.

If you accelerate the ball toward the center of the circle to provide the centripetal acceleration, why doesn't it hit your hand? The answer is that the ball is already moving at a high speed. The force, and therefore the acceleration, that you provide always acts at right angles to the velocity.

Remember: You always have to accelerate an object toward the center of the circle to keep it moving in circular motion. So can you find the magnitude of the acceleration you create?

No doubt. If an object is moving in uniform circular motion at speed v and radius r, you can find the magnitude of the centripetal acceleration with the following equation:

$$a_c = \frac{v^2}{r}$$

For a practical example, imagine you're driving around curves at a high speed. For any constant speed, you can see from the equation $a_c = v^2/r$ that the centripetal acceleration is inversely proportional to the radius of the curve. In other words, on tighter curves (as the radius decreases), your car needs to provide a greater centripetal acceleration (the acceleration increases).

Example

Q. Given that the moon goes around Earth about every 27.3 days and that its distance from the center of Earth is 3.85×10^8 meters, what is the moon's centripetal acceleration?

A. The correct answer is 2.7×10^{-3} meters per second^2.

   1. Start with this equation:

   $$a_c = \frac{v^2}{r}$$

   2. Find the speed of the moon. It goes $2\pi r$ in 27.3 days, so convert 27.3 to seconds:

   $$27.3 \ \text{days} \times \frac{24 \ \text{hr}}{\text{day}} \times \frac{60 \ \text{min}}{\text{hr}} \times \frac{60 \ \text{s}}{\text{min}} = 2.36 \times 10^6 \ \text{s}$$

   3. Therefore, the speed of the moon is

   $$\frac{2\pi r}{T} = \frac{2\pi \left(3.85 \times 10^8 \ \text{m}\right)}{2.36 \times 10^6 \ \text{s}} = 1{,}024 \ \text{m/s}$$

   4. Plug in the numbers:

   $$a_c = \frac{v^2}{r} = \frac{\left(1{,}024 \ \text{m/s}\right)^2}{3.85 \times 10^8 \ \text{m}} = 2.7 \times 10^{-3} \ \text{m/s}^2$$

Practice Questions

   1. The tips of a helicopter's blades are moving at 300 meters per second and have a radius of 7.0 meters. What is the centripetal acceleration of those tips?

   2. Your ball on a string is revolving around in a circle. If it's going 60 miles per hour at a radius of 2.0 meters, what is its centripetal acceleration?

Practice Answers

1. **1.3×10^4 m/s^2.** Use this equation:

$$a_c = \frac{v^2}{r}$$

Plug in the numbers:

$$a_c = \frac{v^2}{r} = \frac{(300 \text{ m/s})^2}{7.0 \text{ m}} = 1.3 \times 10^4 \text{ m/s}^2$$

2. **360 m/s^2.** Use this equation:

$$a_c = \frac{v^2}{r}$$

Convert the 60 miles per hour speed to meters per second:

$$\frac{60 \text{ mi}}{\text{hr}} \times \frac{1.609 \text{ km}}{\text{mi}} \times \frac{1,000 \text{ m}}{\text{km}} \times \frac{1 \text{ hr}}{3,600 \text{ s}} = 26.8 \text{ m/s}$$

Plug in the numbers:

$$a_c = \frac{v^2}{r} = \frac{(26.8 \text{ m/s})^2}{2.0 \text{ m}} = 360 \text{ m/s}^2$$

GETTING ANGULAR WITH DISPLACEMENT, VELOCITY, AND ACCELERATION

For objects moving in a circle, you can work with acceleration and velocity using the horizontal and vertical components, just as in previous chapters on motion. But when objects are undergoing rotational motion, using angular variables instead makes a lot of sense. With these variables, instead of specifying the horizontal and vertical components, you specify the radius and the angle of rotation.

In this section, you discover the angular equivalents of displacement, velocity, and acceleration. You can apply these variables to rotating objects and objects moving in a circle.

Measuring angles in radians

Remember: The natural unit of measurement of angles is the radian, not the degree. A full circle is made up of 2π radians, which is also $360°$, so $360° = 2\pi$ radians. If you travel in a full circle, you go 360°, or 2π radians. (If an object rotates one revolution, then the angle has magnitude of 2π radians. Therefore, sometimes instead of *radians per second,* you see *revolutions per second.*) A half-circle is π radians, and a quarter-circle is $\pi/2$ radians.

The radian is a natural measure of angle because a circular arc that has a length of one radius extends an angle of 1 radian (see Figure 7-3). So if you know the radius and the angle that an object has moved through in radians, you can easily find the distance that the object has moved in proportion to the radius. If the object moves θ radians in a circle of radius r, then the object travels a distance of θr along the circle.

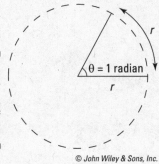

Figure 7-3:
A circular arc extends an angle of one radian.

© John Wiley & Sons, Inc.

This idea is useful in relating the angular velocity to the speed of an object moving in a circle. In addition, you can see why a full circle has an angle of 2π radians: You know that the circumference of a circle is $2\pi r$ and that to go the whole way around the 360° of a circle, you need to travel 2π times the radius. Therefore, there are 2π radians to 360°.

Tip: How do you convert from degrees to radians and back again? Because 360° = 2π radians (or 2 multiplied by 3.14, the rounded version of pi), you have an easy calculation.

Example

Q. Convert 180° into radians.

A. The correct answer is π.

1. Use the conversion factor $\pi/180°$.

2. Plug in the numbers:

$$180°\left(\frac{\pi}{180°}\right) = \pi$$

Practice Questions

1. What is 23° in radians?

2. What is π/16 in degrees?

Practice Answers

1. **0.40 radians.** Use the conversion factor π/180°.

Plug in the numbers:

$$23° \left(\frac{\pi}{180°} \right) = 0.13\pi = 0.40 \text{ radians}$$

2. **11.2°.** Use the conversion factor 180°/π.

Plug in the numbers:

$$\frac{\pi 180°}{16\pi} = 11.2°$$

Relating linear and angular motion

Tip: The fact that you can think of the angle, *θ,* in rotational motion just as you think of the displacement, *s,* in linear motion is great, because it means you have an angular counterpart for many of the linear motion equations (see Chapter 3). Here are the variable substitutions you make to get the angular motion formulas:

- **Displacement:** Instead of *s,* which you use in linear travel, use *θ,* the angular displacement; *θ* is measured in radians.

- **Velocity:** In place of the velocity, *v,* use the angular velocity, *ω;* angular velocity is the number of radians covered per second.

- **Acceleration:** Instead of acceleration, *a,* use the *angular acceleration, α;* the unit for angular acceleration is radians per second².

Table 7-1 compares the formulas for both linear and angular motion.

Table 7-1   Linear and Angular Motion Formulas

Type of Formula	Linear	Angular
Velocity	$v = \dfrac{\Delta s}{\Delta t}$	$\omega = \dfrac{\Delta \theta}{\Delta t}$
Acceleration	$a = \dfrac{\Delta v}{\Delta t}$	$\alpha = \dfrac{\Delta \omega}{\Delta t}$
Displacement	$s = v_i t + \dfrac{1}{2} a t^2$	$\theta = \omega_i t + \dfrac{1}{2} \alpha t^2$
Motion with time canceled out	$v_f{}^2 - v_i{}^2 = 2as$	$\omega_f{}^2 - \omega_i{}^2 = 2\alpha\theta$

GETTING INTO ANGULAR VELOCITY

The symbol for angular velocity is ω, so you can write the equation for angular velocity this way (this is actually a vector equation, of course, but we don't get into the vector nature of angular motion until Chapter 11, so we look at this equation in scalar terms):

$$\omega = \frac{\Delta \theta}{\Delta t}$$

Figure 7-4 shows a line sweeping around in a circle. At a particular moment, it's at angle θ, and if it took time t to get there, its angular velocity is $\omega = \theta/t$.

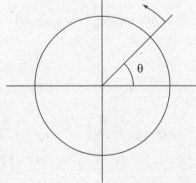

Figure 7-4:
Angular
velocity in a
circle.

© John Wiley & Sons, Inc.

So if the line in Figure 7-5 completes a full circle in 1.0 seconds, its angular velocity is $2\pi/1.0$ s $= 2\pi$ radians/s (because there are 2π radians in a complete circle). Technically speaking, radian isn't a physical unit of measure (it's a ratio), so the angular velocity can also be written 2π s^{-1}.

Given the angular velocity, you also can find the angle swept through in a number of seconds:

$$\Delta \theta = \omega \cdot \Delta t$$

Example

Q. The moon goes around Earth in about 27.3 days. What is its angular velocity?

A. The correct answer is 2.66×10^{-6} radians per second.

1. Convert 27.3 days to seconds:

$$27.3 \ \text{days} \times \frac{24 \ \text{hr}}{\text{day}} \times \frac{60 \ \text{min}}{\text{hr}} \times \frac{60 \ \text{s}}{\text{min}} = 2.36 \times 10^{6} \ \text{s}$$

2. Use the equation for angular velocity:

$$\omega = \frac{\Delta\theta}{\Delta t}$$

3. Plug in the numbers:

$$\omega = \frac{\Delta\theta}{\Delta t} = \frac{2\pi}{2.36 \times 10^{6} \ \text{s}} = 2.66 \times 10^{-6} \ \text{radians/s}$$

Practice Questions

1. You have a toy plane on a string that goes around three complete circles in 9.0 seconds. What is its angular velocity?

2. A satellite is orbiting Earth at 8.7×10^{-4} radians per second. How long will it take to circle the entire world?

Practice Answers

1. **2.1 radians/s.** Use the equation for the angular velocity:

$$\omega = \frac{\Delta\theta}{\Delta t}$$

Plug in the numbers:

$$\omega = \frac{\Delta\theta}{\Delta t} = \frac{(3)(2\pi)}{9.0 \ \text{s}} = \left(\frac{2}{3}\right)\pi \ \text{radians/s} = 2.1 \ \text{radians/s}$$

2. **120 min.** Start with the equation for the angular velocity:

$$\omega = \frac{\Delta\theta}{\Delta t}$$

Solve for Δt:

$$\Delta t = \frac{\Delta \theta}{\omega}$$

Plug in the numbers:

$$\Delta t = \frac{\Delta \theta}{\omega} = \frac{2\pi}{8.7 \times 10^{-4} \text{ radians/s}} = 7,200 \text{ s}$$

That's about 120 minutes.

WHIPPING AROUND WITH ANGULAR ACCELERATION

Just as with linear motion, you can have acceleration when you're dealing with angular motion. For example, the line in Figure 7-5 may be sweeping around the circle faster and faster, which means it's accelerating.

In linear motion, the following is the equation for *acceleration,* the rate at which the object's velocity is changing:

$$a = \frac{\Delta v}{\Delta t}$$

As with all the equations of motion, you need only to substitute the correct angular quantities for the linear ones. In this case, v becomes ω. So the angular acceleration is $\Delta\omega/\Delta t$.

The symbol for linear acceleration is a, and the symbol for angular acceleration is α, which makes the equation for angular acceleration as follows:

$$\alpha = \frac{\Delta \omega}{\Delta t}$$

The unit for angular acceleration is radians per second[2] (or, technically, just seconds^{-2}).

Example

Q. Your toy plane on a string accelerates from $\omega = 2.1$ radians per second to 3.1 radians per second in 1.0 seconds. What is its angular acceleration?

A. The correct answer is 1.0 radians per second^2.

1. Use the equation for angular acceleration:

$$\alpha = \frac{\Delta \omega}{\Delta t}$$

2. Plug in the numbers:

$$\alpha = \frac{\Delta \omega}{\Delta t} = \frac{(3.1 \text{ radians/s} - 2.1 \text{ radians/s})}{1.0 \text{ s}} = 1.0 \text{ radians/s}^2$$

Practice Questions

1. You're square dancing, turning your partner around at 1.0 radians per second. Then you speed up for 0.50 seconds at an angular acceleration of 10.0 radians per second². What is your partner's final angular speed?

2. You're trying a new yoga move, and, starting your arm at rest, you accelerate it at 15 radians per second² over 1.0 second. What's your arm's final angular velocity?

Practice Answers

1. **6.0 radians/s.** Use the equation for angular acceleration:

$$\alpha = \frac{\Delta\omega}{\Delta t}$$

Solve for $\Delta\omega$:

$$\Delta\omega = \alpha\Delta t$$

Plug in the numbers:

$$\Delta\omega = \alpha\,\Delta t = (10.0 \text{ radians/s}^2)(0.50\,\text{s}) = 5.0 \text{ radians/s}$$

Add $\Delta\omega$ to the initial angular velocity, ω_i:

$$1.0\,\text{radian/s} + 5.0 \text{ radians/s} = 6.0 \text{ radians/s}$$

2. **15 radians/s.** Use the equation for angular acceleration:

$$\alpha = \frac{\Delta\omega}{\Delta t}$$

Solve for $\Delta\omega$:

$$\Delta\omega = \alpha\,\Delta t$$

Plug in the numbers:

$$\Delta\omega = \alpha\,\Delta t = (15 \text{ radians/s}^2)(1.0\text{ s}) = 15 \text{ radians/s}$$

Add $\Delta\omega$ to the initial angular velocity, ω_i:

$$0\,\text{radians/s} + 15 \text{ radians/s} = 15 \text{ radians/s}$$

CONNECTING ANGULAR VELOCITY AND ANGULAR ACCELERATION TO ANGLES

You can connect the distance traveled to the original velocity and linear acceleration like this:

$$s = v_i\left(t_f - t_i\right) + \frac{1}{2}a\left(t_f - t_i\right)^2$$

And you can make the substitution from linear to angular motion by putting in the appropriate symbols:

$$\theta = \omega_i\left(t_f - t_i\right) + \frac{1}{2}\alpha\left(t_f - t_i\right)^2$$

Using this equation, you can connect angular velocity, angular acceleration, and time to the angle.

Example

Q. A marble is rolling around a circular track at 6.0 radians per second and then accelerates at 1.0 radian per second². How many radians has it gone through in 1 minute?

A. The correct answer is 2,200 radians.

1. Use this equation:

$$\theta = \omega_i\left(t_f - t_i\right) + \frac{1}{2}\alpha\left(t_f - t_i\right)^2$$

2. Plug in the numbers:

$$\theta = \omega_i\left(t_f - t_i\right) + \frac{1}{2}\alpha\left(t_f - t_i\right)^2$$

$$= \left(6.0 \text{ radians/s}\right)\left(60 \text{ s}\right) + \frac{1}{2}\left(1.0 \text{ radians/s}^2\right)\left(60 \text{ s}\right)^2$$

$$= 360 \text{ radians} + 1{,}800 \text{ radians}$$

$$= 2{,}160 \text{ radians, which rounds to 2,200 radians}$$

Practice Questions

1. Your model globe is spinning at 1.0 radians per second when you give it a push. If you accelerate it at 5.0 radians per second², how many radians has it turned through in 5.0 seconds?

2. A roulette wheel is slowing down, starting from 12.0 radians per second and going through 40.0 radians in 5.0 seconds. What was its angular acceleration?

Practice Answers

1. **68 radians.** Use this equation:

$$\theta = \omega_i \left(t_f - t_i\right) + \frac{1}{2}\alpha\left(t_f - t_i\right)^2$$

Plug in the numbers:

$$\theta = \omega_i \left(t_f - t_i\right) + \frac{1}{2}\alpha\left(t_f - t_i\right)^2$$

$$= \left(1.0 \text{ radians/s}\right)\left(5.0 \text{ s}\right) + \frac{1}{2}\left(5.0 \text{ radians/s}^2\right)\left(5.0 \text{ s}\right)^2$$

$$= 5.0 \text{ radians} + 62.5 \text{ radians} = 67.5 \text{ radians},$$

which rounds to 68 radians with significant figures

2. **–1.6 radians/s².** Use this equation:

$$\theta = \omega_i \left(t_f - t_i\right) + \frac{1}{2}\alpha\left(t_f - t_i\right)^2$$

Solve for α, given that $t_o = 0$:

$$\alpha = \frac{2\left(\theta - \omega_i t_f\right)}{t_f^2}$$

Plug in the numbers:

$$\alpha = \frac{2\left(\theta - \omega_i t_f\right)}{t_f^2}$$

$$= \frac{2\left(40 \text{ radians} - \left[12.0 \text{ radians/s}\right]\left[5.0 \text{ s}\right]\right)}{\left(5.0 \text{ s}\right)^2}$$

$$= -1.6 \text{ radians/s}^2$$

CONNECTING ANGULAR ACCELERATION AND ANGLE TO ANGULAR VELOCITY

You can connect angle, angular velocity, and angular acceleration. The corresponding equation for linear motion is

$$v_f^2 - v_i^2 = 2as$$

Substituting ω for v, α for a, and θ for s gives you:

$$\omega_f^2 - \omega_i^2 = 2\alpha$$

Use this equation when you want to relate angle to angular velocity and angular acceleration.

Example

Q. A merry-go-round slows down from 6.5 radians per second to 2.5 radians per second, undergoing an angular acceleration of 1.0 radians per second^2. How many radians does the merry-go-round go through while this is happening?

A. The correct answer is 18 radians.

1. Start with the equation:

$$\omega_f^2 - \omega_i^2 = 2\alpha$$

2. Solve for θ:

$$\theta = \frac{\omega_f^2 - \omega_i^2}{2\alpha}$$

3. Plug in the numbers:

$$\theta = \frac{\omega_f^2 - \omega_i^2}{2\alpha}$$
$$= \frac{(2.5 \text{ radians/s})^2 - (6.5 \text{ radians/s})^2}{2(-1.0 \text{ radians/s}^2)}$$
$$= 18 \text{ radians}$$

Practice Questions

1. A helicopter's blades are speeding up. They go from 60 radians per second to 80 radians per second. If the angular acceleration is 10 radians per second^2, what is the total angle the blades have gone through?

2. Your ball on a string is traveling around in a circle. If it goes from 12 radians per second to 24 radians per second and the angular acceleration is 20 radians per second^2, what is the total angle the ball has gone through during this acceleration?

Practice Answers

1. **140 radians.** Use this equation:

$$\omega_f^2 - \omega_i^2 = 2\alpha$$

Solve for θ:

$$\theta = \frac{\omega_f^2 - \omega_i^2}{2\alpha}$$

Plug in the numbers:

$$\theta = \frac{\omega_f^{\,2} - \omega_i^{\,2}}{2\alpha}$$

$$= \frac{\left(80 \text{ radians/s}\right)^2 - \left(60 \text{ radians/s}\right)^2}{2\left(10 \text{ radians/s}^2\right)}$$

$$= 140 \text{ radians}$$

2.   **11 radians.** Use this equation:

$$\omega_f^{\,2} - \omega_i^{\,2} = 2\alpha$$

Solve for θ:

$$\theta = \frac{\omega_f^{\,2} - \omega_i^{\,2}}{2\alpha}$$

Plug in the numbers:

$$\theta = \frac{\omega_f^{\,2} - \omega_i^{\,2}}{2\alpha}$$

$$= \frac{\left(24 \text{ radians/s}\right)^2 - \left(12 \text{ radians/s}\right)^2}{2\left(20 \text{ radians/s}^2\right)}$$

$$= 11 \text{ radians}$$

SEEKING THE CENTER: CENTRIPETAL FORCE

When you're driving a car around a bend, you create centripetal acceleration by the friction of your tires on the road. How do you know what force you need to create to turn the car at a given speed and turning radius? That depends on the *centripetal force* — the center-seeking, inward force needed to keep an object moving in uniform circular motion.

In this section, you discover how the centripetal force keeps the object moving in a circle and how the details of the circular motion, such as radius and velocity, depend upon the centripetal force.

Looking at the force you need

Centripetal force isn't some new force that appears out of nowhere when an object travels in a circle; it's the force the object *needs* to keep traveling in that circle.

As you know from Newton's first law (see Chapter 5), if there's no net force on a moving object, the object will continue to move uniformly in a straight line. If a force (or a component of a force) acts in the same direction as the object's velocity, then the object begins to speed up, and if the force acts in the opposite direction to the velocity, then the object slows down. However, if the force always acts perpendicularly to the velocity while remaining of constant magnitude, then the magnitude of the velocity (the speed) doesn't change; only its direction does — the object moves in a circle. In this case, the force is called *centripetal force.*

If you're spinning a ball on a string, then the centripetal force comes from the tension in the string. When the moon orbits the Earth, the centripetal force comes from gravity. And when you drive your car in a circle, the centripetal force comes from the friction of the tires against the road. The origin of the force isn't important, only that it remains of constant magnitude and always acts perpendicularly to the velocity, toward the center of the circle.

Seeing how the mass, velocity, and radius affect centripetal force

Because force equals mass times acceleration, $\mathbf{F} = m\mathbf{a},$ and because centripetal acceleration is equal to v^2/r (see the earlier section "Finding the magnitude of the centripetal acceleration"), you can determine the magnitude of the centripetal force needed to keep an object moving in uniform circular motion with the following equation:

$$F_c = \frac{mv^2}{r}$$

This equation tells you the magnitude of the force that you need to move an object of a given mass, m, in a circle at a given radius, r, and speed, v. (Remember that the direction of the force is always toward the center of the circle.)

Think about how force is affected if you change one of the other variables. The equation shows that if you increase mass or speed, you'll need a larger force; if you decrease the radius, you're dividing by a smaller number, so you'll also need a larger force. Here's how these ideas play out in the real world:

- **Increasing mass:** You may have an easy time swinging a golf ball on a string in a circle, but if you replace the golf ball with a cannonball, watch out. You may now have to whip 10 kilograms around on the end of a 1.0-meter string every half-second. As you can tell, you need a heck of a lot more force.

- **Increasing speed:** Not interested in spinning cannonballs? Then imagine you're driving your car around in a circle. If you're going quite slowly around the circle, your tires have no problem generating enough frictional force to keep you going in the circle. But if you go too fast, then your tires can no longer generate the frictional force acting toward the center of the circle, so you start to skid.

- **Decreasing the radius:** You can see the effect of the radius in your car going around in a circle. If you drive your car at a fixed speed in a circle of smaller and smaller radius, eventually your tires won't be able to supply enough centripetal force from the friction, and you'll skid off the circular path.

The fictitious centrifugal force

You've probably heard of centrifugal force and most likely have felt it when a car you were in turned a corner. However, centrifugal force isn't really a force as defined in Newton's laws. It only *appears* to be a force. When you're in a car turning a corner, your body has inertia and is naturally inclined to move at uniform speed in a straight line. But because the car is turning, it feels as though your body is being thrown outward, toward the car door.

Example

Q. The moon goes around Earth about every 27.3 days with a distance from Earth of 3.85×10^8 meters. If the moon's mass is 7.35×10^{22} kilograms, what is the centripetal force that Earth's gravity exerts on it as it orbits Earth?

A. The correct answer is 2.0×10^{20} newtons.

1. Start with this equation:

$$F_c = \frac{mv^2}{r}$$

2. Find the speed of the moon. It goes $2\pi r$ in 27.3 days, so convert 27.3 days to seconds:

$$27.3 \text{ days} \times \frac{24 \text{ hr}}{\text{day}} \times \frac{60 \text{ min}}{\text{hr}} \times \frac{60 \text{ s}}{\text{min}} = 2.36 \times 10^6 \text{ s}$$

3. Therefore, the speed of the moon is

$$v = \frac{2\pi r}{T} = \frac{2\pi \left(3.85 \times 10^8 \text{ m}\right)}{2.36 \times 10^6 \text{ s}} = 1,024 \text{ m/s}$$

4. Plug in the numbers:

$$F_c = \frac{mv^2}{r} = \frac{(7.35 \times 10^{22} \text{ kg})(1,024 \text{ m/s})^2}{3.85 \times 10^8 \text{ m}} = 2.0 \times 10^{20} \text{ N}$$

Practice Questions

1. You're exerting a force on a string to keep a ball on a string going in a circle. If the ball has a mass of 0.10 kilograms and the speed of the ball is 16.0 meters per second at a distance of 2.0 meters, what is the centripetal force you need to apply to keep the ball going in a circle?

2. You have a 1.0-kilogram toy plane on the end of a 10-meter wire, and it's going around at 60.0 meters per second. What is the force you have to apply to the wire to keep the plane going in a circle?

Practice Answers

1. **13 N.** Use the equation for centripetal force:

$$F_c = \frac{mv^2}{r}$$

Plug in the numbers:

$$F_c = \frac{mv^2}{r}$$
$$= \frac{(0.10\ \text{kg})(16.0\ \text{m/s})^2}{(2.0\ \text{m})}$$
$$= 13\ \text{N}$$

2. **360 N.** Use the equation for centripetal force:

$$F_c = \frac{mv^2}{r}$$

Plug in the numbers:

$$F_c = \frac{mv^2}{r}$$
$$= \frac{(1.0\ \text{kg})(60.0\ \text{m/s})^2}{(10\ \text{m})}$$
$$= 360\ \text{N}$$

Negotiating flat curves and banked turns

Imagine that you're driving a car and you come to a curve. On a flat road, the centripetal force you need to negotiate the curve comes from the friction of the tires against the ground. If the surface is covered with a substance such as ice, you have less friction, and you can't turn as safely at high speeds.

To make turns safer, engineers design roads so that curves are banked. With the road at an angle, there's a component of the normal force of the road against your car, toward the center of the circle. This means that you don't require as much friction from your tires to make the turn.

When you're driving on a flat road, friction provides the centripetal force — toward the center of the circle — that allows you to make a turn.

Say you're sitting in the passenger seat of the car, approaching a turn with a level, non-banked road surface. What's the maximum speed the driver can go and still keep you safe? The frictional force needs to supply the centripetal force, so you come up with the following:

$$F_c = \frac{mv^2}{r} = \mu_s mg$$

where m is the mass of the car, v is the velocity, r is the radius, μ_s is the coefficient of static friction (you use the coefficient of static friction if the wheels aren't slipping), and g is the acceleration due to gravity, 9.8 meters per second^2.

If a curve is banked, then a component of the normal force of the road against the car contributes to the centripetal force, and so you can go around the curve much faster. Because you don't have to rely on friction to supply the centripetal force, the question of whether you can safely make the turn no longer depends on road conditions.

Take a look at Figure 7-5, which shows a car banking around a turn. The engineers can make the driving experience enjoyable if they bank the turn so that drivers garner the centripetal force needed to go around the turn entirely by the component of the normal force of the road against the car acting toward the center of the turn's circle. That component is $F_N \sin\theta$ (F_N is the normal force, the upward force perpendicular to the road; see Chapter 6), so

$$F_c = F_N \sin\theta = \frac{mv^2}{r}$$

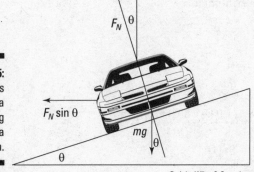

Figure 7-5: The forces acting on a car banking around a turn.

To find the centripetal force, you need the normal force, F_N. If you look at Figure 7-3, you can see that F_N comes from a combination of the centripetal force due to the car's banking around the turn and the car's weight. The purely vertical component of F_N must equal mg, because no other forces are operating vertically, so

$$F_N \cos\theta = mg$$

$$F_N = \frac{mg}{\cos\theta}$$

Plugging this result into the equation for centripetal force gives you

$$F_c = F_N \sin\theta$$

$$\frac{mv^2}{r} = \left(\frac{mg}{\cos\theta}\right)\sin\theta$$

Because $\sin\theta / \cos\theta = \tan\theta$, you can also write this as

$$\frac{mv^2}{r} = mg\tan\theta$$

$$\frac{mv^2}{mgr} = \tan\theta$$

Solve for θ to find the angle of the road. The equation finally breaks down to

$$\theta = \tan^{-1}\left(\frac{v^2}{gr}\right)$$

Tip: You don't have to memorize this result, in case you're panicking — this is the kind of equation used by highway engineers when they have to bank curves (notice that the mass of the car cancels out, meaning that it holds for vehicles regardless of weight). You can always derive this equation from your knowledge of Newton's laws and circular motion.

Example

Q. What should the angle θ be if drivers go around a 200-meter-radius turn at 60.0 miles per hour?

A. The correct answer is 20°.

1. Use this equation:

$$\theta = \tan^{-1}\left(\frac{v^2}{gr}\right)$$

2. Plug in the numbers: 60 miles per hour is about 27 meters per second, and the radius of the turn is 200 meters, so

$$\theta = \tan^{-1}\left(\frac{v^2}{gr}\right)$$

$$\theta = \tan^{-1}\frac{\left(27 \text{ m/s}\right)^2}{\left(9.8 \text{ m/s}^2\right)\left(200 \text{ m}\right)} \approx 20°$$

The designers should bank the turn at about 20° to give drivers a smooth experience.

Practice Questions

1. At what angle should a banked road be tilted if the magnitude of the force exerted by the ground on the car is the same as that exerted by gravity on the car? Give your answer in degrees.

2. What is the minimum velocity a 1,200-kilogram car must travel on an icy, banked road with a radius of curvature of 36 meters and an inclination of 9.0° to prevent it from slipping down the bank if the coefficient of static friction is 0.08?

Practice Answers

1. **0°.** The normal force — the force exerted by a surface on an object touching it — is equal to the force of gravity only when the surface is exactly horizontal. The angle of inclination is thus 0°.

2. **8.1 m/s.** Start by drawing the forces acting on the car:

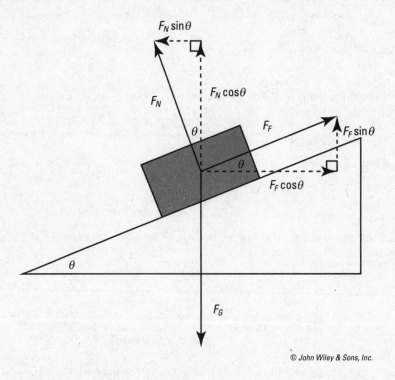

© John Wiley & Sons, Inc.

Because the car isn't accelerating upward or downward, the sum of the vertical forces must equal 0:

$$F_N \cos\theta + F_F \sin\theta - F_G = 0$$

Use $F_F = \mu F_N$ (μ is the coefficient of friction) and $F_G = mg$ to solve for the normal force:

$$F_N \cos\theta + \mu F_N \sin\theta - mg = 0$$
$$F_N \cos\theta + \mu F_N \sin\theta = mg$$
$$F_N(\cos\theta + \mu \sin\theta) = mg$$

$$F_N = \frac{mg}{(\cos\theta + \mu \sin\theta)}$$

The sum of the horizontal forces provides the centripetal force keeping the car in its circular motion. Friction is pointing away from the "center of the circle" and is therefore antiparallel to the centripetal force:

$$F_N \sin\theta - F_F \cos\theta = F_C$$
$$F_N \sin\theta - \mu F_N \cos\theta = F_C$$
$$F_N \left(\sin\theta - \mu \cos\theta\right) = F_C$$

Now use the result you got earlier for the normal force and the centripetal force equation, $F_C = \dfrac{mv^2}{r}$, where m is the mass of the object, v is the object's velocity, and r is the radius (of the curve the object is traversing), to solve the equation.

$$F_N \left(\sin\theta + \mu \cos\theta\right) = F_C$$

$$\frac{mg}{\left(\cos\theta + \mu \sin\theta\right)}\left(\sin\theta - \mu \cos\theta\right) = \frac{mv^2}{r}$$

$$mg\left(\frac{\sin\theta - \mu \cos\theta}{\cos\theta + \mu \sin\theta}\right) = \frac{mv^2}{r}$$

$$g\left(\frac{\sin\theta - \mu \cos\theta}{\cos\theta + \mu \sin\theta}\right) = \frac{v^2}{r}$$

$$\left(9.8 \text{ m/s}^2\right)\left(\frac{\sin 9.0° - 0.08\cos 9.0°}{\cos 9.0° + 0.08\sin 9.0°}\right) = \frac{v^2}{\left(36 \text{ m}\right)}$$

$$(9.8)\left(\frac{0.07742}{1.0002}\right)\text{m}^2/\text{s}^2 = \frac{v^2}{36}$$

$$7.586 \text{ m}^2/\text{s}^2 = \frac{v^2}{36}$$

$$273.1 \text{ m}^2/\text{s}^2 = v^2$$

$$17 \text{ m/s} = v$$

LETTING GRAVITY SUPPLY CENTRIPETAL FORCE

You don't have to tie objects to strings to observe travel in circular motion; larger bodies such as planets move in circular motion, too. Gravity provides the necessary centripetal force.

In this section, you discover Newton's take on the gravitational forc e between two objects, and we show you how his theory relates to 9.8 meters per second², the value experimenters identified as the acceleration due to gravity near the surface of the Earth. Then you put Newton's formula to use in looking at the orbits of satellites.

Using Newton's law of universal gravitation

Remember: Sir Isaac Newton came up with one of the heavyweight laws in physics for you: the *law of universal gravitation*. This law says that every mass exerts an attractive force on every other mass. If the two masses are m_1 and m_2 and the distance between them is r, the magnitude of the force is

$$F = \frac{Gm_1 m_2}{r^2}$$

where G is a constant equal to 6.67×10^{-11} N·m²/kg².

Example

Q. Two people are sitting on a park bench, looking at each other and smiling. If the two love-birds have masses of about 75 kilograms each, what's the force of gravity pulling them together, assuming they start out 0.50 meters away?

A. The correct answer is 1.5×10^{-6} newtons.

Use Newton's law of universal gravitation:

$$F = \frac{Gm_1m_2}{r^2}$$

$$= \frac{\left(6.67 \times 10^{-11} \text{ N} \cdot \text{m}^2/\text{kg}^2\right)\left(75 \text{ kg}\right)\left(75 \text{ kg}\right)}{\left(0.50 \text{ m}\right)^2}$$

$$\approx 1.5 \times 10^{-6} \text{ N}$$

Practice Questions

1. What is the gravitational force between the sun and the Earth? The sun has a mass of about 1.99×10^{30} kilograms, and the Earth has a mass of about 5.98×10^{24} kilograms. A distance of about 1.50×10^{11} meters separates the two bodies.

2. Calculate the force of gravity between Earth and its moon. Approximate the distance between the center of Earth and the center of the moon as 3.80×10^5 kilometers. The moon has a mass of about 7.35×10^{22} kilograms, and the Earth has a mass of about 5.98×10^{24} kilograms.

Practice Answers

1. 3.52×10^{22} N. Plug the numbers into Newton's equation:

$$F = \frac{Gm_1m_2}{r^2}$$

$$= \frac{\left(6.67 \times 10^{-11} \text{ N} \cdot \text{m}^2/\text{kg}^2\right)\left(1.99 \times 10^{30} \text{ kg}\right)\left(5.98 \times 10^{24} \text{ kg}\right)}{\left(1.50 \times 10^{11} \text{ m}\right)^2}$$

$$\approx 3.52 \times 10^{22} \text{ N}$$

2. 2.03×10^{20} N. Convert the Earth to moon distance to meters to match the units in the gravitational constant, G:

$$\left(3.80 \times 10^5 \text{ km}\right)\left(\frac{1 \times 10^3 \text{ m}}{1 \text{ km}}\right) = 3.80 \times 10^8 \text{ m}$$

Plug the numbers into your calculator:

$$F_g = \frac{Gm_1m_2}{r^2}$$

$$= \frac{\left(6.67\times10^{-11}\ \text{N}\cdot\text{m}^2/\text{kg}^2\right)\left(5.98\times10^{24}\ \text{kg}\right)\left(7.35\times10^{22}\ \text{kg}\right)}{\left(3.80\times10^8\ \text{m}\right)^2}$$

$$= \frac{2.93\times10^{37}\ \text{N}}{1.4\times10^{17}}$$

$$= 2.03\times10^{20}\ \text{N}$$

Deriving the force of gravity on the Earth's surface

The equation for the force of gravity — $F = \left(Gm_1m_2\right)/r^2$ — holds true no matter how far apart two masses are. But you also come across a special gravitational case (which most of the work on gravity in this book is about): the force of gravity near the surface of the Earth.

The gravitational force between a mass and the Earth is the object's *weight*. *Mass* is considered a measure of an object's inertia, and its weight is the force exerted on the object in a gravitational field. On the surface of the Earth, the two forces are related by the acceleration due to gravity: $F_g = mg$. Kilograms and slugs are units of mass; newtons and pounds are units of weight.

You can use Newton's law of gravitation to get the acceleration due to gravity, g, on the surface of the Earth just by knowing the gravitational constant G, the radius of the Earth, and the mass of the Earth. The force on an object of mass m_1 near the surface of the Earth is

$$F = m_1g$$

This force is provided by gravity between the object and the Earth, according to Newton's gravity formula, so you can write

$$m_1g = \frac{Gm_1m_2}{r_e^2}$$

The radius of the Earth, r_e, is about 6.38×10^6 meters, and the mass of the Earth is 5.98×10^{24} kilograms. Putting in the numbers, you have

$$m_1g = \frac{\left(6.67\times10^{-11}\ \text{N}\cdot\text{m}^2/\text{kg}^2\right)m_1\left(5.98\times10^{24}\ \text{kg}\right)}{\left(6.38\times10^6\ \text{m}\right)^2}$$

Dividing both sides by m_1 gives you the acceleration due to gravity:

$$g = \frac{\left(6.67\times10^{-11}\ \text{N}\cdot\text{m}^2/\text{kg}^2\right)\left(5.98\times10^{24}\ \text{kg}\right)}{\left(6.38\times10^6\ \text{m}\right)^2}$$

$$\approx 9.8\ \text{m/s}^2$$

Newton's law of gravitation gives you the acceleration due to gravity near the surface of the Earth: 9.8 meters per second[2].

Of course, you can measure g by letting an apple drop and timing it, but what fun is that when you can calculate it in a roundabout way that requires you to first measure the mass of the Earth?

Using the law of gravitation to examine circular orbits

In space, bodies are constantly orbiting other bodies due to gravity. Satellites (including the moon) orbit the Earth, the Earth orbits the sun, the sun orbits around the center of the Milky Way, the Milky Way orbits around the center of its local group of galaxies. This is big-time stuff. In the case of orbital motion, gravity supplies the centripetal force that causes the orbits.

The force of gravity between orbiting bodies is quite a bit different from small-time orbital motion — such as when you have a ball on a string — because for a given distance and two masses, the gravitational force is always going to be the same. You can't increase the force to increase the speed of an orbiting planet as you can with a ball. The following sections examine the speed and period of orbiting bodies in space.

A particular satellite can have only one speed when in orbit around a particular body at a given distance because the force of gravity doesn't change. So what's that speed? You can calculate it with the equations for centripetal force and gravitational force. You know that for a satellite of a particular mass, m_1, to orbit, you need a corresponding centripetal force (see the section "Seeking the Center: Centripetal Force"):

$$F_c = \frac{m_1 v^2}{r}$$

This centripetal force has to come from the force of gravity, so

$$\frac{G m_1 m_2}{r^2} = \frac{m_1 v^2}{r}$$

You can rearrange this equation to get the speed:

$$v = \sqrt{\frac{G m_2}{r}}$$

This equation represents the speed that a satellite at a given radius must have in order to orbit if the orbit is due to gravity. The speed can't vary as long as the satellite has a constant orbital radius — that is, as long as it's going around in circles. This equation holds for any orbiting object where the attraction is the force of gravity, whether it's a human-made satellite orbiting the Earth or the Earth orbiting the sun. If you want to find the speed for satellites that orbit the Earth, for example, you use the mass of the Earth in the equation:

$$v = \sqrt{\frac{G m_E}{r}}$$

Remember: Here are a few details you should note on reviewing the orbiting speed equation:

- **You have to use the distance from the *center* of the Earth, not the distance above Earth's surface, as the radius.** Therefore, the distance you use in the equation is the distance between the two orbiting bodies. In this case, you add the distance from the center of the Earth to the surface of the Earth, 6.38×10^6 meters, to the satellite's height above the Earth.

- **The equation assumes that the satellite is high enough off the ground that it orbits out of the atmosphere.** That assumption isn't really true for artificial satellites; even at 400 miles above the surface of the Earth, satellites do feel air friction. Gradually, the drag of friction brings them lower and lower, and when they hit the atmosphere, they burn up on re-entry. When a satellite is less than 100 miles above the surface, its orbit decays appreciably each time it circles the Earth. (Look out below!)

- **The equation is independent of mass.** If the moon rather than the artificial satellite orbited at 400 miles and you could ignore air friction and collisions with the Earth, it would have to go at the same speed as the satellite in order to preserve its close orbit (which would make for some pretty spectacular moonrises).

You can think of a satellite in motion around the Earth as always falling. The only thing that keeps it from striking the Earth is that its velocity points over the horizon. The satellite *is* falling, but its velocity takes it over the horizon — that is, over the curve of the world as it falls — so it doesn't get any closer to the Earth. (The same is true of the astronauts inside. They only have the appearance of being weightless, but they're continuously falling, too.)

Sometimes it's more important to know the period of an orbit rather than the speed, such as when you're counting on a satellite to come over the horizon before communication can take place. The *period* of a satellite is the time it takes it to make one full orbit around an object. The period of the Earth as it travels around the sun is one year.

If you know the satellite's speed and the radius at which it orbits (see the preceding section), you can figure out its period. The satellite travels around the entire circumference of the circle — which is $2\pi r$ if r is the radius of the orbit — in the period, T. This means the orbital speed must be $2\pi r/T$, giving you

$$\sqrt{\frac{Gm_E}{r}} = \frac{2\pi r}{T}$$

If you solve this for the period of the satellite, you get

$$T = 2\pi\sqrt{\frac{r^3}{Gm_E}}$$

Understanding Kepler's laws of orbiting bodies

Johannes Kepler (1571–1630), a German national born in the Holy Roman Empire, came up with three laws that helped explain a great deal about orbits before Newton came up with his law of universal gravitation. Here are Kepler's laws:

Law 1: Planets orbit in ellipses. An ellipse is a shape like a squashed circle, and the degree of squashedness is called the *eccentricity* of the ellipse. The orbits allowed can have any degree of eccentricity. When the eccentricity is zero, you have a circular orbit.

Law 2: Planets move so that a line between the sun and the planet sweeps out the same area in the same time, independent of where they are in their orbits. This means that when the planet is in that part of its orbit where it is close to the sun, it has to travel faster to sweep out the same area that it does when it's farther away.

Law 3: The square of a planet's orbital period (the time it takes the planet to make one complete orbit) is proportional to its average distance from the sun cubed.

The third law can be derived from Newton's laws. It takes the form of this equation: $r^3 = \left(T^2 Gm_E\right)/4\pi^2$. Although Kepler's third law says that T^2 is proportional to r^3, you can get the exact constant relating these quantities by using Newton's law of gravitation.

Example

Q. Human-made satellites typically orbit at heights of 400 miles from the surface of the Earth (about 640 kilometers, or 6.4×10^5 meters). What's the speed of such a satellite?

A. The correct answer is 7.54×10^3 meters per second or 16,800 miles per hour.

1. All you have to do is put in the numbers into the satellite speed equation:

$$v = \sqrt{\frac{Gm_E}{r}} = \sqrt{\frac{\left(6.67 \times 10^{-11}\ \text{N} \cdot \text{m}^2/\text{kg}^2\right)\left(5.98 \times 10^{24}\ \text{kg}\right)}{\left(6.38 \times 10^6\ \text{m}\right) + \left(6.40 \times 10^5\ \text{m}\right)}} \approx 7.54 \times 10^3\ \text{m/s}$$

2. This converts to about 16,800 miles per hour.

Practice Questions

1. How far above the surface of the Earth is a satellite whose period is the same as the Earth's 24-hour period?

2. If the International Space Station (ISS) is located 420 kilometers above Earth's surface, how many hours does it take to make a complete orbit?

Practice Answers

1. 3.59×10^7 meters. In cases of stationary satellites, the period, T, is 24 hours, or about 86,400 seconds.

   Solve the equation for periods for the radius:

   $$r^3 = \frac{T^2 G m_E}{4\pi^2}$$

   $$r = \left(\frac{T^2 G m_E}{4\pi^2} \right)^{1/3}$$

   Plug in the numbers:

   $$r = \left(\frac{T^2 G m_E}{4\pi^2} \right)^{1/3}$$

   $$= \left(\frac{\left[8.64 \times 10^4 \text{ s} \right]^2 \left[6.67 \times 10^{-11} \text{ N} \cdot \text{m}^2/\text{kg}^2 \right] \left[5.98 \times 10^{24} \text{ kg} \right]}{4\pi^2} \right)^{1/3}$$

   $$\approx \left(7.542 \times 10^{22} \text{ m}^3 \right)^{1/3}$$

   $$\approx 4.23 \times 10^7 \text{ m}$$

   Subtract the Earth's radius of 6.38×10^6 meters to get 3.59×10^7 meters. At this distance, geosynchronous satellites orbit the Earth at the same rate the Earth is turning, which means that they stay put over the same piece of real estate.

2. **1.55.** Use the equation relating orbital period to orbital position, $T^2 = \frac{4\pi^2}{Gm} r^3$.

   Find the distance in meters between the ISS and the center of the Earth:

   $$r = r_{\text{Earth}} + h$$
   $$= 6.38 \times 10^3 \text{ km} + 420 \text{ km}$$
   $$= \left(6.80 \times 10^3 \text{ km} \right) \left(\frac{1 \times 10^3 \text{ m}}{1 \text{ km}} \right) = 6.80 \times 10^6 \text{ m}$$

Substitute that into the orbital equation:

$$T^2 = \frac{4\pi^2}{Gm}r^3$$

$$T = \sqrt{\frac{4\pi^2 r^3}{Gm}}$$

$$= 2\pi\sqrt{\frac{\left(6.80\times10^6\ \text{m}\right)^3}{\left(6.67\times10^{-11}\ \text{N}\cdot\text{m}^2/\text{kg}^2\right)\left(5.98\times10^{24}\ \text{kg}\right)}}$$

$$= 2\pi\sqrt{7.883\times10^5}\ \text{s}$$

$$= 5{,}579\ \text{s}$$

Convert to hours:

$$\left(5{,}579\ \text{s}\right)\left(\frac{1\ \text{min}}{60\ \text{s}}\right)\left(\frac{1\ \text{hr}}{60\ \text{min}}\right) = 1.55\ \text{hr}$$

8

GO WITH THE FLOW: LOOKING AT PRESSURE IN FLUIDS

On a hot summer day, nothing is better than taking a dip in the neighbor's pool. As you execute the perfect swan dive and slip gracefully to the depths, you notice a curious sensation: water pressure. It increases with every foot you're below the surface. You note the rapid increase of that pressure as you go deeper, and you wonder what being miles under the surface of the ocean would feel like. "Hmm," you think. "Just how does water pressure increase with depth?"

This chapter is all about pressures in fluids, and we cover a lot more than pounds per square inch. You also encounter info on Archimedes's principle (which is all about floating and buoyancy), hydraulic machines, fluids moving in pipes, streamlines, and lots more. With all that coming up, it's time to get your feet wet in the physics of fluids.

Remember: Both liquids and gases are considered fluids. A *fluid* is defined as any continuous distribution of matter that can't support a shear stress without moving. If, on the other hand, you *shear* a solid piece of material by applying different forces to different parts of it, then the solid deforms to some degree but eventually finds a balance. For example, if you hold a piece of rubber in one hand and then push the top of it with the other, it bends over, supporting the shear stress you apply to it.

MASS DENSITY: GETTING SOME INSIDE INFORMATION

Density is the ratio of mass to volume. Any solid object that's less dense than water floats. Density is an important property of a fluid because mass is continuously distributed throughout a fluid; the static forces and motions within the fluid depend on the concentration of mass (density) rather than the fluid's overall mass.

Calculating density

Remember: Density (ρ) is mass (m) divided by volume (V), so here's the formula for density:

$$\rho = \frac{m}{V}$$

In the MKS system, the units are kilograms per cubic meter, or kg/m^3.

You can see a sample of the densities of common materials in Table 8-1. Note that ice is less dense than water, so ice floats. Generally, solids and gases expand with temperature and therefore become less dense (you can find out more about the expansion of solids in Chapter 14 and the expansion of gases in Chapter 16). This table includes the density of water at 4°C as a reference point because the density of water varies with temperature. The densities of the gases generally have a stronger dependence on temperature than the solids do, though.

Table 8-1   Densities of Common Materials

Substance	Density (kg/m³)
Gold (near room temperature)	19,300
Mercury (near room temperature)	13,600
Silver (near room temperature)	10,500
Copper (near room temperature)	8,890
Diamond (near room temperature)	3,520
Aluminum (near room temperature)	2,700
Blood (near body temperature)	1,060
Water (4°C)	1,000
Ice (0°C)	917
Oxygen (at 0°C, 101.325 kPa)	1.43
Helium (at 0°C, 101.325 kPa)	0.179

Example

Q. Say that you have a whopper diamond with a volume of 0.0500 cubic meters (that's a cube that's about 1 foot on each side, so it's truly a whopper). You measure its mass as 176.0 kilograms. So what is its density?

A. The correct answer is 3,520 kilograms per cubic meter.

  1. Use this equation:

$$\rho = \frac{m}{V}$$

2. Plug in the numbers:

$$\rho = \frac{m}{V}$$

$$= \frac{176.0 \text{ kg}}{0.0500 \text{ m}^3}$$

$$= 3,520 \text{ kg/m}^3$$

Practice Questions

1. The mass of the Earth is about 5.97×10^{24} kilograms and its average radius is about 6,371 kilometers. What is the average density of the Earth?

2. You make a cake that has a mass of 0.300 kilograms and fits in a cake pan that is 0.30 by 0.10 by 0.06 meters cubed. What is the density of the cake?

Practice Answers

1. **5,510 kg/m³.** Use this equation:

$$\rho = \frac{m}{V}$$

The volume of a sphere is

$$V = \frac{4}{3}\pi r^3$$

Plug in the numbers to find the volume of the Earth:

$$V = \frac{4}{3}\pi r^3$$

$$= \frac{4}{3}\pi (6,371,000 \text{ m})^3$$

$$= 1.083 \times 10^{21} \text{ m}^3$$

Plug your result into the density equation:

$$\rho = \frac{m}{V}$$

$$= \frac{5.97 \times 10^{24} \text{ kg}}{1.083 \times 10^{21} \text{ m}^3}$$

$$= 5,510 \text{ kg/m}^3$$

This is greater than the density of titanium but less than the density of steel.

2. 1.7×10^2 kg/m^3. Use the density equation:

$$\rho = \frac{m}{V}$$

The volume of the cake is

$$V = (0.30 \text{ m})(0.10 \text{ m})(0.06 \text{ m})$$
$$= 0.0018 \text{ m}^3$$

Plug in the numbers:

$$\rho = \frac{m}{V}$$
$$= \frac{0.300 \text{ kg}}{0.0018 \text{ m}^3}$$
$$= 1.7 \times 10^2 \text{ kg/m}^3$$

Comparing densities with specific gravity

A substance's *specific gravity* is the ratio of that substance's density to the density of water at 4°C. Because the density of water at 4°C is 1,000 kilograms per cubic meter, that ratio is easy to find. For example, the density of gold is 19,300 kilograms per cubic meter, so its specific gravity is the following:

$$\text{specific gravity}_{\text{gold}} = \frac{19,300 \text{ kg/m}^3}{1,000 \text{ kg/m}^3} = 19.3$$

Specific gravity has no units, because it's a ratio of density divided by density, so all units cancel out. Therefore, the specific gravity of gold is simply 19.3.

Tip: Anything with a specific gravity greater than 1 sinks in pure water at 4°C, and anything with a specific gravity less than 1 floats. As you'd expect, gold, with a specific gravity of 19.3, sinks. Ice, on the other hand, with a specific gravity of 0.917, floats. So how can a ship, which is made of metal with a specific gravity very much greater than water, float? The ship floats because of the shape of its hull. The ship is mostly hollow and displaces water weighing more than the weight of the ship. Averaged throughout, the ship is less dense than the water overall, so the effective specific gravity of the ship is less than that of water.

Example

Q. The density of gasoline is 721 kilograms per cubic meter. What is its specific gravity?

A. The correct answer is 0.721.

   1. Use the equation for specific gravity:

$$\text{specific gravity} = \frac{\rho}{1{,}000 \text{ kg/m}^3}$$

   2. Plug in the numbers:

$$\text{specific gravity}_{\text{gasoline}} = \frac{\rho_{\text{gasoline}}}{1{,}000 \text{ kg/m}^3}$$

$$= \frac{721 \text{ kg/m}^3}{1{,}000 \text{ kg/m}^3}$$

$$= 0.721$$

Practice Questions

   1. If the specific gravity of cork is 0.240, what is its density?

   2. If the density of table salt is 2,170 kilograms per cubic meter, what is its specific gravity?

Practice Answers

   1. **240 kg/m³.** Use the equation for specific gravity:

$$\text{specific gravity} = \frac{\rho}{1{,}000 \text{ kg/m}^3}$$

Solve the equation for specific gravity for the density of cork:

$$\text{specific gravity}_{\text{cork}} = \frac{\rho_{\text{cork}}}{1{,}000 \text{ kg/m}^3}$$

$$\rho_{\text{cork}} = \text{specific gravity}_{\text{cork}} \times 1{,}000 \text{ kg/m}^3$$

Insert the given value for the specific gravity of cork:

$$\rho_{\text{cork}} = \text{specific gravity}_{\text{cork}} \times 1{,}000 \text{ kg/m}^3$$

$$= 0.240 \times 1{,}000 \text{ kg/m}^3$$

$$= 240 \text{ kg/m}^3$$

2.   **2.17.** Use the equation for specific gravity:

$$\text{specific gravity} = \frac{\rho}{1,000 \text{ kg/m}^3}$$

Plug in the numbers:

$$\text{specific gravity}_{salt} = \frac{\rho_{salt}}{1,000 \text{ kg/m}^3}$$

$$= \frac{2,170 \text{ kg/m}^3}{1,000 \text{ kg/m}^3}$$

$$= 2.17$$

APPLYING PRESSURE

Everyone who's ever talked about car or bicycle tires or blown up a balloon knows about air pressure. And if you've gone swimming under water, you know about water pressure. When you push on something, people say you exert pressure on it.

Remember: In physics terms, *pressure* is force per area — a fact you may already know if you've filled a tire to a certain number of pounds per square inch. The equation for pressure, *P,* is as follows:

$$P = \frac{F}{A}$$

where *F* is force and *A* is area. Note that pressure isn't a vector — it's a scalar (that is, just a number without a direction).

In this section, you look at the units of pressure, see how pressure changes along with depth or altitude, and discover how hydraulic machines work.

Looking at units of pressure

Because pressure is force divided by area, its MKS units are newtons per square meter, or N/m^2. In the foot-pound-second (FPS) system, the units are pounds per square inch, or psi.

Remember: The unit *newtons per square meter* is so common in physics that it has a special name: the *pascal,* which equals 1 newton per square meter. The pascal is abbreviated as Pa.

You don't have to be under water to experience pressure from a fluid. Air exerts pressure, too, due to the weight of the air above you. Here's how much pressure the air exerts on you at sea level:

$$\text{air pressure}_{sea\ level} = 1.013 \times 10^5 \text{ Pa}$$

Remember: The air pressure at sea level is a standard pressure that people refer to as 1 *atmosphere* (abbreviated atm):

$$\text{air pressure}_{\text{sea level}} = 1.013 \times 10^5 \text{ Pa} = 1 \text{ atm}$$

If you convert an atmosphere to pounds per square inch, it's about 14.7 pounds per square inch. That means that 14.7 pounds of force are pressing in on every square inch of your body at sea level.

Your body pushes back with 14.7 pounds per square inch, so you don't feel any pressure on you at all. But if you suddenly got transported to outer space, the inward pressure of the air pushing on you would be gone, and all that would remain would be the 14.7 pounds per square inch your body exerted outward. You wouldn't explode, but your lungs could burst if you tried to hold your breath. The change in pressure could also cause the nitrogen in your blood to form bubbles and give you the bends!

Example

Q. Say you're in your neighbor's pool, waiting near the bottom until your neighbors give up trying to chase you off and go back into the house. You're near the deep end of the pool, and using the handy pressure gauge you always carry, you measure the pressure on the back of your hand as 1.2×10^5 pascals. What force does the water exert on the back of your hand if the back of your hand has an area of about 8.4×10^{-3} square meters?

A. The correct answer is 1.0×10^3 newtons.

1. Solve the pressure equation $P = F / A$ for the force:

$$F = PA$$

2. Plug in the numbers:

$$
\begin{aligned}
F &= PA \\
&= \left(1.2 \times 10^5 \text{ Pa}\right)\left(8.4 \times 10^{-3} \text{ m}^2\right) \\
&= \left(1.2 \times 10^5 \text{ N/m}^2\right)\left(8.4 \times 10^{-3} \text{ m}^2\right) \\
&\approx 1.0 \times 10^3 \text{ N}
\end{aligned}
$$

Practice Questions

1. A book resting on a table exerts a downward force of 5.5 newtons on the table. If the surface area of the book is 0.0050 square meters, what is the pressure applied on the table by the book?

2. What is the magnitude of the force exerted on a piston of cross-sectional area 0.3 square meters by a fluid at a pressure of 1.2 pascals?

Practice Answers

1. 1.1×10^3 Pa. Pressure is the force per unit area: $P = \frac{F}{A}$. Plug in the numbers:

$$P = \frac{F}{A}$$
$$= \frac{5.5\,\text{N}}{5.0\times10^{-3}\,\text{m}^2}$$
$$= 1.1\times10^3\,\text{Pa}$$

2. **0.36 N.** Pressure is the force per unit area: $P = \frac{F}{A}$. Solve for the force:

$$P = \frac{F}{A}$$
$$F = PA$$

Plug the numbers into your calculator:

$$F = PA$$
$$= 1.2\,\text{Pa}\times0.30\,\text{m}^2$$
$$= 0.36\,\text{N}$$

Connecting pressure to changes in depth

You know that pressure increases the farther you go under water, but by how much? As a physicist, you can put some numbers in and get numerical results out. Just what pressure would you expect for a given depth?

Say that you're under water and you're considering the imaginary cube of water you see in Figure 8-1. At the top of the cube, the water pressure is P_1. At the bottom of the cube, it's P_2. The cube has horizontal faces of area A and a height h. First find the forces on the top and bottom of the cube.

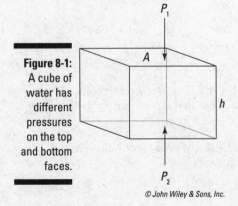

Figure 8-1: A cube of water has different pressures on the top and bottom faces.

© John Wiley & Sons, Inc.

The sum of the forces is the difference between the force on the bottom face of the cube, F_2, and the force on the top face of the cube, F_1:

$$\sum F = F_2 - F_1$$

You can say the force pushing down on the top face is $F_1 = P_1 A$ and that the force pushing on the bottom face is $F_2 = P_2 A$. Therefore, in terms of pressure, the sum of forces is the following:

$$\sum F = P_2 A - P_1 A$$

So what's the net force upward on the cube of water? The upward force must be equal to the weight of the water, mg, where m is the mass of the water and g is the gravitational constant (9.8 meters per second^2). So you have the following equation:

$$P_2 A - P_1 A = mg$$

Hmm. You don't know m, the mass of the water. Can you get the weight of the water in terms of A, the area of the top and bottom faces of the cube? The mass of the water is the density of water, ρ, multiplied by the volume of the cube, which is Ah. So you can replace m with ρAh, which gives you the following equation:

$$P_2 A - P_1 A = \rho gAh$$

Now you're talking. Dividing everything by A gives you the difference in pressures:

$$P_2 - P_1 = \rho gh$$

Remember: If you call the difference in the pressures ΔP, you get the following equation:

$$\Delta P = \rho gh$$

The preceding equation is an important, general result that holds for any fluid: water, air, gasoline, and so on. This equation says that the difference in pressure between two points in a fluid is equal to the fluid's density multiplied by g (the acceleration due to gravity) multiplied by the difference in height between the two points.

Remember: To find the total pressure on something submerged in a liquid, you have to add the pressure due to the liquid to the atmospheric pressure, which is about 14.7 pounds per square inch, or 1.013×10^5 pascals.

Example

Q. How much does the pressure increase for every meter you go under water?

A. The correct answer is 9,800 pascals.

   1. Use the equation for the difference in pressure in a fluid:

$$\Delta P = \rho g h$$

   2. Plug in the numbers:

$$\Delta P = \rho g h$$
$$= \left(1,000 \text{ kg/m}^3\right)\left(9.8 \text{ m/s}^2\right)(1.0 \text{ m})$$
$$= 9,800 \left(\frac{\text{kg} \cdot \text{m}}{\text{s}^2}\right)\left(\frac{1}{\text{m}^2}\right)$$
$$= 9,800 (\text{N})\left(\frac{1}{\text{m}^2}\right)$$
$$= 9,800 \text{ Pa}$$

That works out to be about 1.4 pounds per square inch added pressure for every meter you go down.

Practice Questions

   1. Say that your head is 1.5 meters above your feet. What's the difference in blood pressure between your head and your feet (neglecting the action of the heart)?

   2. A water slide park drills a well to get water. The water in the well is 20 meters down. A pump at the top of the well is pulling up air in the pipe, creating a vacuum with $P = 0$. If the bottom of the well is at atmospheric pressure, will the water reach the top of the well?

Practice Answers

   1. 1.6×10^4 Pa. Use this equation:

$$\Delta P = \rho g h$$

Table 8-1 lists the density, ρ, of blood as 1,060 kilograms per cubic meter. Do the math:

$$\Delta P = \rho g h = \left(1,060 \text{ kg/m}^3\right)\left(9.8 \text{ m/s}^2\right)(1.5 \text{ m}) \approx 1.6 \times 10^4 \text{ Pa}$$

That pressure works out to be slightly less than 2.0 pounds per square inch.

2.   **No.** The difference in pressure is $\Delta P = \rho g h$, so when the pump pulls the water up as far as it's going to go, $\rho g h$ of the column of water in the pipe equals atmospheric pressure or 1.01×10^5 pascals:

$$\rho g h = 1.01 \times 10^5 \text{ Pa}$$

Solve for h to find how far the water can rise:

$$h = \frac{1.01 \times 10^5 \text{ Pa}}{\rho g}$$

Plug in the numbers:

$$h = \frac{1.01 \times 10^5 \text{ Pa}}{\left(1{,}000 \text{ kg/m}^3\right)\left(9.8 \text{ m/s}^2\right)} \approx 10.3 \text{ m}$$

Therefore, the maximum height you can pump water out of a well with the pump at the top of the well is 10.3 meters. But in this case, the well is 20 meters deep.

Hydraulic machines: Passing on pressure with Pascal's principle

Pascal's principle says that given a fluid in a totally enclosed system, a change in pressure at one point in the fluid is transmitted to all points in the fluid as well as to the enclosing walls. In other words, if you have a fluid enclosed in a pipe (with no air bubbles) and change the pressure in the fluid at one end of the pipe, the pressure changes all throughout the pipe to match.

The fact that pressure inside an enclosed system is the same (neglecting gravitational differences) has an interesting consequence. Because $P = F/A$, you get the following equation for force:

$$F = PA$$

So if the *pressure* is the same everywhere in an enclosed system but the *areas* you consider are different, can you get different forces?

To make this question clearer, look at Figure 8-2, which shows a system of enclosed fluid with two hydraulic pistons, one with a piston head of area A_1 and one with a piston head of area A_2. You apply a force of F_1 on the smaller piston. What is the force on the other piston, F_2?

Pressure at each point is F/A. According to Pascal's principle, the pressure is the same everywhere inside the fluid, so $F_1/A_1 = F_2/A_2$:

$$\frac{F_1}{A_1} = \frac{F_2}{A_2}$$

Solving for F_2 gives you the force at Point 2:

$$F_2 = \frac{F_1 A_2}{A_1}$$

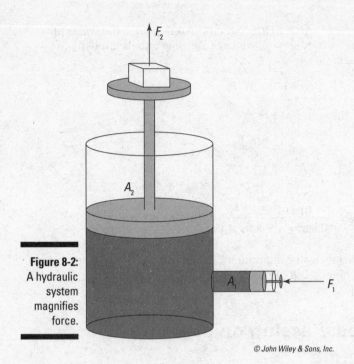

Figure 8-2:
A hydraulic
system
magnifies
force.

© John Wiley & Sons, Inc.

Cool. That means that you can develop a huge force from a small force if the ratio of the piston sizes is big. For example, say the area of Piston 2 is bigger than Piston 1 by a factor of 100. Does that mean that any force you apply to Piston 1 will be multiplied by 100 times on Piston 2?

Yes, indeed — that's how hydraulic equipment works. By using a small piston at one end and a large piston at the other, you can create huge forces. Backhoes and other hydraulic machines, such as garbage trucks and hydraulic lifts, use Pascal's principle to function.

What's the catch here? If you push on Piston 1 and get 100 times the force on Piston 2, you seem to be getting something for nothing. The catch is that you have to push the smaller piston 100 times as far as the second piston will move.

Example

Q. You jump on a water bed and apply a force of 610 newtons over an area of 0.0030 square meters. Your friend is sitting on the other end of the water bed and feels an upward force of 2,500 newtons. What surface area does your friend cover?

A. The correct answer is 1.2×10^{-2} square meters.

1. Use Pascal's principle: $\dfrac{F_1}{A_1} = \dfrac{F_2}{A_2}$.

2. Solve for A_2:

$$\frac{F_1}{A_1} = \frac{F_2}{A_2}$$

$$A_2 = F_2 \frac{A_1}{F_1}$$

3. Plug into your calculator:

$$A_2 = F_2 \frac{A_1}{F_1}$$

$$= 2,500 \text{ N} \times \frac{0.0030 \text{ m}^2}{610 \text{ N}}$$

$$= 1.2 \times 10^{-2} \text{ m}^2$$

Practice Questions

1. In a hydraulic system, a piston with a cross-sectional area of 21 square centimeters pushes on an incompressible liquid with a force of 38 newtons. The far end of the hydraulic pipe connects to a second piston with a cross-sectional surface area of 100.0 square centimeters. What is the force on the second piston?

2. You want to inflate your bike tire to a pressure of 700,000 pascals. If the radius of your bike pump is 1.5 centimeters, what force do you need to apply to the handle of the bike pump to inflate your tire?

Practice Answers

1. 1.8×10^2 N. Use Pascal's principle: $\frac{F_1}{A_1} = \frac{F_2}{A_2}$. Solve for F_2:

$$\frac{F_1}{A_1} = \frac{F_2}{A_2}$$

$$F_2 = A_2 \frac{F_1}{A_1}$$

Insert the given values to find the force on the second piston:

$$F_2 = A_2 \frac{F_1}{A_1}$$

$$= 100 \text{ cm}^2 \times \frac{38 \text{ N}}{21 \text{ cm}^2}$$

$$= 1.8 \times 10^2 \text{ N}$$

2. 4.9×10^2 N. Use Pascal's principle: $\dfrac{F_1}{A_1} = \dfrac{F_2}{A_2}$. Pressure is force per unit area, so rewrite this equation as $\dfrac{F_1}{A_1} = P_2$, where $P_2 = 700{,}000$ pascals.

The cross-sectional area of the piston in the bike pump is $A_1 = \pi r_1^2$.

Solve for the force F_1:

$$\frac{F_1}{A_1} = P_2$$
$$F_1 = P_2 A_1$$
$$= P_2 \pi r_1^2$$

Plug in the numbers:

$$F_1 = P_2 \pi r_1^2$$
$$= 700{,}000 \,\text{Pa} \times \pi \left(0.015 \,\text{m}\right)^2$$
$$= 4.9 \times 10^2 \,\text{N}$$

BUOYANCY: FLOAT YOUR BOAT WITH ARCHIMEDES'S PRINCIPLE

Archimedes's principle says that any fluid exerts a buoyant force on an object wholly or partially submerged in it, and the magnitude of the buoyant force equals the weight of the fluid displaced by the object. An object that's less dense than water floats because the water it displaces weighs more than the object does. Therefore, as the object pushes down, the water pushes back up more strongly.

If you've ever tried to push a beach ball under water, you've felt this principle in action. As you push the ball down, it pushes back up. In fact, a big beach ball can be tough to hold under water. As a physicist in a bathing suit, you may wonder, "What's happening here?"

What is the buoyant force, F_b, the water exerts on the beach ball? To make this problem easier, you decide to consider the beach ball as a cube of height h and horizontal face with area A. So the buoyant force on the cubic beach ball is equal to the force at the bottom of the beach ball minus the force at the top:

$$F_{buoyancy} = F_{bottom} - F_{top}$$

And because $F = PA$, you can work pressure into the equation with a simple substitution:

$$F_{buoyancy} = \left(P_{bottom} - P_{top}\right) A$$

You can also write the change in pressure, $P_{bottom} - P_{top}$, as ΔP:

$$F_{buoyancy} = \Delta P A$$

The change in pressure equals ρgh, so replace ΔP:

$$F_{buoyancy} = \rho ghA$$

Note that hA is the volume of the cube. Multiplying volume, V, by density, ρ, gives you the mass of the water displaced by the cube, m, so you can replace ρhA with m:

$$F_{buoyancy} = mg$$

You should recognize mg (mass times acceleration due to gravity) as the expression for weight, so the force of buoyancy is equal to the weight of the water displaced by the object you're submerging:

$$F_{buoyancy} = W_{water\ displaced}$$

That equation turns out to be Archimedes's principle.

Example

Q. The Acme Raft Company built the raft shown in the following figure. The density of the wood used in its rafts is 550 kilograms per cubic meter, and the raft is 0.20 meters high. How much of the raft will be under water when it's launched?

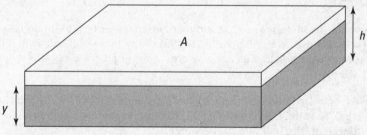

© John Wiley & Sons, Inc.

A. The correct answer is 0.11 meters.

1. If the raft floats, then the weight of the raft must equal the buoyant force the water exerts on the raft.

2. The weight of the raft is equal to the following:

$$W_{raft} = \rho_{raft} Ahg$$

where h is the height of the raft and A is horizontal surface area.

3. If the bottom of the raft is a distance y under water, then the submerged volume of the raft is Ay. That makes the mass of the water displaced by the raft equal to the following:

$$m_{water\ displaced} = \rho_{water} Ay$$

4. The weight of the displaced water is just its mass multiplied by g, or

$$W_{water\ displaced} = \rho_{water} Ayg$$

5. Set the values for raft weight and water weight equal to each other:

$$\rho_{raft} Ahg = \rho_{water} Ayg$$

6. Solve for y:

$$\rho_{raft} Ahg = \rho_{water} Ayg$$
$$\rho_{raft} h = \rho_{water} y$$
$$y = \frac{\rho_{raft} h}{\rho_{water}}$$

7. Plug in the numbers:

$$y = \frac{550\ \text{kg/m}^3}{1{,}000\ \text{kg/m}^3}(0.2\ \text{m})$$
$$= 0.11\ \text{m}$$

Practice Questions

1. Legend says that Archimedes submerged a crown in water to find its density. He then measured the volume of water displaced to find the volume of the crown. If he measured 1.2 liters of water displaced and the crown weighed 3.3 kilograms, what was the density of the crown?

2. You plunge a basketball beneath the surface of a swimming pool until half the volume of the basketball is submerged. If the basketball has a radius of 12 centimeters, what is the buoyant force on the ball due to the water?

Practice Answers

1. $2.8 \times 10^3\ \text{kg/m}^3$. The volume of the crown is the volume of water displaced:

$$V = 1.2\ \cancel{L} \times \frac{1\ \text{m}^3}{1{,}000\ \cancel{L}}$$
$$= 0.0012\ \text{m}^3$$

The mass of the crown is m = 3.3 kilograms, so the density of the crown is

$$\rho = \frac{m}{V}$$
$$= \frac{3.3\ \text{kg}}{0.0012\ \text{m}^3}$$
$$= 2.8 \times 10^3\ \text{kg/m}^3$$

2. **35 N.** The buoyant force is the mass of the water displaced multiplied by the acceleration due to gravity: $F_{\text{buoyancy}} = m_{\text{water displaced}}g$.

The volume of water displaced is half the volume of the basketball:

$$V_{\text{water displaced}} = \frac{1}{2}V_{\text{basketball}}$$

$$= \frac{1}{2}\left(\frac{4}{3}\pi r^3\right)$$

$$= \frac{2}{3}\pi r^3$$

Here, r = 12 centimeters. In meters, the radius is

$$r = 12 \text{ cm} \times \frac{1\text{ m}}{100\text{ cm}}$$

$$= 0.12\text{ m}$$

Using the equation for density, the mass of water displaced is $m_{\text{water displaced}} = \rho_{\text{water}}V_{\text{water displaced}}$. The buoyancy force is

$$F_{\text{buoyancy}} = m_{\text{water displaced}}g$$

$$= \rho_{\text{water}}V_{\text{water displaced}}g$$

$$= \rho_{\text{water}} \times \frac{2}{3}\pi r^3 \times g$$

$$= 1{,}000 \text{ kg/m}^3 \times \frac{2}{3}\pi\left(0.12\text{ m}\right)^3 \times 9.8\text{ m/s}^2$$

$$= 35\text{ N}$$

FLUID DYNAMICS: GOING WITH FLUIDS IN MOTION

Fluids move according to simple laws that are consistent with Newton's laws of motion. But even though these laws are simple, the range of possible fluid flows is enormous! As you can see all around you, fluids can do all kinds of motions: They can swirl like in a hurricane, they can be in steady uniform flows like when a tap is running, and they can tumble in the most complicated patterns like the steam rising from a boiling kettle. All these different kinds of flow can be characterized with certain properties, which is the subject of this section.

Characterizing the type of flow

Fluid flow has all kinds of aspects — it can be steady or unsteady, compressible or incompressible, and more. Some of these characteristics reflect properties of the liquid or gas itself, and others focus on how the fluid is moving. This section looks at the possibilities.

Remember: Fluid flow can actually get very complex when it becomes turbulent. Physicists haven't developed any elegant equations to describe turbulence because how turbulence works depends on the individual system — whether you have water cascading through a pipe or air streaming out of a jet engine. Usually, you have to resort to computers to handle problems that involve fluid turbulence.

Evenness: Steady or unsteady flow

Fluid flow can be steady or unsteady, depending on the fluid's velocity:

- **Steady:** In steady fluid flow, the velocity of the fluid is constant at any point.
- **Unsteady:** When the flow is unsteady, the fluid's velocity can differ between any two points.

For example, suppose that you're sitting by the side of a stream and note that the water flow isn't steady: You see eddies and backwash and all kinds of swirling. Imagine velocity vectors for a hundred points in the water, and you get a good picture of unsteady flow — the velocity vectors can be pointing all over the map, although the velocity vectors generally follow the stream's overall average flow. (Sometimes, in a complex flow, physicists divide the flow into a sum of a smooth average flow and complicated fluctuations, but you don't need to do that here.)

Squeezability: Compressible or incompressible flow

Remember: Fluid flow can be *compressible* or *incompressible,* depending on whether you can easily compress the fluid. Liquids are usually nearly impossible to compress, whereas gases (also considered a fluid) are very compressible.

A hydraulic system works only because liquids are incompressible — that is, when you increase the pressure in one location in the hydraulic system, the pressure increases to match everywhere in the whole system (for details, see the earlier section "Hydraulic machines: Passing on pressure with Pascal's principle"). Gases, on the other hand, are very compressible — even when your bike tire is stretched to its limit, you can still pump more air into it by pushing down on the plunger and squeezing it in. The laws of how gases behave when compressing and expanding in different situations can be found in Chapter 16.

Thickness: Viscous or nonviscous flow

Liquid flow can be *viscous* or *nonviscous,* depending on how resistant the fluid is to flow. *Viscosity* is a measure of the thickness of a fluid, and very gloppy fluids such as motor oil or shampoo are called *viscous fluids.*

Remember: Viscosity is actually a measure of friction in the fluid. When a fluid flows, the layers of fluid rub against one another, and in very viscous fluids, the friction is so great that the layers of flow pull against one other and hamper that flow.

Viscosity usually varies with temperature, because when the molecules of a fluid are moving faster (when the fluid is warmer), the molecules can more easily slide over each other. So when you pour pancake syrup, for example, you may notice that it's very thick in the bottle, but the syrup becomes quite runny when it spreads over the warm pancakes and heats up.

Spinning: Rotational or irrotational flow

Fluid flow can be rotational or irrotational. If, as you travel in a closed loop, you add up all the components of the fluid velocity vectors along your path and the end result isn't zero, then the flow is *rotational.*

Tip: To test whether a flow has a rotational component, you can put a small object in the flow and let the flow carry it. If the small object spins, the flow is rotational; if the object doesn't spin, the flow is irrotational.

For example, look at the water flowing in a brook. It eddies around stones, curling around obstacles. At such locations, the water flow has a rotational component.

Some flows that you may think are rotational are actually irrotational. For example, away from the center, a vortex is actually an irrotational flow! You can see this if you look at the water draining from your bathtub. If you place a small floating object in the flow, it goes around the plug hole, but it doesn't spin about itself; therefore, the flow is irrotational.

On the other hand, flows that have no apparent rotation can actually be rotational. Take a shear flow, for example. In a *shear flow,* all the fluid is moving in the same direction, but the fluid is moving faster on one side. Suppose that the fluid is moving faster on the left than on the right. The fluid isn't moving in a circle at all, but if you place a small floating object in this flow, the flow on the left side of the object is slightly faster, so the object begins to spin. The flow is rotational.

Picturing flow with streamlines

A handy way of visualizing the flow of a fluid is through streamlines. You draw a fluid's *streamline* so that a tangent to the streamline at any point is parallel to the fluid's velocity at that point. In other words, a streamline follows the fluid flow.

You can see an example in Figure 8-3, where the streamline is the darker line in the middle of the fluid flow.

If you plot streamlines for any flow, you get an immediate picture of how that fluid is flowing. You can have as many or as few streamlines as you need to get an accurate picture of fluid flow.

When fluid flow is turbulent, streamlines can become all mixed up. That's why dealing with turbulent flows in a precise, mathematical way is difficult.

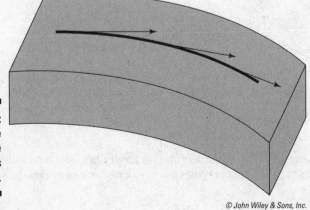

Figure 8-3:
A streamline shows the directions of flow.

GETTING UP TO SPEED ON FLOW AND PRESSURE

However complicated a fluid flow may seem, fluids do obey some simple laws that can be expressed in equations. This section introduces the equation that describes the continuity of fluid flow (the result of the fact that matter is neither created nor destroyed) and the relation between speed and pressure. You also take a look at some of the consequences of these relations.

The equation of continuity: Relating pipe size and flow rates

If a fluid is flowing at a certain speed at a certain point in a system of pipes, you can predict what its speed will be at another point using the *equation of continuity*. Because the mass of the fluid is neither created nor destroyed, if mass moves away from one place at a certain rate, it must therefore move to the neighboring place at the same rate. With this idea expressed as an equation, you can find out how the speed changes in a narrowing pipe, for example.

Conserving mass with the equation of continuity

The equation of continuity comes from the idea that no mass disappears when fluid is flowing. In other words, the fluid you get out equals what you put in. You can find the equation of continuity by mixing a little geometry with the physics formulas for mass (which remains constant), density, and speed.

Imagine a cube of fluid flowing in a pipe with the rest of the fluid, as Figure 8-4 shows. The cube has an area A perpendicular to the fluid flow and has a length h along the fluid flow.

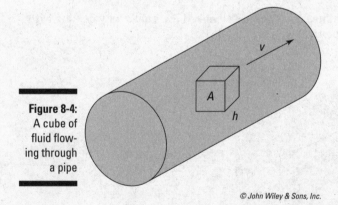

Figure 8-4:
A cube of fluid flowing through a pipe

© John Wiley & Sons, Inc.

Now say that the pipe narrows so much that the cube no longer fits. The boundary of the cube is going to change shape. What can you say stays constant between the original cube and the deformed cube? The mass of the fluid inside the box shape will stay constant because no fluid flows through the boundary. Therefore, you can say that

$$m_1 = m_2$$

where m_1 is the mass of fluid in the first cube and m_2 is the mass of the fluid in the deformed cube later on, where the pipe narrows.

If you instead look at the situation in terms of density, ρ, the mass of the fluid in the cube is the density multiplied by the volume of the cube, which is Ah. So you can restate the equation as

$$\rho_1 A_1 h_1 = \rho_2 A_2 h_2$$

where A_1 is the area of the front face of the cube originally, h_1 is the original length of the cube, and so on.

Now say that you're measuring the amount of mass going by in time t to get the flow rate, so you divide the equation you just got by a time interval, t:

$$\rho_1 A_1 \frac{h_1}{t} = \rho_2 A_2 \frac{h_2}{t}$$

The length of the cube passing you by in time, t, gives you the speed of the fluid at that location, so h/t becomes the speed of the fluid at that location. Substituting v, the speed, for h/t, you get the following equation:

$$\rho_1 A_1 v_1 = \rho_2 A_2 v_2$$

And this quantity, ρAv, is called the *mass flow rate* — it's the mass of fluid that passes by a certain point per second. The MKS units of the mass flow rate are kilograms per second, or kg/s.

Remember: The mass flow rate has the same value at every point in a fluid conduit that has a single entry and single exit point. The mass flow rate at any two points along the conduit can be related like this, with the *equation of continuity:*

$$\rho_1 A_1 v_1 = \rho_2 A_2 v_2$$

Incompressible liquids: Changing the pipe size to change the flow rate

Because liquids are virtually incompressible, the density doesn't change along the flow. Therefore, if you come to a location where the same amount of water must squeeze through a smaller space than before, the water's velocity has to change — it goes faster. Just think of what you do to get water to squirt faster from a garden hose: You put your thumb over most of the end of the hose, forcing the water to squirt through a smaller area.

You can show this idea mathematically using the equation of continuity. For incompressible liquids, density must be the same at Point 1 and Point 2. Because $\rho_1 = \rho_2$, you have this equation, where ρ is the shared density:

$$\rho A_1 v_1 = \rho A_2 v_2$$

Dividing by ρ gives you

$$A_1v_1 = A_2v_2$$

where Av is called the *volume flow rate*, whose symbol is Q. So at any two points along the flow of an incompressible liquid, the following is true:

$$Q_1 = Q_2$$

Example

Q. Say that you're playing with a fire hose, and you note that water is spraying from the hose at 7.7 meters per second. The cross-sectional area of the nozzle is 4.0×10^{-4} square meters. What is the speed of the water leaving the hydrant when it first enters the fire hose, which has a cross-sectional area of 1.0×10^{-2} square meters?

A. The correct answer is 0.31 meters per second.

    1. Use the fact that the volume rate of flow of an incompressible liquid is the same at any point along its flow:

$$A_1v_1 = A_2v_2$$

    where A_1 is the cross-sectional area of the hose, v_1 is the speed of the water as it enters the hose, A_2 is the cross-sectional area of the nozzle, and v_2 is the speed at which water leaves the hose.

    2. Solve for v_1, the speed of the water as it enters the hose:

$$v_1 = \frac{A_2v_2}{A_1}$$

    3. Plug in the numbers:

$$v_1 = \frac{A_2v_2}{A_1} = \frac{\left(4.0 \times 10^{-4} \text{ m}^2\right)\left(7.7 \text{ m/s}\right)}{\left(1.0 \times 10^{-2} \text{ m}^2\right)} \approx 0.31 \text{ m/s}$$

Practice Questions

    1. Oil passing through a pipeline at 0.22 meters per second encounters a pipeline section where the radius is reduced by half. What is the speed of the oil in this section of pipeline?

    2. If the volume flow rate through a pipe is 0.039 cubic meters per second and the pipe diameter is 12 centimeters, what is the speed of the fluid flow through the pipe?

Practice Answers

1. **0.88 m/s.** Use the equation of continuity: $A_1 v_1 = A_2 v_2$. In terms of the radius, the cross-sectional area of the pipeline is $A = \pi r^2$. Use this in the continuity equation:

    $$A_1 v_1 = A_2 v_2$$
    $$\pi r_1^2 v_1 = \pi r_2^2 v_2$$
    $$r_1^2 v_1 = r_2^2 v_2$$

    You know that $r_2 = \frac{1}{2} r_1$ so you can solve the continuity equation for the speed v_2:

    $$r_1^2 v_1 = r_2^2 v_2$$
    $$\cancel{r_1^2} v_1 = \left(\frac{1}{2}\cancel{r_1}\right)^2 v_2$$
    $$v_2 = 4 v_1$$
    $$= 4 \times 0.22 \text{ m/s}$$
    $$= 0.88 \text{ m/s}$$

2. **3.5 m/s.** The volume flow rate is $Q = Av$, where $Q = 0.039$ cubic meters per second. You know the diameter of the pipe, so you can calculate the area A:

    $$A = \pi r^2$$
    $$= \pi \left(\frac{d}{2}\right)^2$$
    $$= \frac{\pi d^2}{4}$$

    where

    $$d = 12 \text{ cm} \times \frac{1 \text{ m}}{100 \text{ cm}}$$
    $$= 0.12 \text{ m}$$

    Solve the equation for volume flow rate for the speed v of the fluid:

    $$Q = Av$$
    $$= \frac{\pi d^2}{4} v$$
    $$v = \frac{4Q}{\pi d^2}$$
    $$= \frac{4 \times 0.039 \text{ m}^3/\text{s}}{\pi (0.12 \text{ m})^2}$$
    $$= 3.5 \text{ m/s}$$

Bernoulli's equation: Relating speed and pressure

Now you come to the powerhouse of fluid flow — Daniel Bernoulli's equation, which lets you relate pressure, fluid speed, and height. Using Bernoulli's equation, you can find the difference in fluid pressure between two points if you know the fluid's speed and height at those two points.

Remember: Bernoulli's equation relates a moving fluid's pressure, density, speed, and height from Point 1 to Point 2 in this way:

$$P_1 + \frac{1}{2}\rho v_1^2 + \rho g y_1 = P_2 + \frac{1}{2}\rho v_2^2 + \rho g y_2$$

Here's what the variables stand for in this equation (where the subscripts indicate whether you're talking about Point 1 or Point 2):

- P is the pressure of the fluid.
- ρ is the fluid's density.
- g is the acceleration due to gravity.
- v is the fluid's speed.
- y is the height of the fluid.

The equation assumes that you're working with the steady flow of an incompressible, irrotational, nonviscous fluid (see the earlier section "Characterizing the type of flow" for details).

Getting a lift

You can easily demonstrate Bernoulli's principle at home. All you need are two pieces of regular copy paper. Hold the two pieces of paper by the very top so that they dangle freely; then hold them face to face at a separation of a few inches. The air pressure between the two pieces of paper is the same as the air pressure on the other side, so they hang there without moving. Now here comes the cool part. If you blow between the two pieces of paper, what do you think will happen? Most of your friends would probably say that the two pieces of paper will drift apart. But if you try it, you find that the two pieces of paper actually move together! You can work out why this happens because you know Bernoulli's principle.

When you blow between the two pieces of paper, the air increases its speed and reduces its pressure simultaneously. Because the pressure between the pieces of paper is now less than the pressure outside, the pieces of paper move together.

And the fun doesn't end there: If you amaze your friends with the paper trick, you can further boggle their minds by using the principle to explain how airplanes fly. The cross-section of an airplane wing has a kind of swept dome shape to it. Because of this particular shape, air moving toward the wing diverges at the leading edge. Some air goes over the wing, and some air goes under the wing before rejoining at the trailing edge of the wing. The wing is designed so the air traveling over the wing moves faster than the air traveling below the wing. And as you have just demonstrated with two pieces of paper (according to Bernoulli's principle), faster air has a lower pressure. So the pressure of the air below the wing is greater than the pressure above the wing. This pressure difference provides the lift required for the airplane to fly.

Remember: One thing you can take immediately from this equation is what's called *Bernoulli's principle,* which says that increasing a fluid's speed can lead to a decrease in pressure.

Tip: Together, the equation of continuity and Bernoulli's equation allow you to relate the pressure in pipes to their changes in diameter. You often use the equation of continuity, which tells you that a particular volume of a liquid flows at a constant rate, to find the speeds you use in Bernoulli's equation, which relates speed to pressure.

Example

Q. On the operating table lies a very important person who has an aneurism in the aorta, the principal artery leading from the heart. An aneurysm is an enlargement in a blood vessel where the walls have weakened.

The normal speed of blood through a person's aorta is 0.40 meters per second. The density of blood is 1,060 kilograms per cubic meter. How much higher is the pressure in the aneurysm than the normal aorta if the cross-sectional area of the aneurysm is twice the cross-sectional area of the normal aorta?

A. The correct answer is 64 pascals.

1. Use Bernoulli's equation here because it relates pressure and velocity:

$$P_1 + \frac{1}{2}\rho v_1^2 + \rho g y_1 = P_2 + \frac{1}{2}\rho v_2^2 + \rho g y_2$$

2. Because the patient is lying on the operating table, which means that $y_1 = y_2$, Bernoulli's equation becomes the following:

$$P_1 + \frac{1}{2}\rho v_1^2 = P_2 + \frac{1}{2}\rho v_2^2$$

3. Rearrange the equation to solve for the difference in pressure between the aneurysm and normal aorta $\Delta P = P_2 - P_1$:

$$P_2 - P_1 = \frac{1}{2}\rho v_1^2 - \frac{1}{2}\rho v_2^2$$

$$\Delta P = \frac{1}{2}\rho v_1^2 - \frac{1}{2}\rho v_2^2$$

4. Use the equation of continuity to find the speed of the blood in the aneurysm:

$$\rho_1 A_1 v_1 = \rho_2 A_2 v_2$$

5. Because the density of blood is the same at Point 1 and Point 2, in the normal aorta and inside the aneurysm, divide out the density to get

$$A_1 v_1 = A_2 v_2$$

6.   Solving for v_2:

$$v_2 = \frac{A_1 v_1}{A_2}$$

7.   Plug in the numbers:

$$v_2 = \frac{A_1 (0.4 \text{ m/s})}{2.0 A_1} = 0.2 \text{ m/s}$$

8.   Plug your result into the equation for the difference in pressure from Step 3:

$$\Delta P = \frac{1}{2}\rho v_1{}^2 - \frac{1}{2}\rho v_2{}^2$$

$$= \frac{1}{2}\left(1,060 \text{ kg/m}^3\right)\left(0.4 \text{ m/s}\right)^2 - \frac{1}{2}\left(1,060 \text{ kg/m}^3\right)\left(0.2 \text{ m/s}\right)^2$$

$$\approx 64 \text{ Pa}$$

The pressure is about 64 pascals or about 0.01 pounds per square inch higher in the aneurysm.

Practice Questions

1.   A dam holds back the water in a lake. If the dam has a small hole 1.4 meters below the surface of the lake, at what speed does water exit the hole?

2.   A hose lying on the ground has water coming out of it at a speed of 5.4 meters per second. You lift the nozzle of the hose to a height of 1.3 meters above the ground. At what speed does the water now come out of the hose?

Practice Answers

1.   **5.2 m/s.** Use Bernoulli's equation: $P_1 + \frac{1}{2}\rho v_1^2 + \rho g y_1 = P_2 + \frac{1}{2}\rho v_2^2 + \rho g y_2$.

Let Point 1 be on the surface of the lake and Point 2 be at the outlet of the hole in the dam. The pressure at each point is just atmospheric pressure, so $P_1 = P_2$.

The hole is 1.4 meters below the lake, so $y_1 - y_2 = 1.4$ meters. Because the hole is "small," the level of the lake doesn't change much as water leaks out of the hole, so $v_1 = 0$ meters per second. Solve for the speed of the fluid at Point 2:

$$P_1 + \frac{1}{2}\rho v_1^2 + \rho g y_1 = P_2 + \frac{1}{2}\rho v_2^2 + \rho g y_2$$

$$\rho g\left(y_1 - y_2\right) = \frac{1}{2}\rho v_2^2$$

$$v_2 = \pm\sqrt{2g\left(y_1 - y_2\right)}$$

$$= \pm\sqrt{2 \times 9.8 \text{ m/s}^2 \times 1.4 \text{ m}}$$

$$= \pm 5.2 \text{ m/s}$$

Because you're interested in the speed of the water, which is a positive quantity, use the plus sign in the equation. Thus, the speed of the water coming out of the hole is 5.2 meters per second.

2. **1.9 m/s.** Use Bernoulli's equation: $P_1 + \frac{1}{2}\rho v_1^2 + \rho g y_1 = P_2 + \frac{1}{2}\rho v_2^2 + \rho g y_2$.

Let Point 1 be on the ground and Point 2 be at 1.3 meters above the ground. At both points, the pressure is atmospheric pressure, so $P_1 = P_2 = 101,000$ pascals. The difference in heights between Points 1 and 2 is $y_2 - y_1 = 1.3$ meters.

Using these equations, solve Bernoulli's equation for the speed v_2:

$$P_1 + \frac{1}{2}\rho v_1^2 + \rho g y_1 = P_2 + \frac{1}{2}\rho v_2^2 + \rho g y_2$$

$$\frac{1}{2}\rho v_1^2 + \rho g y_1 - \rho g y_2 = \frac{1}{2}\rho v_2^2$$

$$v_2^2 = v_1^2 + 2 g y_1 - 2 g y_2$$

$$v_2 = \pm\sqrt{v_1^2 + 2 g (y_1 - y_2)}$$

$$= \pm\sqrt{(5.4 \text{ m/s})^2 + 2 \times 9.8 \text{ m/s}^2 \times (-1.3 \text{ m})}$$

$$= 1.9 \text{ m/s}$$

PART III

working

ENERGETICALLY

WEB EXTRA *If you're feeling energetic, take a look at www.dummies.com/extras/ucanphysics1 for a free article on energy diagrams.*

IN THIS PART . . .

- Work through problems about kinetic energy, potential energy, and power.
- Predict what will happen when objects collide using momentum.
- Take angular motion and torque for a spin.
- Get going with moments of inertia, rotational kinetic energy, and angular momentum.
- Solve problems of simple harmonic motion and springs.

9

GETTING SOME WORK
OUT OF PHYSICS

IN THIS CHAPTER

- Working through forces
- Evaluating kinetic and potential energy
- Walking the path of conservative and nonconservative forces
- Accounting for mechanical energy and power in work

You know all about work; it's what you do when you have to do physics problems. You sit down with your calculator, you sweat a little, and you get through it. You've done your work. Unfortunately, that doesn't count as work in physics terms.

In physics, *work* is done when a force moves an object through a displacement. That may not be your boss's idea of work, but it gets the job done in physics. Along with the basics of work, we use this chapter to introduce kinetic and potential energy, look at conservative and nonconservative forces, and examine mechanical energy and power.

LOOKING FOR WORK

Holding heavy objects — like, say, a set of exercise weights — up in the air seems to take a lot of work. In physics terms, however, that isn't true. Even though holding up weights may take a lot of biological work, no mechanical work takes place if the weights aren't moving.

Remember: In physics, mechanical *work* is performed on an object when a force moves the object through a displacement. When the force is constant and the angle between the force and the displacement is θ, then the work done is given by $W = Fs\cos\theta$. In layman's terms, if you push a 1,000-pound hockey puck for some distance, physics says that the work you do is the component of the force you apply in the direction of travel multiplied by the distance you go.

To get a picture of the full work spectrum, you need to look across different systems of measurement. After you have the measurement units down, you can look at practical working examples, such as pushing and dragging. You can also figure out what negative work means.

Working on measurement systems

Work is a scalar, not a vector; therefore, it has only a magnitude, not a direction (more on scalars and vectors in Chapter 4). Because work is force times distance, $Fs \cos \theta$, it has the unit newton-meter (N·m or N-m) in the MKS system (see Chapter 2 for info on systems of measurement).

Remember: Mechanical work done by a net force is equivalent to a transfer of energy (this is called the *work-energy theorem*), which has units called *joules*. Because of this, work and energy have the same units. For conversion purposes, 1 newton of force applied through a distance of 1 meter (where the force is applied along the line of the displacement) is equivalent to 1 joule, or 1 J, of work. (In the foot-pound-second system, work has the unit *foot-pound*. You can also discuss energy and work in terms of kilowatt-hours, which you may be familiar with from electric bills; 1 kilowatt-hour [kWh] = 3.6×10^6 joules.)

Working with force

Motion is a requirement of work. For work to be done, a net force has to move an object through a displacement. Work is a product of force and displacement.

The angle between the force and the displacement affects how much work gets done. If you apply force at an angle instead of parallel to the direction of motion, you have to supply more force to perform the same amount of work. If the force moving the object has a component in the same direction as the motion, the work that force does on the object is positive. If the force moving the object has a component in the opposite direction of the motion, the work done by that force on the object is negative. Here are some examples and practice questions for you to work through.

Examples

Q. Say that you're pushing a huge gold ingot 3 kilometers to your home, as the following figure shows. Suppose that the kinetic coefficient of friction (see Chapter 6), μ_k, between the ingot and the ground is 0.250 and that the ingot has a mass of 1,000.0 kilograms. How much work do you have to do to get it home?

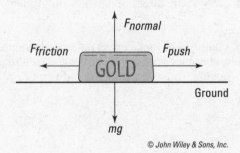

© John Wiley & Sons, Inc.

A. The correct answer is 7.35×10^6 joules.

1. Start with this equation for the force of friction:

$$F_F = \mu_k F_N$$

2. Because the road is flat, the magnitude of the normal force, F_N, is just mg (mass times the acceleration due to gravity on the surface of the Earth). That means that

$$F_F = \mu_k mg$$

3. Plug in the numbers:

$$F_F = \mu_k mg$$
$$= (0.250)(1,000 \text{ kg})(9.8 \text{ m/s}^2)$$
$$= 2,450 \text{ N}$$

4. Use the formula for work:

$$W = F_{push}\, s \cos\theta$$

5. Because you're pushing the ingot with a force that's parallel to the ground, the angle between F_{push} and s is $0°$, and $\cos 0° = 1$. You do the least amount of work if you push the ingot at a constant velocity, which means that $F_{push} = F_F$.

6. Plug in the numbers:

$$W = F_{push}\, s \cos\theta = (2,450 \text{ N})(3,000 \text{ m})(1) = 7.35 \times 10^6 \text{ J}$$

Q. You find another huge gold ingot 3.0 kilometers from your home. This time you decide to drag it along with rope that's at an angle of $10°$ from the ground, as the following figure shows. Suppose that the kinetic coefficient of friction (see Chapter 6), μ_k, between the ingot and the ground is 0.250 and that the ingot has a mass of 1,000.0 kilograms. How much work do you have to do to get it home?

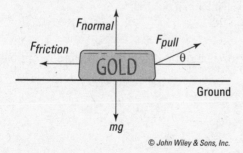

© John Wiley & Sons, Inc.

A. The correct answer is 7.0×10^6 joules.

1. The normal force with the ground is given by the weight of the ingot minus the upward component of the force you supply. Therefore, the force of friction is given by

$$F_{friction} = \mu\left(mg - F_{pull} \sin\theta\right)$$

2. The horizontal component of your force equals the force of friction:

$$F_{pull} \cos\theta = F_{friction}$$

3. Plug in the frictional force:

$$F_{pull} \cos\theta = \mu\left(mg - F_{pull}\sin\theta\right)$$

4. Solve for F_{pull}, and plug in the numbers:

$$F_{pull} = \frac{\mu mg}{\cos\theta + \mu\sin\theta}$$

$$= \frac{(0.25)(1,000 \text{ kg})(9.8 \text{ m/s}^2)}{\cos 10° + (0.25)\sin 10°} \approx 2,380 \text{ N}$$

5. Use the equation for work:

$$W = F_{pull}\,s\cos\theta$$

$$= (2,380 \text{ N})(3,000 \text{ m})(\cos 10°)$$

$$\approx 7.0 \times 10^6 \text{ J}$$

Q. You and your friends try to lift the huge gold ingot to take it inside your home. You manage to lift the gold ingot 25 centimeters, before realizing that it's too heavy. You then very slowly set the ingot down. If the ingot has a mass of 1,000.0 kilograms, how much work did you do setting the ingot down?

A. The correct answer is -2.4×10^3 joules.

1. The force you exerted on the ingot is equal to the weight of the ingot, or 9,800 newtons.

2. Use the equation for work:

$$W = F_{pull}\,s\cos\theta$$

$$= (9,800 \text{ N})(0.25 \text{ m})(\cos 180°)$$

$$\approx -2.4 \times 10^3 \text{ J}$$

You did negative work, because the force and the displacement were in opposite directions.

Practice Questions

1. You're pulling a chest of drawers, applying a force of 60.0 newtons at an angle of 60.0°. How much work do you do pulling it over 10.0 meters?

2. You're dragging a sled on a rope at 30°, applying a force of 20.0 newtons. How much work do you do over 1.0 kilometers?

3. You're pushing a box of dishes across the kitchen floor, using 100.0 joules to move it 10.0 meters. If you apply the force at 60.0°, what is the force you used?

4. You're pushing an out-of-gas car down the road, applying a force of 800.0 newtons. How much work have you done in moving the car 10.0 meters?

5. A construction foreman exerts 1,300 newtons of force trying to move a 1,200-kilogram block of concrete. How many joules of work does the foreman perform?

Practice Answers

1. **300 J.** Use the equation $W = Fs \cos \theta$.

   Plug in the numbers:

   $$W = Fs \cos \theta = (60.0 \text{ N})(10.0 \text{ m})(0.5) = 300 \text{ J}$$

2. **1.7 × 10⁴ J.** Use the equation $W = Fs \cos \theta$.

   Plug in the numbers:

   $$W = Fs \cos \theta = (20 \text{ N})(1,000 \text{ m})(0.866) = 1.7 \times 10^4 \text{ J}$$

3. **20 N.** Use the equation $W = Fs \cos \theta$.

   Solve for force:

   $$F = \frac{W}{s \cos \theta}$$

   Plug in the numbers:

   $$F = \frac{W}{s \cos \theta} = \frac{100.0 \text{ J}}{(10.0 \text{ m})(0.5)} = 20 \text{ N}$$

4. **8,000 J.** Use the equation $W = Fs \cos \theta$.

Plug in the numbers:

$$W = Fs \cos \theta = (800.0 \text{ N})(10.0 \text{ m})(1) = 8,000 \text{ J}$$

5. **0 J.** Although the foreman exerted a force, the concrete didn't move, which means that no distance was traversed. Because work is the product of force and distance, if the distance is 0, then the work done must also be 0.

MAKING A MOVE: KINETIC ENERGY

Remember: When you start pushing or pulling a stationary object with a constant force, it starts to move if the force you exert is greater than the net forces resisting the movement, such as friction and gravity. If the object starts to move at some speed, it will acquire kinetic energy. *Kinetic energy* is the energy an object has because of its motion. *Energy* is the ability to do work.

You know the ins and outs of kinetic energy. So how do you calculate it?

The work-energy theorem: Turning work into kinetic energy

A force acting on an object that undergoes a displacement does work on the object. If this force is a *net* force that accelerates the object (according to Newton's second law — see Chapter 5), then the velocity changes due to the acceleration (see Chapter 3). The change in velocity means that there is a change in the kinetic energy of the object.

Remember: The change in kinetic energy of the object is equal to the work done by the net force acting on it. This is a very important principle called the *work-energy theorem.*

After you know how work relates to kinetic energy, you're ready to take a look at how kinetic energy relates to the speed and mass of the object.

The equation to find kinetic energy, *KE,* is as follows, where *m* is mass and *v* is velocity:

$$KE = \frac{1}{2}mv^2$$

Using a little math, you can show that work required to accelerate the object from rest is also equal to $(1/2)mv^2$. Say, for example, that you apply a force to a model airplane to get it flying and that the plane is accelerating. Here's the equation for net force:

$$F = ma$$

The work done on the plane, which becomes its kinetic energy, equals the following:

$$W = Fs\cos\theta$$

Net force, F, equals mass times acceleration. Assume that you're pushing in the same direction that the plane is going; in this case, $\cos 0° = 1$, so

$$W = Fs = mas$$

You can tie this equation to the final and original velocity of the object. Use the equation $v_f^2 - v_i^2 = 2as$ (see Chapter 3), where v_f equals final velocity and v_i equals initial velocity. Solving for a gives you

$$a = \frac{v_f^2 - v_i^2}{2s}$$

If you plug this value of a into the equation for work, $W = mas$, you get the following:

$$W = \frac{1}{2}m\left(v_f^2 - v_i^2\right)$$

If the initial velocity is 0, you get

$$W = \frac{1}{2}mv_f^2$$

This is the work that you put into accelerating the model plane — that is, into the plane's motion — and that work becomes the plane's kinetic energy, KE:

$$KE = \frac{1}{2}mv_f^2$$

This is just the work-energy theorem stated as an equation.

Using the kinetic energy equation

You normally use the kinetic energy equation to find the kinetic energy of an object when you know its mass and velocity. The equation to find kinetic energy is

$$KE = \frac{1}{2}mv^2$$

That's the energy of motion.

Example

Q. Say that you push a spaceship, mass 1,000.0 kilograms, by applying a force of 1.00×10^4 newtons for 1.00 meters. How fast does the spaceship end up traveling?

A. The correct answer is 4.47 meters per second.

1. Find the total work done:

$$(1.00 \times 10^4 \text{ N})(1.00 \text{ m}) = 1.00 \times 10^4 \text{ J}$$

2. The work done on the spaceship goes into its kinetic energy, and its initial kinetic energy is 0, so its final kinetic energy is 1.0×10^4 joules.

3. Use the equation for kinetic energy and solve for v:

$$KE = \frac{1}{2}mv^2$$
$$1.00 \times 10^4 \text{ J} = \frac{1}{2}(1,000.0 \text{ kg})v^2$$
$$v^2 = \frac{(1.00 \times 10^4 \text{ J})}{500.0 \text{ kg}}$$
$$v^2 = 20.0 \text{ m}^2/\text{s}^2$$
$$v = \sqrt{20.0} \text{ m/s} = 4.47 \text{ m/s}$$

The final speed is 4.47 meters per second.

Practice Questions

1. You're applying 600.0 newtons of force to a car with a mass of 1,000.0 kilograms. You're traveling a distance of 100.0 meters on a frictionless, icy road. What is the car's final speed?

2. You're applying 3,000.0 newtons of force to a hockey puck with a mass of 0.10 kilograms. It travels over a distance of 0.10 meters on a frictionless ice rink. What is the puck's final speed?

Practice Answers

1. **11 m/s.** Find the total work done:

$$(600 \text{ N})(100 \text{ m}) = 6.0 \times 10^4 \text{ J}$$

The work done on the car goes into its kinetic energy, so its final kinetic energy is 6.0×10^4 joules.

Use the equation for kinetic energy and solve for v:

$$KE = \frac{1}{2}mv^2$$

$$v^2 = \frac{2}{m}(KE)$$

Plug in the numbers:

$$v^2 = \frac{2(6 \times 10^4 \text{ J})}{1{,}000 \text{ kg}} = 120 \text{ m}^2/\text{s}^2$$

The final speed = $120^{1/2}$ m/s = 11 m/s.

2. **77 m/s.** Find the total work done:

$$(3{,}000\,\text{N})(0.10\,\text{m}) = 300\,\text{J}$$

The work done on the puck goes into its kinetic energy, so its final kinetic energy is 300 joules.

Use the equation for kinetic energy and solve for v:

$$KE = \frac{1}{2}mv^2$$

$$v^2 = \frac{2}{m}(KE)$$

Plug in the numbers:

$$v^2 = \frac{2(300 \text{ J})}{0.10 \text{ kg}} = 6{,}000 \text{ m}^2/\text{s}^2$$

The final speed = $6{,}000^{1/2}$ = 77 m/s.

Calculating changes in kinetic energy by using net force

In everyday life, multiple forces act on an object, and you have to take them into account. If you want to find the change in an object's kinetic energy, you have to consider only the work done by the net force. In other words, you convert only the work done by the net force into kinetic energy.

Example

Q. In the following figure, a 1,000.0-kilogram safe full of gold bars is sliding down a 3.0-meter ramp that meets the horizontal at an angle of 23°. The coefficient of kinetic friction is 0.15. What will the safe's speed be when it reaches the bottom of the ramp?

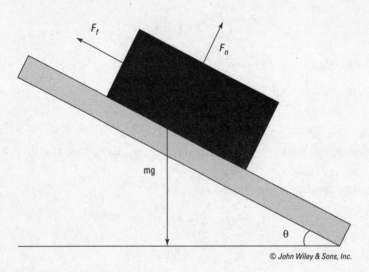

© John Wiley & Sons, Inc.

A. The correct answer is 3.8 meters per second.

1. Determine the net force acting on the safe. The component of the safe's weight acting along the ramp is

$$F = mg \sin\theta$$

2. You know that the normal force is

$$F_n = mg \cos\theta$$

That means that the kinetic force of friction is

$$F_f = \mu_k F_n = \mu_k mg \cos\theta$$

3. The net force accelerating the safe down the ramp, F_{net}, is

$$F_{net} = mg \sin\theta - F_f = mg \sin\theta - \mu_k mg \cos\theta$$

4. Plug in the numbers:

$$F_{net} = (1,000 \text{ kg})(9.8 \text{ m/s}^2)(\sin 23°) - (0.15)(1,000 \text{ kg})(9.8 \text{ m/s}^2)(\cos 23°)$$
$$= 3,800 \text{ N} - 1,350 \text{ N}$$
$$= 2,450 \text{ N}$$

5. The net force, 2,450 newtons, acts over the entire 3.0-meter ramp, so the work done by this force is

$$W = F_{net}s = (2,450 \text{ N})(3.0 \text{ m}) = 7,350 \text{ J}$$

6. The 7,350 joules of work done on the safe by the net force goes into the safe's kinetic energy. You find the safe's kinetic energy by using this equation:

$$W = F_{net}s = 7,350 \text{ J} = \frac{1}{2}mv^2$$

7. Solve for v^2:

$$v^2 = 14.7 \text{ m}^2/\text{s}^2, \text{ so } v = 3.8 \text{ m/s}$$

Practice Questions

1. An 80.0-kilogram person trips and slides down a 20.0-meter toboggan run of 27°. If the coefficient of kinetic friction is 0.10, what is the person's speed at the bottom of the run?

2. You're going down a 100.0-meter ski jump at an angle of 40.0°. Your mass (including skis!) is 90.0 kilograms, and the coefficient of kinetic friction is 0.050. What is your speed at the bottom of the jump?

Practice Answers

1. **11.8 m/s.** Find the net force on the sliding person. The component of the person's weight acting along the run is

$$F = mg \sin\theta$$

The normal force on the person is

$$F_n = mg \cos\theta$$

which means that the kinetic force of friction is

$$F_f = \mu_k F_n = \mu_k mg \cos\theta$$

The net force accelerating the person down the ramp, F_{net}, is

$$F_{net} = mg \sin\theta - F_f = mg \sin\theta - \mu_k mg \cos\theta$$

Plug in the numbers:

$$F_{net} = (80.0 \text{ kg})(9.8 \text{ m/s}^2)(\sin\ 27°) - (0.10)(80.0 \text{ kg})(9.8 \text{ m/s}^2)(\cos\ 27°)$$
$$= 350 \text{ N} - 70 \text{ N}$$
$$= 280 \text{ N}$$

The net force acts over the 20.0-meter toboggan ramp, so the work done by this force is

$$W = F_{net}s = (280 \text{ N})(20.0 \text{ m}) = 5,600 \text{ J}$$

The 5,600 joules of work done on the person goes into the person's kinetic energy. Find the person's kinetic energy:

$$W = F_{net} = 5,600 \text{ J} = \frac{1}{2}mv^2$$

Solve:

$$5,600 \text{ J} = \frac{1}{2}mv^2$$
$$5,600 \text{ J} = \frac{1}{2}(80.0 \text{ kg})v^2$$
$$v^2 = 140 \text{ m}^2/\text{s}^2$$
$$v = 11.8 \text{ m/s}$$

2. **35 m/s.** Find the net force on yourself to start. The component of your weight acting along the ski jump is

$$F = mg\sin\theta$$

The normal force on yourself is

$$F_n = mg\cos\theta$$

which means the kinetic force of friction is

$$F_f = \mu_k F_n = \mu_k mg\cos\theta$$

The net force accelerating you down the ski jump, F_{net}, is

$$F_{net} = mg\sin\theta - F_f = mg\sin\theta - \mu_k mg\cos\theta$$

Plug in the numbers:

$$F_{net} = (90.0 \text{ kg})(9.8 \text{ m/s}^2)(\sin\ 40°) - (0.05)(90.0 \text{ kg})(9.8 \text{ m/s}^2)(\cos\ 40°)$$
$$= 570 \text{ N} - 34 \text{ N}$$
$$= 540 \text{ N}$$

This net force acts over the 100-meter jump, so the work done by this force is

$$W = F_{net}s = (540\,\text{N})(100.0\,\text{m}) = 5.4 \times 10^4\,\text{J}$$

That work goes into your kinetic energy:

$$W = F_{net}s = 5.4 \times 10^4\ \text{J} = \frac{1}{2}mv^2$$

Solve:

$$5.4 \times 10^4\ \text{J} = \frac{1}{2}mv^2$$
$$5.4 \times 10^4\ \text{J} = \frac{1}{2}(90.0\ \text{kg})v^2$$
$$v^2 = 1{,}200\ \text{m}^2/\text{s}^2$$
$$v = 35\ \text{m/s}$$

ENERGY IN THE BANK: POTENTIAL ENERGY

There's more to motion than kinetic energy — an object can also have *potential energy,* which is stored energy or the energy an object has because of its position. The energy is called *potential* because it can be converted to kinetic energy or other forms of energy, such as thermal energy.

Objects can have potential energy from different sources. To give an object potential energy, all you need to do is perform work on an object against a force, such as when you pull back on an object connected to a spring. Gravity is a very common source of potential energy in physics problems.

Suppose that you have the job of taking your little cousin Jackie to the park, and you put the little tyke on the slide. Jackie starts at rest and then accelerates, ending up with quite a bit of speed at the bottom of the slide. You sense physics at work here. Taking out your clipboard, you put Jackie higher up the slide and let go, watching carefully. Sure enough, Jackie ends up going even faster at the bottom of the slide. You decide to move Jackie even higher up. (Suddenly, Jackie's mother shows up and grabs him from you. That's enough physics for one day.)

What was happening on the slide? Where did Jackie's kinetic energy come from? It ultimately came from the work you did lifting Jackie against the force of gravity. Jackie sits at rest at the bottom of the slide, so he has no kinetic energy. If you lift him to the top of the slide and hold him, he waits for the next trip down the slide, so he has no motion and no kinetic energy. However, you did work lifting him up against the force of gravity, so he has potential energy. As Jackie slides down the (frictionless) slide, gravity turns your work — and Jackie's potential energy — into kinetic energy.

To new heights: Gravitational potential energy

How much work do you do when you lift an object against the force of gravity? Suppose that you want to store a cannonball on an upper shelf at height h above where the cannonball is now. The work you do is

$$W = Fs\cos\theta$$

In this case, F equals the force required to overcome gravity, s equals distance, and θ is the angle between them. The gravitational force on an object is mg (mass times the acceleration due to gravity, 9.8 meters per second^2), and when you lift the cannonball straight up, $\theta = 0°$, so

$$W = Fs\cos\theta = mgh$$

The variable h here is the distance you lift the cannonball. To lift the ball, you have to do a certain amount of work, or m times g times h. The cannonball is stationary when you put it on the shelf, so it has no kinetic energy. However, it does have potential energy, which is the work you put into the ball to lift it to its present position.

If the cannonball rolls to the edge of the shelf and falls off, how much kinetic energy would it have just before it strikes the ground (which is where it started when you first lifted it)? It would have mgh joules of kinetic energy at that point. The ball's potential energy, which came from the work you put in lifting it, changes to kinetic energy thanks to the fall.

Remember: In general, you can say that if you have an object of mass m near the surface of the Earth (where the acceleration due to gravity is g), at a height h, then the potential energy of that mass compared to what it'd be at height 0 (where $h = 0$ at some reference height) is

$$PE = mgh$$

And if you move an object vertically against the force of gravity from height h_i to height h_f, its change in potential energy is

$$\Delta PE = mg\left(h_f - h_i\right)$$

The work you perform on the object changes its potential energy.

E

Example

Q. During a basketball game, a 1.0-kilogram ball gets thrown vertically in the air. It's momentarily stationary at a height of 5.0 meters above the floor and then falls back down. What is the ball's speed when it hits the floor?

A. The correct answer is 9.9 meters per second.

   1. Find out how much potential energy the ball has when it starts to fall. Use this equation for potential energy:

$$PE = mgh$$

2. This potential energy will be converted into kinetic energy, so you expand the equation like so:

$$PE = mgh = KE = \frac{1}{2}mv^2$$

3. Solve for v^2:

$$v^2 = 2gh$$

4. Solve for v:

$$v = \sqrt{2gh}$$

5. Plug in the numbers:

$$v = \sqrt{2gh}$$
$$= \sqrt{2(9.8 \text{ m/s}^2)(5.0 \text{ kg})} = 9.9 \text{ m/s}$$

Practice Questions

1. You jump out of an airplane at 2,000 feet and fall 1,000 feet before opening your parachute. What is your speed in feet per second (neglecting air resistance) when you open your chute?

2. If you're in an airplane at 30,000.0 feet, what is your potential energy if you have a mass of 70.0 kilograms?

Practice Answers

1. **254 ft/s.** Figure out how much potential energy you have when you start to fall. That potential energy is

$$PE = mgh_i$$

Find out how much potential energy is converted into kinetic energy when you open your parachute. The change in potential energy, ΔPE, is

$$\Delta PE = mgh_i - mgh_f$$

This potential energy will be converted into kinetic energy, so you have

$$\Delta PE = mgh_i - mgh_f = KE = \frac{1}{2}mv^2$$

Note that in the FPI system of units, $g = 32.2 \text{ ft/s}^2$.

Solve:

$$v^2 = 2g\left(h_i - h_f\right)$$
$$v^2 = 2\left(32.2 \text{ ft/s}^2\right)\left(2,000 \text{ ft} - 1,000 \text{ ft}\right)$$
$$v^2 = 64,400 \text{ ft}^2/\text{s}^2$$
$$v = 254 \text{ ft/s}$$

2. **6.3×10^6 J.** The equation for potential energy is

$$PE = mgh$$

Convert 30,000 feet into meters, using the fact that 1 foot = 0.305 meters:

$$30,000 \text{ ft} \times \frac{0.305 \text{ m}}{\text{ft}} = 9,150 \text{ m}$$

Plug in the numbers:

$$PE = mgh = \left(70.0 \text{ kg}\right)\left(9.8 \text{ m/s}^2\right)\left(9,150 \text{ m}\right) = 6.3 \times 10^6 \text{ J}$$

CHOOSE YOUR PATH: CONSERVATIVE VERSUS NONCONSERVATIVE FORCES

The work a *conservative force* does on an object is path-independent; the actual path taken by the object makes no difference. Fifty meters up in the air has the same gravitational potential energy whether you get there by taking the steps or by hopping on a Ferris wheel. That's different from the force of friction, which dissipates kinetic energy as heat. When friction is involved, the path you take does matter — a longer path will dissipate more kinetic energy than a short one. For that reason, friction is a *nonconservative force*.

For example, suppose that you and some buddies arrive at Mt. Newton, a majestic peak that rises h meters into the air. You can take two ways up — the quick way or the scenic route. Your friends drive up the quick route, and you drive up the scenic way, taking time out to have a picnic and to solve a few physics problems. They greet you at the top by saying, "Guess what — our potential energy compared to before is mgh greater."

"Mine, too," you say, looking out over the view. You pull out this equation (originally presented in the section "To new heights: Gravitational potential energy," earlier in this chapter):

$$\Delta PE = mg\left(h_f - h_i\right)$$

This equation basically states that the actual path you take when going vertically from h_i to h_f doesn't matter. All that matters is your beginning height compared to your ending height. Because the path taken by the object against gravity doesn't matter, gravity is a conservative force.

Here's another way of looking at conservative and nonconservative forces. Say that you're vacationing in the Alps and that your hotel is at the top of Mt. Newton. You spend the whole day driving around — down to a lake one minute, to the top of a higher peak the next. At the end of the day, you end up back at the same location: your hotel on top of Mt. Newton.

What's the change in your gravitational potential energy? In other words, how much net work did gravity perform on you during the day? Gravity is a conservative force, so the change in your gravitational potential energy is 0. Because you've experienced no net change in your gravitational potential energy, gravity did no net work on you during the day.

The road exerted a normal force on your car as you drove around (see Chapter 6), but that force was always perpendicular to the road, so it didn't do any work, either.

Conservative forces are easier to work with in physics because they don't "leak" energy as you move around a path — if you end up in the same place, you have the same amount of energy. If you have to deal with nonconservative forces such as friction, including air friction, the situation is different. If you're dragging something over a field carpeted with sandpaper, for example, the force of friction does different amounts of work on you depending on your path. A path that's twice as long will involve twice as much work to overcome friction.

KEEPING THE ENERGY UP: THE CONSERVATION OF MECHANICAL ENERGY

Mechanical energy is the sum of potential and kinetic energy, or the energy acquired by an object upon which work is done. The *conservation of mechanical energy,* which occurs in the absence of nonconservative forces (see the preceding section), makes your life much easier when solving physics problems, because the sum of kinetic energy and potential energy stays the same.

In this section, you examine the different forms of mechanical energy: kinetic and potential. You also find out how to relate the kinetic energy to the object's motion, how potential energy arises from the forces acting on the object, and how you can calculate the potential energy for the particular case of gravitational forces. And last, we explain how you can use mechanical energy to make calculations easier.

Shifting between kinetic and potential energy

Imagine a roller coaster car traveling along a straight stretch of track. The car has mechanical energy because of its motion: *kinetic energy.* Imagine that the track has a hill and that the car has just enough energy to get to the top before it descends the other side, back down to a straight and level track (see Figure 9-1). What happens? Well, at the top of the hill, the car is pretty much stationary, so where has all its kinetic energy gone? The answer is that it has been converted to *potential energy*. As the car begins its descent on the other side of the hill, the potential energy begins to be converted back to kinetic energy, and the car gathers speed until it reaches the bottom of the hill. Back at the bottom, all the potential energy the car had at the top of the hill has been converted back into kinetic energy.

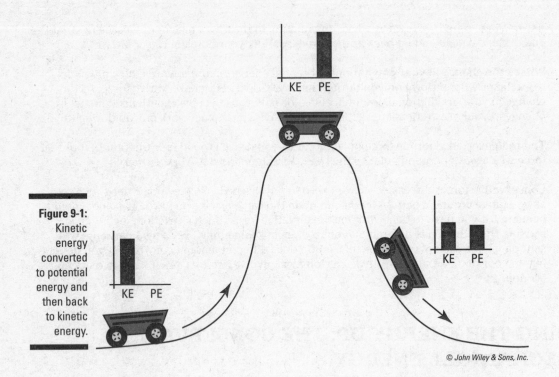

Figure 9-1:
Kinetic
energy
converted
to potential
energy and
then back
to kinetic
energy.

An object's potential energy derives from work done by forces, and a label for a particular potential energy comes from the forces that are its source. For example, the roller coaster has potential energy because of the gravitational forces acting on it, so this is often called *gravitational potential energy*. For more on potential energy, see the section "Energy in the Bank: Potential Energy," earlier in this chapter.

Remember: The roller coaster car's total mechanical energy, which is the sum of its kinetic and potential energies, remains constant at all points of the track. The combination of the kinetic and potential energies does vary, however. When no work is done on an object, its mechanical energy remains constant, whatever motions it may undergo.

Say, for example, that you see a roller coaster at two different points on a track — Point 1 and Point 2 — so that the coaster is at two different heights and two different speeds at those points. Because mechanical energy is the sum of the potential energy $(\text{mass} \times \text{gravity} \times \text{height})$ and kinetic energy $\left[(1/2)\text{mass} \times \text{velocity}^2 \right]$, the total mechanical energy at Point 1 is

$$ME_1 = mgh_1 + \frac{1}{2}mv_1^2$$

At Point 2, the total mechanical energy is

$$ME_2 = mgh_2 + \frac{1}{2}mv_2^2$$

What's the difference between ME_2 and ME_1? If there's no friction (or another non-conservative force), then $ME_1 = ME_2$, or

$$mgh_1 + \frac{1}{2}mv_1^2 = mgh_2 + \frac{1}{2}mv_2^2$$

These equations represent the *principle of conservation of mechanical energy.* The principle says that if the net work done by nonconservative forces is zero, the total mechanical energy of an object is conserved; that is, it doesn't change. (If, on the other hand, friction or another nonconservative force is present, the difference between ME_2 and ME_1 is equal to the net work the nonconservative forces do: $ME_2 - ME_1 = W_{nc}$.)

Tip: Another way of rattling off the principle of conservation of mechanical energy is that at Point 1 and Point 2,

$$PE_1 + KE_1 = PE_2 + KE_2$$

You can simplify that mouthful to the following:

$$ME_1 = ME_2$$

where *ME* is the total mechanical energy at any one point. In other words, an object always has the same amount of energy as long as the net work done by nonconservative forces is zero.

The mechanical-energy balance: Finding velocity and height

Breaking apart the equation for mechanical energy into potential and kinetic energy at two different points — $mgh_1 + (1/2)mv_1^2 = mgh_2 + (1/2)mv_2^2$ — allows you to solve for individual variables, such as velocity and height. You can use the principle of conservation of mechanical energy to find an object's final speed. Check out the following examples.

Examples

Q. You strap yourself into the Physics Park's new Bullet Blaster III coaster. The crew closes the hatch, and you're off down the totally frictionless track. What is your speed when you get to the bottom of the 400-meter drop?

A. The correct answer is 90 meters per second.

1. Use conservation of mechanical energy:

$$mgh_1 + \frac{1}{2}mv_1^2 = mgh_2 + \frac{1}{2}mv_2^2$$

2.   Your initial velocity is 0 and your final height is 0, so two of the terms drop out when you plug in the numbers. Divide both sides by m to get

$$mgh_1 = \frac{1}{2}mv_2^{\,2}$$

$$gh_1 = \frac{1}{2}v_2^{\,2}$$

3.   Solve for v_2 by rearranging the terms and taking the square root of both sides:

$$v_2 = \sqrt{2gh_1}$$

4.   Plug in the numbers:

$$v_2 = \sqrt{2gh_1} = \sqrt{2\left(9.8 \text{ m/s}^2\right)\left(400 \text{ m}\right)} \approx 89 \text{ m/s}$$

Round to the correct number of significant figures to get 90 meters per second.

Q.   A 2-kilogram ball is tossed into the air with a velocity of 14 meters per second. What is its height when its velocity is zero?

A.   The correct answer is 10 meters.

1.   The ball's kinetic energy is converted into potential energy:

$$\frac{1}{2}mv_i^2 = mgh_f$$

where m is the mass, v_i is the initial velocity, g is the acceleration due to gravity near Earth's surface, and h_f is the final height (in this case, when the velocity equals 0).

2.   Substitute the correct values after canceling the m from both sides to solve for h_f:

$$\frac{1}{2}mv_i^2 = mgh_f$$

$$\frac{1}{2}v_i^2 = gh_f$$

$$\frac{1}{2}\left(14 \text{ m/s}\right)^2 = \left(9.8 \text{ m/s}^2\right)h_f$$

$$98 \text{ m}^2/\text{s}^2 = \left(9.8 \text{ m/s}^2\right)h_f$$

$$10 \text{ m} = h_f$$

Practice Questions

1.   Tarzan is swinging on a vine over a crocodile-infested river at a speed of 13.0 meters per second. He needs to reach the opposite river bank 9.0 meters above his present position to be safe. Will he make it?

2.   At the top of a frictionless ramp angled at 30° to the horizontal, a cart has a total mechanical energy of 50 joules. Halfway down the ramp, how many joules of mechanical energy does the cart have?

Practice Answers

1. **No.** Use conservation of mechanical energy:

$$mgh_1 + \frac{1}{2}mv_1{}^2 = mgh_2 + \frac{1}{2}mv_2{}^2$$

At Tarzan's maximum height at the end of the swing, his speed, v_2, will be 0 meters per second, assuming $h_1 = 0$ meters. Plug in the values for v_2 and h_1:

$$\frac{1}{2}v_1{}^2 = gh_2$$

Solve for h_2:

$$h_2 = \frac{v_1{}^2}{2g}$$

$$= \frac{(13.0 \text{ m/s})^2}{2(9.8 \text{ m/s}^2)} \approx 8.6 \text{ m}$$

Tarzan will come up 0.4 meters short of the 9.0 meters he needs to be safe, so he needs some help.

2. **50 J.** Although the amounts of kinetic and potential energy change individually along the ramp, their sum — the total mechanical energy — always remains constant in the absence of frictional forces, which is the case here.

POWERING UP: THE RATE OF DOING WORK

Sometimes, it isn't just the amount of work you do but the rate at which you do work that's important. The concept of power gives you an idea of how much work you can expect in a certain amount of time.

Remember: *Power* in physics is the amount of work done divided by the time it takes, or the *rate* of work. Here's what that looks like in equation form:

$$P = \frac{W}{t}$$

Assume that you have two speedboats of equal mass, and you want to know which one will get you up to a speed of 120 miles per hour faster. Ignoring silly details like friction, you'll need the same amount of work to get up to that speed, but how long will it take? If one boat takes three weeks to get you up to 120 miles per hour, that may not be the one you take to the races. In other words, the amount of work you do in a certain amount of time can make a big difference.

If the work done at any one instant varies, you may want to work out the average work done over the time, t. An average quantity in physics is often written with a bar over it, as in the following equation for average power:

$$\bar{P} = \frac{\bar{W}}{t}$$

This section covers what units you're dealing with and the various ways to find power.

Using common units of power

Power is work or energy divided by time, so power has the units of joules per second, which is called the *watt* — a familiar term for just about anybody who uses anything electrical. You abbreviate a watt as simply W, so a 100-watt light bulb converts 100 joules of electrical energy into light and heat every second.

Tip: You occasionally run across symbol conflicts in physics, such as the W for watts and the *W* for work. This conflict isn't serious, however, because one symbol is for units (watts) and one is for a concept (work). Capitalization is standard, so be sure to pay attention to units versus concepts.

Note that because work and time are scalar quantities (they have no direction), power is a scalar as well.

Other units of power include foot-pounds per second (ft·lbs/s) and horsepower (hp). One hp = 550 ft·lbs/s = 745.7 W.

Doing alternate calculations of power

Because work equals force times distance, you can write the equation for power the following way, assuming that the force acts along the direction of travel:

$$P = \frac{W}{t} = \frac{Fs}{t}$$

where s is the distance traveled. However, the object's speed, v, is just s divided by t, so the equation breaks down to

$$P = \frac{W}{t} = \frac{Fs}{t} = Fv$$

That's an interesting result — power equals force times speed? Yep, that's what it says. However, because you often have to account for acceleration when you apply a force, you usually write the equation in terms of average power and average speed:

$$\bar{P} = F\bar{v}$$

Example

Q. You're riding a toboggan down an icy run to a frozen lake, and you accelerate the 80.0-kilogram combination of you and the toboggan from 1.0 meters per second to 2.0 meters per second in 2.0 seconds. How much power does that require?

A. The correct answer is 60 watts.

1. Assuming that there's no friction on the ice, you use this equation for the total work:

$$W = \frac{1}{2}mv_2{}^2 - \frac{1}{2}mv_1{}^2$$

2. Plug in the numbers:

$$W = \frac{1}{2}mv_2{}^2 - \frac{1}{2}mv_1{}^2$$
$$= \frac{1}{2}(80.0 \text{ kg})(2.0 \text{ m/s})^2 - \frac{1}{2}(80.0 \text{ kg})(1.0 \text{ m/s})^2$$
$$= 120 \text{ J}$$

3. Because this work is done in 2.0 seconds, the power involved is

$$P = \frac{120 \text{ J}}{2.0 \text{ s}} = 60 \text{ W}$$

Practice Questions

1. A 1,000.0-kilogram car accelerates from 88 meters per second to 100.0 meters per second in 30.0 seconds. How much power does that require?

2. A 60.0-kilogram person is running and accelerates from 5.0 meters per second to 7.0 meters per second in 2.0 seconds. How much power does that require?

3. A 120-kilogram linebacker accelerates from 5.0 meters per second to 10.0 meters per second in 1.0 second. How much power does that require?

4. You're driving a snowmobile that accelerates from 10 meters per second to 20 meters per second over a time interval of 10.0 seconds. If you and the snowmobile together have a mass of 500 kilograms, how much power is used?

Practice Answers

1. **3.8×10^4 W.** The equation for power is

$$P = \frac{W}{t}$$

The amount of work done is the difference in kinetic energy:

$$W = \frac{m}{2}\left(v_2{}^2 - v_1{}^2\right)$$

Therefore, the power is

$$P = \frac{W}{t} = \frac{m}{2t}\left(v_2{}^2 - v_1{}^2\right)$$

Plug in the numbers:

$$
\begin{aligned}
P = \frac{W}{t} &= \frac{m}{2t}\left(v_2{}^2 - v_1{}^2\right) \\
&= \frac{1{,}000.0 \text{ kg}}{2(30 \text{ s})}\left(\left[100 \text{ m/s}\right]^2 - \left[88 \text{ m/s}\right]^2\right) \\
&= 3.8 \times 10^4 \text{ W}
\end{aligned}
$$

2. **360 W.** The equation for power is

$$P = \frac{W}{t}$$

The amount of work done is the difference in kinetic energy:

$$W = \frac{m}{2}\left(v_2{}^2 - v_1{}^2\right)$$

Therefore, the power is

$$P = \frac{W}{t} = \frac{m}{2t}\left(v_2{}^2 - v_1{}^2\right)$$

Plug in the numbers:

$$
\begin{aligned}
P = \frac{W}{t} &= \frac{m}{2t}\left(v_2{}^2 - v_1{}^2\right) \\
&= \frac{60.0 \text{ kg}}{2(2.0 \text{ s})}\left(\left[7.0 \text{ m/s}\right]^2 - \left[5.0 \text{ m/s}\right]^2\right) \\
&= 360 \text{ W}
\end{aligned}
$$

3. **4,500 W.** The equation for power is

$$P = \frac{W}{t}$$

The amount of work done is the difference in kinetic energy:

$$W = \frac{m}{2}\left(v_2^{\,2} - v_1^{\,2}\right)$$

Therefore, the power is

$$P = \frac{W}{t} = \frac{m}{2t}\left(v_2^{\,2} - v_1^{\,2}\right)$$

Plug in the numbers:

$$P = \frac{W}{t} = \frac{m}{2t}\left(v_2^{\,2} - v_1^{\,2}\right)$$
$$= \frac{120 \text{ kg}}{2(1.0 \text{ s})}\left(\left[10.0 \text{ m/s}\right]^2 - \left[5.0 \text{ m/s}\right]^2\right)$$
$$= 4,500 \text{ W}$$

4. **8 × 10³ W.** The equation for power is

$$P = \frac{W}{t}$$

The amount of work done is the difference in kinetic energy:

$$W = \frac{m}{2}\left(v_2^{\,2} - v_1^{\,2}\right)$$

Therefore, the power is

$$P = \frac{W}{t} = \frac{m}{2t}\left(v_2^{\,2} - v_1^{\,2}\right)$$

Plug in the numbers:

$$P = \frac{W}{t} = \frac{m}{2t}\left(v_2^{\,2} - v_1^{\,2}\right)$$
$$= \frac{500 \text{ kg}}{2(10.0 \text{ s})}\left(\left[20 \text{ m/s}\right]^2 - \left[10 \text{ m/s}\right]^2\right)$$
$$= 7.5 \times 10^3 \text{ W}$$

Round to the correct number of significant digits.

10

PUTTING OBJECTS IN MOTION: MOMENTUM AND IMPULSE

Both momentum and impulse are very important to *kinematics,* or the study of objects in motion. After you have these topics under your belt, you can start talking about what happens when objects collide (hopefully not your car or bike). Sometimes they bounce off each other (like when you hit a tennis ball with a racket), and sometimes they stick together (like when a dart hits a dart board). With the knowledge of impulse and momentum you pick up in this chapter, you can handle either case.

LOOKING AT THE IMPACT OF IMPULSE

In physics terms, impulse tells you how much the momentum of an object will change when a force is applied for a certain amount of time. Say, for example, that you're shooting pool. Instinctively, you know how hard to tap each ball to get the results you want. The nine ball in the corner pocket? No problem — tap it and there it goes. The three ball bouncing off the side cushion into the other corner pocket? Another tap, this time a little stronger.

The taps you apply are called *impulses.* Take a look at what happens on a microscopic scale, millisecond by millisecond, as you tap a pool ball. The force you apply with your cue appears in Figure 10-1. The tip of each cue has a cushion, so the impact of the cue is spread out over a few milliseconds as the cushion squashes slightly. The impact lasts from the time when the cue touches the ball, t_i, to the time when the ball loses contact with the cue, t_f. As you can see from Figure 10-1, the force exerted on the ball changes during that time; in fact, it changes drastically, and figuring out what the force was doing at any 1 millisecond would be hard without some fancy equipment.

Because the pool ball doesn't come with any fancy equipment, you have to do what physicists normally do, which is to talk in terms of the average force over time. You can see what that average force looks like in Figure 10-2. Speaking as a physicist, you say that the impulse — or the tap — that the pool cue provides is the average force multiplied by the time that you apply the force.

Figure 10-1:
Examining
force versus
time gives
you the
impulse you
apply on
objects.

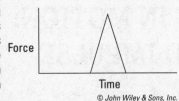

© John Wiley & Sons, Inc.

Here's the equation for impulse:

$$\text{Impulse} = \mathbf{J} = \overline{\mathbf{F}}\Delta t$$

Note that this equation is a vector equation, meaning it deals with both direction and magnitude (see Chapter 4). Impulse, **J**, is a vector, and it's in the same direction as the average force (which itself may be a net vector sum of other forces).

Figure 10-2:
The average
force over a
time interval
depends on
the values
the force
has over
that time.

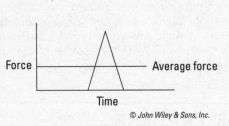

© John Wiley & Sons, Inc.

Remember: You get impulse by multiplying a quantity with units of newtons by a quantity with units of seconds, so the units of impulse are *newton-seconds* (N-s) in the MKS system and *pound-seconds* in the FPS system. (See Chapter 2 for details on systems of measurement.)

Example

Q. Suppose that you're playing pool and hit a pool ball for 10.0 milliseconds with a force of 20.0 newtons. What impulse did you impart to the pool ball?

A. The correct answer is 0.200 newton-seconds, in the direction of the force.

   1. Use the equation $\text{Impulse} = \overline{\mathbf{F}}t$.

   2. Plug in the numbers:

   $$\text{Impulse} = \overline{\mathbf{F}}t = (20.0 \text{ N})(1.0 \times 10^{-2} \text{ s}) = 0.200 \text{ N-s in the direction of the force}$$

Practice Questions

1. You're disgusted with your computer and give it a whack. If your hand is in contact with the computer for 100.0 milliseconds with a force of 100.0 newtons, what impulse do you impart to the computer?

2. You're standing under the eaves of your house when a huge icicle breaks off and hits you, imparting a force of 300.0 newtons for 0.10 seconds. What was the impulse?

Practice Answers

1. **10 N-s.** Use the equation Impulse $= \bar{\mathbf{F}}t$.

Plug in the numbers:

$$\text{Impulse} = \bar{\mathbf{F}}t = (100.0 \text{ N})(0.1 \text{ s}) = 10 \text{ N-s}$$

2. **30 N-s.** Use the equation Impulse $= \bar{\mathbf{F}}t$.

Plug in the numbers:

$$\text{Impulse} = \bar{\mathbf{F}}t = (300.0 \text{ N})(0.10 \text{ s}) = 30 \text{ N-s}$$

GATHERING MOMENTUM

In physics terms, momentum is proportional to both mass and velocity, and to make your job easy, physics defines *momentum* as the product of mass times velocity. Momentum is a big concept both in introductory physics and in some advanced topics such as high-energy particle physics, where the components of atoms zoom around at high speeds. When the particles collide, you can often predict what will happen based on your knowledge of momentum.

Even if you're unfamiliar with the physics of momentum, you're already familiar with the general idea. Catching a runaway car going down a steep hill is a problem because of its momentum. If a car without any brakes is speeding toward you at 40 miles per hour, trying to stop it by standing in its way may not be a great idea. The car has a lot of momentum, and bringing it to a stop requires plenty of effort. Now think of an oil tanker. Its engines aren't strong enough to make it turn or stop on a dime. Therefore, an oil tanker can take 20 miles or more to come to a stop, all because of the ship's momentum.

Remember: The more mass that's moving, the more momentum the mass has. The greater the magnitude of its velocity (think of an even faster oil tanker), the more momentum it has. The symbol for momentum is **p,** so you can say that

$$\mathbf{p} = m\mathbf{v}$$

Momentum is a vector quantity, meaning that it has a magnitude and a direction (see Chapter 4). The magnitude is in the same direction as the velocity — all you have to do to get the momentum of an object is to multiply its mass by its velocity. Because you multiply mass by velocity, the units for momentum are kilogram-meters per second (kg·m/s) in the MKS system.

Example

Q. Suppose that you're in an 800.0-kilogram race car going 200.0 miles an hour due east. If you have a mass of 60.0 kilograms, what is the total momentum?

A. The correct answer is 7.7×10^4 kilogram-meters per second, due east.

1. Use the equation $\mathbf{p} = m\mathbf{v}$.

2. Plug in the numbers, after first converting 200 miles per hour to 89.4 meters per second:

$$\mathbf{p} = m\mathbf{v} = (800.0\,\text{kg} + 60.0\,\text{kg})(89.4\,\text{m/s}) = 7.7 \times 10^4 \ \text{kg} \cdot \text{m/s, due east}$$

Practice Questions

1. You're falling out of an airplane, and before opening your parachute, you hit a speed of 100.0 meters per second. What is your momentum if you have a mass of 80.0 kilograms?

2. You're pushing an 800.0-kilogram car down the road, and it's going at 6.0 meters per second west. How much momentum does it have?

Practice Answers

1. **8,000 kg·m/s downward.** Use the equation $\mathbf{p} = m\mathbf{v}$.

Plug in the numbers:

$$\mathbf{p} = m\mathbf{v} = (80.0\ \text{kg})(100.0\ \text{m/s}) = 8,000\ \text{kg} \cdot \text{m/s downward}$$

2. **4,800 kg·m/s west.** Use the equation $\mathbf{p} = m\mathbf{v}$.

Plug in the numbers:

$$\mathbf{p} = m\mathbf{v} = (800.0\ \text{kg})(6.0\ \text{m/s}) = 4,800\ \text{kg} \cdot \text{m/s west}$$

THE IMPULSE-MOMENTUM THEOREM: RELATING IMPULSE AND MOMENTUM

You can connect the impulse you give to an object — like striking a pool ball with a cue — with the object's change in momentum; all you need is a little algebra and the process you explore in this section, called the *impulse-momentum theorem*. What makes the connection easy is that you can play with the equations for impulse and momentum to simplify them so you can relate the two topics. What equations does physics have in its arsenal to connect these two? Relating force and velocity is a start. For example, force equals mass times acceleration (see Chapter 5), and the definition of average acceleration is

$$\bar{\mathbf{a}} = \frac{\Delta \mathbf{v}}{\Delta t}$$

where $\mathbf{v}$ stands for velocity and t stands for time. Now you may realize that if you multiply that equation by the mass, you get force, which brings you closer to working with impulse:

$$\mathbf{F} = m\mathbf{a} = m\left(\frac{\Delta \mathbf{v}}{\Delta t}\right)$$

Now you have force in the equation. To get impulse, multiply the force equation by Δt, the time over which you apply the force:

$$\mathbf{F}\Delta t = m\left(\frac{\Delta \mathbf{v}}{\Delta t}\right)\Delta t = m\Delta \mathbf{v}$$

Take a look at the final expression, $m\Delta \mathbf{v}$. Because momentum equals $m\mathbf{v}$ (see the section "Gathering Momentum," earlier in this chapter), this is just the difference in the object's initial and final momentum: $\mathbf{p}_f - \mathbf{p}_i = \Delta \mathbf{p}$. Therefore, you can put that into the equation:

$$\mathbf{F}\Delta t = \Delta \mathbf{p}$$

Now take a look at the term on the left, $\mathbf{F}\Delta t$. That's the impulse, $\mathbf{J}$ (see the section "Looking at the Impact of Impulse," earlier in this chapter), or the force applied to the object multiplied by the time that force was applied. Therefore, you can write this equation as

$$\mathbf{J} = \mathbf{F}\Delta t = \Delta \mathbf{p}$$

Remember: Getting rid of everything in the middle finally gives you the impulse-momentum theorem, which says that impulse equals change in momentum:

$$\mathbf{J} = \Delta \mathbf{p}$$

The rest of this section provides some examples so you can practice this equation. But before you set off, think about what the formula means for the relation among impulse, force, and momentum. The impulse-momentum theorem defines a very simple relation between the impulse and momentum, namely that impulse is equal to the change in momentum. You can also see how a constant force applied over a time is equal to an impulse that is given by the force multiplied by the time:

$$\mathbf{J} = \mathbf{F}\Delta t$$

Last, you can tie the force and momentum together through the impulse, which gives you

$$\mathbf{F}\Delta t = \Delta \mathbf{p}$$

The meaning of this relation may become clearer if you divide both sides by Δt:

$$\mathbf{F} = \frac{\Delta \mathbf{p}}{\Delta t}$$

So you see that the force is given by the rate of change of momentum. This is a whole new way of thinking about force! Wherever you see momentum changing with time, you know a force is acting, and if calculating the momentum is easier, it can lead to an easier way of calculating the force. Check out the following examples.

Examples

Q. If you hit a stationary 160.0-gram pool ball with a force of 100.0 newtons for 0.10 seconds, what is its final speed?

A. The correct answer is 63 meters per second.

1. Use the equation $\mathbf{F}t = \Delta \mathbf{p}$.

2. Find the impulse first:

$$\text{Impulse} = Ft = (100.0 \text{ N})(0.10 \text{ s}) = 10 \text{ N-s}$$

3. That impulse is the change in the pool ball's momentum:

$$\Delta p = (100.0 \text{ N})(0.10 \text{ s}) = 10 \text{ kg} \cdot \text{m/s}$$

4. Now the pool ball's initial momentum was 0, so the final momentum is

$$p = 0 + \Delta p = 10 \text{ kg} \cdot \text{m/s}$$

5. Momentum equals mass times velocity, so to find speed (the magnitude of velocity), you can solve for v, like so:

$$v = \frac{p}{m}$$
$$= \frac{10 \text{ kg} \cdot \text{m/s}}{0.16 \text{ kg}} = 63 \text{ m/s}$$

Q. The handy rain gauge on your umbrella's top tells you that 100.0 grams of water are hitting the umbrella each second at an average speed of 10.0 meters per second. If the umbrella has a total mass of 1.00 kilograms, what force do you need to hold it upright in the rain?

A. The correct answer is 11 newtons.

   1. You need to apply a force to counteract the weight of the umbrella plus the force of the rain on the umbrella.

   2. The weight of the umbrella is just the mass times the acceleration due to gravity: $(F_G = mg)$, or $(1.0 \text{ kg})(9.8 \text{ m/s}^2) = 9.8$ N.

   3. When that rain hits your umbrella, the water comes to rest, so the vertical component of the change in momentum of rain every second is

$$\Delta p = m\Delta v$$

   4. Plug in numbers to get the change in momentum:

$$\Delta p = m\Delta v = (0.100 \text{ kg})(10 \text{ m/s}) = 1.0 \text{ kg} \cdot \text{m/s}$$

   5. Find the force by using the impulse-momentum theorem:

$$F_{rain}\Delta t = \Delta p$$

   6. Divide both sides by Δt to solve for the force, F:

$$F_{rain} = \frac{\Delta p}{\Delta t}$$

   7. Plug in Δp and set Δt to 1.0 second to get the force of the rain:

$$F_{rain} = \frac{\Delta p}{\Delta t} = \frac{1.0 \text{ kg} \cdot \text{m/s}}{1.0 \text{ s}} = 1.0 \text{ kg} \cdot \text{m/s}^2 = 1.0 \text{ N}$$

   8. Add the 1.0 newtons of force of the falling rain to the 9.8 newtons of the umbrella's weight for a total of 11 newtons, taking into account significant digits.

Practice Questions

   1. You hit a hockey puck, mass 170 grams, with a force of 150 newtons for 0.10 seconds. If it started at rest, what is its final speed?

   2. You're standing on an ice rink when another skater hits you, imparting a force of 200.0 newtons for 0.20 seconds. If you have a mass of 90.0 kilograms, what is your final speed?

   3. You kick a 450-gram soccer ball with a force of 400.0 newtons for 0.20 seconds. What is its final speed?

   4. You hit a 150-gram baseball with a force of 400.0 newtons for 0.10 seconds. The baseball was traveling toward you at 40 meters per second. What is its final speed?

Practice Answers

1. **88 m/s.** Use the equation $Ft = \Delta \mathbf{p}$.

    Find the impulse:

    $$\text{Impulse} = Ft = (150 \text{ N})(0.10 \text{ s}) = 15 \text{ N-s}$$

    Because the initial momentum of the puck was 0, that impulse becomes the puck's new momentum:

    $$p = (150 \text{ N})(0.10 \text{ s}) = 15 \text{ kg} \cdot \text{m/s}$$

    Momentum equals mass times velocity, so to find speed (the magnitude of velocity), you solve for v:

    $$
    \begin{aligned}
    v &= \frac{p}{m} \\
    &= \frac{15 \text{ kg} \cdot \text{m/s}}{0.17 \text{ kg}} \\
    &= 88 \text{ m/s}
    \end{aligned}
    $$

2. **0.44 m/s.** Use the equation $Ft = \Delta \mathbf{p}$.

    Find the impulse:

    $$\text{Impulse} = Ft = (200.0 \text{ N})(0.20 \text{ s}) = 40 \text{ N-s}$$

    Because your initial momentum was 0, that impulse becomes your new momentum:

    $$p = (200.0 \text{ N})(0.20 \text{ s}) = 40 \text{ kg} \cdot \text{m/s}$$

    Momentum equals mass times velocity, so to find speed (the magnitude of velocity), you solve for v:

    $$
    \begin{aligned}
    v &= \frac{p}{m} \\
    &= \frac{40 \text{ kg} \cdot \text{m/s}}{90.0 \text{ kg}} \\
    &= 0.44 \text{ m/s}
    \end{aligned}
    $$

3. **178 m/s.** Use the equation $Ft = \Delta \mathbf{p}$.

    Find the impulse:

    $$\text{Impulse} = Ft = (400.0 \text{ N})(0.20 \text{ s}) = 80 \text{ N-s}$$

    That impulse becomes the soccer ball's new momentum:

    $$p = (400.0 \text{ N})(0.20 \text{ s}) = 80 \text{ kg} \cdot \text{m/s}$$

Momentum equals mass times velocity, so to find speed (the magnitude of velocity), you solve for v:

$$v = \frac{p}{m}$$
$$= \frac{80 \text{ kg} \cdot \text{m/s}}{0.45 \text{ kg}}$$
$$= 178 \text{ m/s}$$

4.   **227 m/s.** Use the equation $\mathbf{F}t = \Delta\mathbf{p}$.

Find the impulse first:

$$\text{Impulse} = \mathbf{F}t = (400.0 \text{ N})(0.10 \text{ s}) = 40 \text{ N-s}$$

That impulse becomes the ball's change in momentum:

$$\Delta p = (400.0 \text{ N})(0.10 \text{ s}) = 40.0 \text{ kg} \cdot \text{m/s}$$

The baseball started with a momentum of (0.150 kg)(40 m/s) = 6 kg·m/s (toward you), so its final momentum will be 40 kg·m/s – 6 kg·m/s = 34 kg·m/s (away from you).

Momentum equals mass times velocity, so to find speed (the magnitude of velocity), you solve for v:

$$v = \frac{p}{m}$$
$$= \frac{34 \text{ kg} \cdot \text{m/s}}{0.150 \text{ kg}}$$
$$= 227 \text{ m/s}$$

WHEN OBJECTS GO BONK: CONSERVING MOMENTUM

The *principle of conservation of momentum* states that when you have an isolated system with no external forces, the initial total momentum of objects before a collision equals the final total momentum of the objects after the collision. In other words, $\Sigma\mathbf{p}_i = \Sigma\mathbf{p}_f$.

You may have a hard time dealing with the physics of impulses because of the short times and the irregular forces. But with the principle of conservation, items that are hard to measure — for example, the force and time involved in an impulse — are out of the equation altogether. Thus, this simple principle may be the most useful idea we provide in this chapter.

You can derive the principle of conservation of momentum from Newton's laws, what you know about impulse, and a little algebra.

Say that two careless space pilots are zooming toward the scene of an interplanetary crime. In their eagerness to get to the scene first, they collide. During the collision, the average force the second ship exerts on the first ship is $\mathbf{F}_{12}$. By the impulse-momentum theorem (see the earlier section "The Impulse-Momentum Theorem: Relating Impulse and Momentum"), you know the following for the first ship:

$$\mathbf{F}_{12}\Delta t = \Delta\mathbf{p}_1 = m_1\Delta\mathbf{v}_1 = m_1\left(\mathbf{v}_{f1} - \mathbf{v}_{i1}\right)$$

And if the average force exerted on the second ship by the first ship is $\mathbf{F}_{21}$, you also know that

$$\mathbf{F}_{21}\Delta t = \Delta\mathbf{p}_2 = m_2\Delta\mathbf{v}_2 = m_2\left(\mathbf{v}_{f2} - \mathbf{v}_{i2}\right)$$

Now you add these two equations together, which gives you the resulting equation:

$$\mathbf{F}_{12}\Delta t + \mathbf{F}_{21}\Delta t = m_1\left(\mathbf{v}_{f1} - \mathbf{v}_{i1}\right) + m_2\left(\mathbf{v}_{f2} - \mathbf{v}_{i2}\right)$$

Distribute the mass terms and rearrange the terms on the right until you get the following:

$$\mathbf{F}_{12}\Delta t + \mathbf{F}_{21}\Delta t = m_1\mathbf{v}_{f1} - m_1\mathbf{v}_{i1} + m_2\mathbf{v}_{f2} - m_2\mathbf{v}_{i2}$$
$$\mathbf{F}_{12}\Delta t + \mathbf{F}_{21}\Delta t = \left(m_1\mathbf{v}_{f1} + m_2\mathbf{v}_{f2}\right) - \left(m_1\mathbf{v}_{i1} + m_2\mathbf{v}_{i2}\right)$$

This is an interesting result, because $m_1\mathbf{v}_{1i} + m_2\mathbf{v}_{2i}$ is the *initial total momentum* of the two rocket ships $\left(\mathbf{p}_{i1} + \mathbf{p}_{i2}\right)$ and $m_1\mathbf{v}_{f1} + m_2\mathbf{v}_{f1}$ is the *final total momentum* $\left(\mathbf{p}_{f1} + \mathbf{p}_{f2}\right)$ of the two rocket ships. Therefore, you can write this equation as follows:

$$\mathbf{F}_{12}\Delta t + \mathbf{F}_{21}\Delta t = \left(\mathbf{p}_{f1} + \mathbf{p}_{f2}\right) - \left(\mathbf{p}_{i1} + \mathbf{p}_{i2}\right)$$

If you write the initial total momentum as $\mathbf{p}_f$ and the final total momentum as $\mathbf{p}_i$, the equation becomes

$$\mathbf{F}_{12}\Delta t + \mathbf{F}_{21}\Delta t = \mathbf{p}_f - \mathbf{p}_i$$

Where do you go from here? Both terms on the left include Δt, so you can rewrite $\mathbf{F}_{12}\Delta t + \mathbf{F}_{21}\Delta t$ as the sum of the forces involved, $\Sigma\mathbf{F}$, multiplied by the change in time:

$$\sum\mathbf{F}\Delta t = \mathbf{p}_f - \mathbf{p}_i$$

Remember: If you're working with what's called an *isolated* or *closed system*, you have no external forces to deal with. Such is the case in space. If two rocket ships collide in space, there are no external forces that matter, so by Newton's third law (see Chapter 5), $\mathbf{F}_{12}\Delta t = -\mathbf{F}_{21}\Delta t$. In other words, when you have a closed system, you get the following:

$$\sum\mathbf{F}\Delta t = \mathbf{p}_f - \mathbf{p}_i$$
$$0 = \mathbf{p}_f - \mathbf{p}_i$$

This converts to

$$\mathbf{p}_f = \mathbf{p}_i$$

Remember: The equation $\mathbf{p}_f = \mathbf{p}_i$ says that when you have an isolated system with no external forces, the initial total momentum before a collision equals the final total momentum after a collision, giving you the principle of conservation of momentum.

Example

Q. A pool ball with a mass of 170 grams and a speed of 30 meters per second hits another pool ball that's at rest. If the first pool ball ends up going in the same direction with a speed of 10 meters per second, what is the new speed of the second pool ball?

A. The correct answer is 20 meters per second.

1. Use the equation $\mathbf{p}_i = \mathbf{p}_f$.

2. All travel is in the same direction in this question, so you can treat it as a scalar equation. Find the original total momentum:

$$p_i = (0.17 \text{ kg})(30 \text{ m/s}) = 5.1 \text{ kg} \cdot \text{m/s}$$

3. The original total momentum equals the total final momentum, which is given by this expression, where p_{2f} is the final momentum of the second ball:

$$p_i = 5.1 \text{ kg} \cdot \text{m/s} = p_f = (0.17 \text{ kg})(10 \text{ m/s}) + p_{2f}$$

4. Solve for p_{2f}:

$$p_{2f} = 5.1 \text{ kg} \cdot \text{m/s} - 1.7 \text{ kg} \cdot \text{m/s} = 3.4 \text{ kg} \cdot \text{m/s}$$

5. Divide p_{2f} by the pool ball's mass to find its speed:

$$\frac{p_{2f}}{m} = \frac{3.4 \text{ kg} \cdot \text{m/s}}{0.17 \text{ kg}} = 20 \text{ m/s}$$

Practice Questions

1. You're driving a bumper car at an amusement park circus at 18 meters per second, and you hit another car that's at rest. If you end up going at 6.0 meters per second, what is the final speed of the other car, given that both cars have 100.0-kilogram mass, you have 80.0-kilogram mass, and the other person has a mass of 70.0 kilograms?

2. On the athletic field, a golf ball with a mass of 0.20 kilograms and speed of 100.0 meters per second hits a soccer ball at rest that has a mass of 0.45 kilograms. If the golf ball ends up at rest, what is the soccer ball's final speed, given that it travels in the same direction as the golf ball was originally traveling?

Practice Answers

1. **13 m/s.** Use the equation $\mathbf{p}_o = \mathbf{p}_f$.

    The bumper cars travel in the same direction here, so you can treat this as a scalar equation. Find the original total momentum:

    $$p_i = (18 \text{ m/s})(100.0 \text{ kg} + 80.0 \text{ kg}) = 3,240 \text{ kg} \cdot \text{m/s}$$

    This equals the total final momentum, which is given by this expression, where p_{2f} is the final momentum of the second bumper car:

    $$p_i = 3,240 \text{ kg} \cdot \text{m/s} = p_f = (100.0 \text{ kg} + 80.0 \text{ kg})(6.0 \text{ m/s}) + p_{2f}$$

    Solve for p_{2f}:

    $$p_{2f} = 3,240 \text{ kg} \cdot \text{m/s} - 1,080 \text{ kg} \cdot \text{m/s} = 2,160 \text{ kg} \cdot \text{m/s}$$

    Divide p_{2f} by the bumper car and person's mass to find its speed:

    $$\frac{p_{2f}}{m} = \frac{2,160 \text{ kg} \cdot \text{m/s}}{100.0 \text{ kg} + 70.0 \text{ kg}}$$
    $$= 12.7 \text{ m/s}$$

    which rounds to 13 with significant figures.

2. **44 m/s.** Use the equation $p_i = p_f$.

    Find the original total momentum:

    $$p_i = (0.20 \text{ kg})(100.0 \text{ m/s}) = 20 \text{ kg} \cdot \text{m/s}$$

    This equals the total final momentum, which is given by this expression, where p_{2f} is the final momentum of the soccer ball:

    $$p_i = 20 \text{ kg} \cdot \text{m/s} = p_f = p_{2f}$$

    Solve for p_{2f}:

    $$p_{2f} = 20 \text{ kg} \cdot \text{m/s}$$

    Divide p_{2f} by the soccer ball's mass to find its speed:

    $$\frac{p_{2f}}{m} = \frac{20 \text{ kg} \cdot \text{m/s}}{0.45 \text{ kg}} = 44 \text{ m/s}$$

WHEN WORLDS (OR CARS) COLLIDE: ELASTIC AND INELASTIC COLLISIONS

Examining collisions in physics can be pretty entertaining, especially because the principle of conservation of momentum makes your job so easy (see the earlier section "When Objects Go Bonk: Conserving Momentum"). But when you're dealing with collisions, there's often more to the story than impulse and momentum. Sometimes, kinetic energy is also conserved, which gives you the extra edge you need to figure out what happens in all kinds of collisions, even across two dimensions.

Collisions are important in many physics problems. Two cars collide, for example, and you need to find the final velocity of the two when they stick together. You may even run into a case where two railway cars going at different velocities collide and couple together, and you need to determine the final velocity of the two cars.

But what if you have a more general case where the two objects don't stick together? Say, for example, you have two pool balls that hit each other at different speeds and at different angles and bounce off with different speeds and different angles. How the heck do you handle that situation? You have a way to handle these collisions, but you need more than just the principle of conservation of momentum gives you. In this section, we explain the difference between elastic and inelastic collisions and then do a few elastic-collision problems.

Determining whether a collision is elastic

Remember: When bodies collide in the real world, they sometimes squash and deform to some degree. The energy to perform the deformation comes from the objects' original kinetic energy. In other cases, friction turns some of the kinetic energy into heat. Physicists classify collisions in *closed systems* (where the net forces add up to zero) based on whether colliding objects lose kinetic energy to some other form of energy:

- **Elastic collision:** In an elastic collision, the total kinetic energy in the system is the same before and after the collision. If losses to heat and deformation are much smaller than the other energies involved, such as when two pool balls collide and go their separate ways, you can generally ignore the losses and say that kinetic energy was conserved.

- **Inelastic collision:** In an inelastic collision, the collision changes the total kinetic energy in a closed system. In this case, friction, deformation, or some other process transforms the kinetic energy. If you can observe appreciable energy losses due to nonconservative forces (such as friction), kinetic energy isn't conserved.

  You see inelastic collisions when objects stick together after colliding, such as when two cars crash and weld themselves into one. However, objects don't need to stick together in an inelastic collision; all that has to happen is the loss of some kinetic energy. For example, if you smash your car into a car and deform it, the collision is inelastic, even if you can drive away after the accident.

Remember: Regardless of whether a collision is elastic or inelastic, momentum is *always* the same before and after the collision, as long as you have a closed system.

Colliding elastically along a line

In elastic collisions, kinetic energy and momentum are conserved. What do elastic collisions look like? In general, there's no deformation of any of the objects from an elastic collision. Here are the equations for the conservation of these factors:

$$KE_i = KE_f$$

$$\mathbf{p}_i = \mathbf{p}_f$$

Check out this idea in action: Suppose that you're in a car when you hit the car in front of you (elastically — no deformation of bumpers is involved), which started at rest. You know that momentum is always conserved, and you know that the car in front of you was stopped when you hit it, so if your car is Car 1 and the other one is Car 2, you get this equation:

$$m_1\mathbf{v}_{f1} + m_2\mathbf{v}_{f2} = m_1\mathbf{v}_{i1}$$

This equation can't tell you what $\mathbf{v}_{f1}$ and $\mathbf{v}_{f2}$ are, because there are two unknowns and only one equation. You can't solve for either $\mathbf{v}_{f1}$ or $\mathbf{v}_{f2}$ exactly in this case, even if you know the masses and $\mathbf{v}_{o1}$. So to solve for both final speeds, you need another equation to constrain what's going on here. That means using the conservation of kinetic energy. The collision was an elastic one, so kinetic energy was indeed conserved. That means that

$$\frac{1}{2}m_1\mathbf{v}_{f1}^2 + \frac{1}{2}m_2\mathbf{v}_{f2}^2 = \frac{1}{2}m_1\mathbf{v}_{i1}^2$$

With two equations and two unknowns, $\mathbf{v}_{f1}$ and $\mathbf{v}_{f2}$, you can solve for those unknowns in terms of the masses and $\mathbf{v}_{i1}$.

You probably won't be asked to solve questions of this kind on physics tests because, in addition to being two simultaneous equations, the second equation has a lot of squared velocities in it. But it's one you may see in homework. When you do the math, you get

$$v_{f1} = \frac{(m_1 - m_2)v_{i1}}{(m_1 + m_2)}$$

and

$$v_{f2} = \frac{2 \cdot m_1 \cdot v_{i1}}{(m_1 + m_2)}$$

This is a more substantial result than you get from problems that just use the conservation of momentum; in such problems, you can solve for only one final speed. Here, using both the conservation of momentum and kinetic energy, you can solve for both objects' final speeds. However, remember that the formulas in this section work only in the special case of an elastics collision. Also, the formulas would need to be modified if the initial velocity of the second object wasn't zero.

Example

Q. You're in a car that hits the at-rest car in front of you. If you and your car's mass is 1,000.0 kilograms, the mass of the car and driver ahead of you is 900.0 kilograms, and if you started at 44 meters per second, what are the final speeds of the two cars? Assume that the collision is elastic and all the action happens in the same line as your original direction of travel.

A. The correct answer is that your car moves 2.3 meters per second, and the other car moves 46 meters per second.

1. You know that this collision is elastic and the second car starts at rest, so you can use the equations given earlier in this section. Use this equation to find the final speed of your car:

$$v_{f1} = \frac{(m_1 - m_2)v_{i1}}{(m_1 + m_2)}$$

2. Plug in the numbers:

$$v_{f1} = \frac{(m_1 - m_2)v_{i1}}{(m_1 + m_2)} = \frac{(100.0 \text{ kg})(44 \text{ m/s})}{1,900.0 \text{ kg}} = 2.3 \text{ m/s}$$

3. Use this equation to find the final speed of the other car:

$$v_{f2} = \frac{2m_1 v_{i1}}{(m_1 + m_2)}$$

4. Plug in the numbers:

$$v_{f2} = \frac{2m_1 v_{i1}}{(m_1 + m_2)} = \frac{2(1,000.0 \text{ kg})(44 \text{ m/s})}{(1,900.0 \text{ kg})} = 46 \text{ m/s}$$

Practice Questions

1. A 160-gram hockey puck traveling at 60.0 meters per second hits a stationary puck with the same mass. What are the final speeds of the pucks, given that the collision is elastic and that all motion takes place along the same line?

2. You're driving a bumper car at 23 meters per second, and you hit another bumper car that's at rest. If you and your car have a mass of 300 kilograms, and the mass of the other car and driver is 240 kilograms, what are the final speeds of the cars?

Practice Answers

1. **First puck: 0 m/s; second puck: 60 m/s.** Use this equation to find the final speed of the first puck:

$$v_{f1} = \frac{(m_1 - m_2)v_{i1}}{(m_1 + m_2)}$$

Substituting the numbers gives you:

$$v_{f1} = \frac{(m_1 - m_2)v_{i1}}{(m_1 + m_2)} = \frac{(160 \text{ g} - 160 \text{ g})(60 \text{ m/s})}{(160 \text{ g} + 160 \text{ g})} = 0 \text{ m/s}$$

Use this equation to find the final speed of the second puck:

$$v_{f2} = \frac{2m_1 v_{i1}}{(m_1 + m_2)}$$

Putting in the numbers gives you:

$$v_{f2} = \frac{2m_1 v_{i1}}{(m_1 + m_2)} = \frac{2(160 \text{ g})(60 \text{ m/s})}{(160 \text{ g} + 160 \text{ g})} = 60 \text{ m/s}$$

Note that when the masses are the same, the first puck stops, and the second puck takes off with the same speed as the first puck had.

2. **You: 2.6 m/s; the other car: 26 m/s.** Use this equation to find the final speed of your car:

$$v_{f1} = \frac{(m_1 - m_2)v_{i1}}{(m_1 + m_2)}$$

Plug in the numbers:

$$v_{f1} = \frac{(m_1 - m_2)v_{i1}}{(m_1 + m_2)} = \frac{(300 \text{ kg} - 240 \text{ kg})(23 \text{ m/s})}{(300 \text{ kg} + 240 \text{ kg})} = 2.6 \text{ m/s}$$

Use this equation to find the final speed of the second car:

$$v_{f2} = \frac{2m_1 v_{i1}}{(m_1 + m_2)}$$

Plug in the numbers:

$$v_{f2} = \frac{2m_1 v_{i1}}{(m_1 + m_2)} = \frac{2(300 \text{ kg})(23 \text{ m/s})}{(300 \text{ kg} + 240 \text{ kg})} = 26 \text{ m/s}$$

Colliding in two dimensions

Collisions can take place in two dimensions. For example, soccer balls can move any which way on a soccer field, not just along a single line. Soccer balls can end up going north or south, east or west, or a combination of those. So you have to be prepared to handle collisions in two dimensions.

Example

Q. In the following figure, there's been an accident at an Italian restaurant, and two meatballs are colliding. Assuming that v_{i1} = 10.0 m/s, v_{i2} = 5.0 m/s, v_{f2} = 6.0 m/s, and the masses of the meatballs are equal, what are θ and v_{f1}?

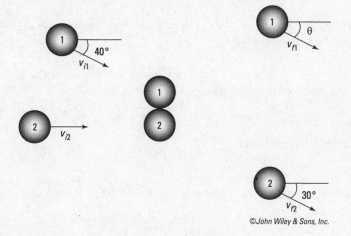

©John Wiley & Sons, Inc.

A. The correct answer is θ = 24° and v_{f1} = 8.2 m/s.

1. You can't assume that these meatballs conserve kinetic energy when they collide because the meatballs probably deform from the collision. However, momentum is conserved. In fact, momentum is conserved in both the x and y directions, which means

$$p_{fx} = p_{ix}$$

and

$$p_{fy} = p_{iy}$$

2. Here's what the original momentum in the x direction was:

$$p_{fx} = p_{ix} = m_1 v_{i1} \cos 40° + m_2 v_{i2}$$

3. Momentum is conserved in the x direction, so you get

$$p_{fx} = p_{ix} = m_1 v_{i1} \cos 40° + m_2 v_{i2} = m_1 v_{f1x} + m_2 v_{f2} \cos 30°$$

4. Which means that

$$m_1 v_{f1x} = m_1 v_{i1} \cos 40° + m_2 v_{i2} - m_2 v_{f2} \cos 30°$$

5. Divide by m_1:

$$v_{f1x} = \frac{m_1 v_{i1} \cos 40° + m_2 v_{i2} - m_2 v_{f2} \cos 30°}{m_1}$$

And because $m_1 = m_2$, this becomes

$$v_{f1x} = v_{i1} \cos 40° + v_{i2} - v_{f2} \cos 30°$$

6. Plug in the numbers:

$$\begin{aligned} v_{f1x} &= v_{i1} \cos 40° + v_{i2} - v_{f2} \cos 30° \\ &= (10 \text{ m/s})(0.766) + 5.0 \text{ m/s} - (6.0 \text{ m/s})(0.866) \\ &= 7.5 \text{ m/s} \end{aligned}$$

7. Now for the y direction. Here's what the original momentum in the y direction looks like (in the downward direction):

$$p_{fy} = p_{iy} = m_1 v_{i1} \sin 40°$$

8. Set that equal to the final momentum in the y direction:

$$p_{fy} = p_{iy} = m_1 v_{i1} \sin 40° = m_1 v_{f1y} + m_2 v_{f2} \sin 30°$$

That equation turns into

$$m_1 v_{f1y} = m_1 v_{i1} \sin 40° - m_2 v_{f2} \sin 30°$$

9. Solve for the final velocity component of Meatball 1's y velocity:

$$v_{f1y} = \frac{m_1 v_{i1} \sin 40° - m_2 v_{f2} \sin 30°}{m_1}$$

10. Because the two masses are equal, this becomes

$$v_{f1y} = v_{i1} \sin 40° - v_{f2} \sin 30°$$

11. Plug in the numbers:

$$\begin{aligned} v_{f1y} &= v_{i1} \sin 40° - v_{f2} \sin 30° \\ &= (10 \text{ m/s})(0.642) - (6.0 \text{ m/s})(0.5) \\ &= 3.4 \text{ m/s} \end{aligned}$$

12. So you end up with

$$v_{f1x} = 7.5 \text{ m/s (to the right)}$$
$$v_{f1y} = 3.4 \text{ m/s (downward)}$$

That means that the angle θ is

$$\theta = \tan^{-1}\left(\frac{3.4 \text{ m/s}}{7.5 \text{ m/s}}\right)$$
$$= \tan^{-1}(0.45)$$
$$= 24°$$

And the magnitude of v_{f1} is

$$v_{f1} = \sqrt{v_{f1x}^2 + v_{f1y}^2}$$
$$= \sqrt{(7.5 \text{ m/s})^2 + (3.4 \text{ m/s})^2}$$
$$= 8.2 \text{ m/s}$$

Practice Questions

1. Assume that the two objects in the figure shown with the preceding example question are hockey pucks of equal mass. Assuming that v_{i1} = 15 m/s, v_{i2} = 7.0 m/s, and v_{f2} = 7.0 m/s, what are θ and v_{f1}, assuming that momentum is conserved but kinetic energy is not?

2. Assume that the two objects in the following figure are tennis balls of equal mass. Assuming that v_{i1} = 12 m/s, v_{i2} = 8.0 m/s, and v_{f2} = 6.0 m/s, what are θ and v_{f1}, assuming that momentum is conserved but kinetic energy is not?

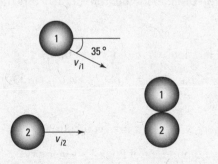

© John Wiley & Sons, Inc.

Practice Answers

1.   **14 m/s, 26°.** Momentum is conserved in this collision. In fact, momentum is conserved in both the x and y directions, which means the following are true:

$$p_{fx} = p_{ix}$$
$$p_{fy} = p_{iy}$$

The original momentum in the x direction was

$$p_{fx} = p_{ix} = m_1 v_{i1} \cos 40° + m_2 v_{i2}$$

Momentum is conserved in the x direction, so

$$p_{fx} = p_{ix} = m_1 v_{i1} \cos 40° + m_2 v_{i2} = m_1 v_{f1x} + m_2 v_{f2} \cos 30°$$

Solving for $m_1 v_{flx}$ gives you

$$m_1 v_{f1x} = m_1 v_{i1} \cos 40° + m_2 v_{i2} - m_2 v_{f2} \cos 30°$$

Divide by m_1:

$$v_{f1x} = \frac{m_1 v_{i1} \cos\ 40°\ + m_2 v_{i2} - m_2 v_{f2} \cos\ 30°}{m_1}$$

Because $m_1 = m_2$, that equation becomes

$$v_{f1x} = v_{i1} \cos 40° + v_{i2} - v_{f2} \cos 30°$$

Plug in the numbers:

$$
\begin{aligned}
v_{f1x} &= v_{i1} \cos\ 40°\ + v_{i2} - v_{f2} \cos\ 30° \\
&= (15 \text{ m/s})(0.766) + 7.0 \text{ m/s} - (7.0 \text{ m/s})(0.866) \\
&= 12.4 \text{ m/s}
\end{aligned}
$$

Now for the y direction. The original momentum in the y direction was

$$p_{fy} = p_{iy} = m_1 v_{i1} \sin 40°$$

Set that equal to the final momentum in the y direction:

$$p_{fy} = p_{iy} = m_1 v_{i1} \sin 40° = m_1 v_{f1y} + m_2 v_{f2} \sin 30°$$

Which turns into

$$m_1 v_{f1y} = m_1 v_{i1} \sin 40° - m_2 v_{f2} \sin 30°$$

Solve for the final velocity component of Puck 1's y velocity:

$$v_{f1y} = \frac{m_1 v_{i1} \sin\ 40° - m_2 v_{f2} \sin 30°}{m_1}$$

Because the two masses are equal, the equation becomes

$$v_{f1y} = v_{i1} \sin 40° - v_{f2} \sin 30°$$

Plug in the numbers:

$$v_{f1y} = v_{i1} \sin\ 40° - v_{f2} \sin\ 30°$$
$$= (15 \text{ m/s})(0.642) - (7.0 \text{ m/s})(0.5)$$
$$= 6.1 \text{ m/s}$$

So you end up with

$$v_{f1x} = 12.4 \text{ m/s}$$
$$v_{f1y} = 6.1 \text{ m/s}$$

That means the angle θ is

$$\theta = \tan^{-1}\left(\frac{6.1 \text{ m/s}}{12.4 \text{ m/s}}\right)$$
$$= \tan^{-1}(0.49) = 26°$$

And the magnitude of v_{f1} is

$$v_{f1} = \sqrt{v_{f1x}^2 + v_{f1y}^2}$$
$$= \sqrt{(12.4 \text{ m/s})^2 + (6.1 \text{ m/s})^2}$$
$$= 14 \text{ m/s}$$

2.   **14 m/s, 12°.** In this situation, momentum is conserved in both the x and y directions, so the following are true:

$$p_{fx} = p_{ix}$$
$$p_{fy} = p_{iy}$$

The original momentum in the x direction was

$$p_{fx} = p_{ix} = m_1 v_{i1} \cos 35° + m_2 v_{i2}$$

Momentum is conserved in the x direction, so

$$p_{fx} = p_{ix} = m_1 v_{i1} \cos 35° + m_2 v_{i2} = m_1 v_{f1x} + m_2 v_{f2} \cos 42°$$

Which means

$$m_1 v_{f1x} = m_1 v_{i1} \cos 35° + m_2 v_{i2} - m_2 v_{f2} \cos 42°$$

Divide by m_1:

$$v_{f1x} = \frac{m_1 v_{i1} \cos\ 35° + m_2 v_{i2} - m_2 v_{f2} \cos\ 42°}{m_1}$$

Because $m_1 = m_2$, this becomes

$$v_{f1x} = v_{i1} \cos 35° + v_{i2} - v_{f2} \cos 42°$$

Plug in the numbers:

$$\begin{aligned} v_{f1x} &= v_{i1} \cos 35° + v_{i2} - v_{f2} \cos 42° \\ &= (12 \text{ m/s})(0.82) + 8.0 \text{ m/s} - (6.0 \text{ m/s})(0.73) \\ &= 13.4 \text{ m/s} \end{aligned}$$

Now for the y direction. The original momentum in the y direction was

$$p_{fy} = p_{iy} = m_1 v_{i1} \sin 35°$$

Set that equal to the final momentum in the y direction:

$$p_{fy} = p_{iy} = m_1 v_{i1} \sin 35° = m_1 v_{f1y} + m_2 v_{f2} \sin 42°$$

Solving for $m_1 v_{f1y}$ gives you

$$m_1 v_{f1y} = m_1 v_{i1} \sin 35° - m_2 v_{f2} \sin 42°$$

Solve for the final velocity component of Tennis Ball 1's y velocity:

$$v_{f1y} = \frac{m_1 v_{i1} \sin 35° - m_2 v_{f2} \sin 42°}{m_1}$$

Because the two masses are equal, the equation becomes

$$v_{f1y} = v_{i1} \sin 35° - v_{f2} \sin 42°$$

Plug in the numbers:

$$\begin{aligned} v_{f1y} &= v_{i1} \sin 35° - v_{f2} \sin 42° \\ &= (12 \text{ m/s})(0.57) - (6.0 \text{ m/s})(0.67) \\ &= 2.9 \text{ m/s} \end{aligned}$$

So you end up with

$$v_{f1x} = 13.4 \text{ m/s}$$
$$v_{f1y} = 2.9 \text{ m/s}$$

Which means the angle θ is

$$\begin{aligned} \theta &= \tan^{-1}\left(\frac{2.9 \text{ m/s}}{13.4 \text{ m/s}} \right) \\ &= \tan^{-1}(0.22) = 12° \end{aligned}$$

And the magnitude of v_{f1} is

$$\begin{aligned} v_{f1} &= \sqrt{v_{f1x}^2 + v_{f1y}^2} \\ &= \sqrt{(13.4 \text{ m/s})^2 + (2.9 \text{ m/s})^2} \\ &= 14 \text{ m/s} \end{aligned}$$

11

WINDING UP WITH ANGULAR MOTION

IN THIS CHAPTER

- Changing gears from linear motion to rotational motion
- Calculating tangential speed and acceleration
- Understanding angular acceleration and velocity
- Identifying the torque involved in rotational motion
- Maintaining rotational equilibrium

This chapter is the first of two on handling objects that rotate, from space stations to marbles. Rotation is what makes the world go round — literally — and if you know how to handle linear motion and Newton's laws (see the first two parts of the book if you don't), the rotational equivalents we present in this chapter and in Chapter 12 are pieces of cake. And if you don't have a grasp on linear motion, no worries. You can get a firm grip on the basics of rotation here. You see all kinds of rotational ideas in this chapter: angular acceleration, tangential speed and acceleration, torque, and more. Kinetics deals not only with the motions of objects but also the forces behind those motions. Rotational kinetics deals with rotational motions and the forces behind them (torque). But enough spinning the wheels. Read on!

GOING FROM LINEAR TO ROTATIONAL MOTION

You need to change equations when you go from linear motion to rotational motion. Here are the angular equivalents (or analogs) for the linear motion equations:

	Linear	*Angular*
Velocity	$v = \dfrac{\Delta s}{\Delta t}$	$\omega = \dfrac{\Delta \theta}{\Delta t}$
Acceleration	$a = \dfrac{\Delta v}{\Delta t}$	$\alpha = \dfrac{\Delta \omega}{\Delta t}$
Displacement	$s = v_i \Delta t + \dfrac{1}{2} a t^2$	$\theta = \omega_i \Delta t + \dfrac{1}{2} \alpha t^2$
Motion with time canceled out	$v_f{}^2 - v_i{}^2 = 2as$	$\omega_f{}^2 - \omega_i{}^2 = 2\alpha\theta$

In all these equations, t stands for time, Δ means "change in," $_f$ means final, and $_i$ means initial. In the linear equations, v is velocity, s is displacement, and a is acceleration. In the angular equations, ω is angular velocity (measured in radians per second), θ is angular displacement in radians, and α is angular acceleration (in radians per second^2).

You know that the quantities displacement, velocity, and acceleration are all vectors; well, their angular equivalents are vectors, too. First, consider angular displacement, $\Delta\theta$ — this is a measure of the angle through which an object has rotated. The magnitude tells you the size of the angle of the rotation, and the direction is parallel to the axis of the rotation. Similarly, angular velocity, ω, has a magnitude equal to the angular speed and a direction that defines the axis of rotation. The angular acceleration, α, has a magnitude equal to the rate at which the angular velocity is changing; it's also directed along the axis of rotation.

If you consider only motion in a plane, then you have only one possible direction for the axis of rotation: perpendicular to the plane. In this case, these vector quantities have only one component — this vector component is just a number, and the sign of the number indicates all you need to know about the direction. For example, a positive angular displacement may be a clockwise rotation, and a negative angular displacement may be a counterclockwise rotation.

Just as the magnitude of the velocity is the speed, the magnitude of the angular velocity is the angular speed. Just as the magnitude of a displacement is a distance, the magnitude of an angular displacement is an angle — that is, the magnitude of the vector quantity is a scalar quantity.

Note: In the next section, we begin by looking at the motion in a plane considering only the single component of the vectors — which are scalar numbers (we identify the vector with its single component). So for that section, the quantities, $\Delta\theta$, ω, and α don't appear in bold type because they represent the single component of a rotation in a plane. In the section "Applying Vectors to Rotation," we take a closer look at the vector nature of the angular displacement, velocity, and acceleration.

UNDERSTANDING TANGENTIAL MOTION

Tangential motion is motion that's perpendicular to *radial motion,* or motion along a radius. Given a central point, vectors in the surrounding space can be broken into two components: *radial direction,* which points directly away from the center of the circle, and *tangential direction,* which follows the circle and is directed perpendicular to the radial direction. Motion in the tangential direction is referred to as *tangential motion.*

Tip: You can tie angular quantities such as angular displacement (θ), angular velocity (ω), and angular acceleration (α) to their associated tangential quantities. All you have to do is multiply by the radius, using these equations:

- $s = r\theta$
- $v = r\omega$
- $a = r\alpha$

Warning: These equations rely on using radians as the measure of angles; they don't work if you try to use degrees.

Say you're riding a motorcycle, for example, and the wheels' angular speed is $\omega = 21.5\pi$ radians per second. What does this mean in terms of your motorcycle's speed? To determine your motorcycle's velocity, you need to relate angular velocity, ω, to linear velocity, v. The following sections explain how you can make such relations.

Finding tangential velocity

At any point on a circle, you can pick two special directions: The direction that points directly away from the center of the circle (along the radius) is called the *radial* direction, and the direction that's perpendicular to this is called the *tangential* direction.

When an object moves in a circle, you can think of its *instantaneous velocity* (the velocity at a given point in time) at any particular point on the circle as an arrow drawn from that point and directed in the tangential direction. For this reason, this velocity is called the *tangential velocity*. The magnitude of the tangential velocity is the *tangential speed,* which is simply the speed of an object moving in a circle.

Remember: Given an angular velocity of magnitude ω, the tangential velocity at any radius is of magnitude $r\omega$. The idea that the tangential velocity increases as the radius increases makes sense, because given a rotating wheel, you'd expect a point at radius r to be going faster than a point closer to the hub of the wheel.

Take a look at Figure 11-1, which shows a ball tied to a string. The ball is whipping around with angular velocity of magnitude ω.

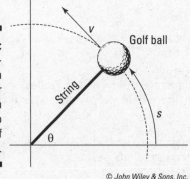

Figure 11-1:
A ball in circular motion has angular speed with respect to the radius of the circle.

© John Wiley & Sons, Inc.

Tip: You can easily find the magnitude of the ball's velocity, v, if you measure the angles in radians. A circle has 2π radians; the complete distance around a circle — its circumference — is $2\pi r$, where r is the circle's radius. In general, therefore, you can connect an angle measured in radians with the distance you cover along the circle, s, like this:

$$s = r\theta$$

where r is the radius of the circle. Now, you can say that $v = s/t$, where v is magnitude of the velocity, s is the distance, and t is time. You can substitute for s to get

$$v = \frac{s}{t} = \frac{r\theta}{t}$$

Because $\omega = \theta/t$, you can say that

$$v = \frac{s}{t} = \frac{r\theta}{t} = r\omega$$

In other words,

$$v = r\omega$$

Example

Q. The wheels of a motorcycle are turning with an angular velocity of 21.5π radians per second What is the motorcycle's speed if the radius of one of the motorcycle's wheels is 40 centimeters?

A. The correct answer is 27 meters per second.

1. Use the equation $v = r\omega$.

2. Plug in the numbers:

$$v = r\omega = (0.40 \text{ m/s})21.5\pi \approx 27 \text{ m/s}$$

Practice Questions

1. If a satellite is orbiting Earth, which has an average radius of 3,960 miles, at an altitude of 150 miles and an angular speed of 1.17×10^{-3} radians per second, what is the satellite's tangential speed in miles per hour?

2. You're flying a toy plane on a string, and it's going around at 20.0 miles per hour, 100.0 feet from you. What is its angular speed in radians per second?

Practice Answers

1. **17,300 mph.** Use the equation $v = r\omega$.

Convert 1.17×10^{-3} radians per second into radians per hour:

$$\frac{1.17 \times 10^{-3} \text{ radians}}{\cancel{s}} \times \frac{60 \; \cancel{s}}{\cancel{min}} \times \frac{60 \; \cancel{min}}{hr} = 4.21 \text{ radians/hr}$$

The radius at which the satellite orbits is 150 miles added to the radius of Earth, which is about 3,960 miles, making that radius 4,110 miles. That makes the satellite's tangential speed

$$v = r\omega = (4{,}110 \text{ miles})(4.21 \text{ radians/hr}) = 17{,}300 \text{ mph}$$

2.   **0.29 radians/s.** Use the equation $v = r\omega$.

Solve for ω:

$$\omega = \frac{v}{r}$$

Convert 20.0 miles per hour to feet per hour:

$$\frac{20.0 \text{ mi}}{\text{hr}} \times \frac{5{,}280 \text{ ft}}{\text{mi}} = 1.06 \times 10^5 \text{ ft/hr}$$

Convert from feet per hour to feet per second:

$$\frac{1.06 \times 10^5 \text{ ft}}{\text{hr}} \times \frac{1 \text{ hr}}{60 \text{ min}} \times \frac{1 \text{ min}}{60 \text{ s}} = 29 \text{ ft/s}$$

Solve for ω:

$$\omega = \frac{v}{r} = \frac{29 \text{ ft/s}}{100.0 \text{ ft}} = 0.29 \text{ radians/s}$$

Finding tangential acceleration

Tangential acceleration is a measure of how the tangential velocity of a point at a certain radius changes with time. Tangential acceleration is just like linear acceleration (see Chapter 3), but it's particular to the tangential direction, which is relevant to circular motion. Here, you look at the magnitude of the angular acceleration, α, which tells you how the speed of the object in the tangential direction is changing.

For example, when you start a lawn mower, a point on the tip of one of its blades starts at a tangential velocity of zero and ends up with a tangential velocity with a pretty large magnitude. So how do you determine the point's tangential acceleration? You can use the following equation from Chapter 3, which relates velocity to acceleration (where Δv is the change in velocity and Δt is the change in time) to relate tangential quantities like tangential velocity to angular quantities such as angular velocity:

$$a = \frac{\Delta v}{\Delta t}$$

Tangential velocity, v, equals $r\omega$ (as you see in the preceding section), so you can plug in this information:

$$a = \frac{\Delta (r\omega)}{\Delta t}$$

Because the radius is constant here, the equation becomes

$$a = \frac{r\Delta\omega}{\Delta t}$$

However, $\Delta\omega/\Delta t = \alpha$, the angular acceleration, so the equation becomes

$$a = r\alpha$$

Translated into layman's terms, this says tangential acceleration equals angular acceleration multiplied by the radius.

Example

Q. A set of helicopter blades has a radius of 4.3 meters. If a point on the tip of one blade starts at 0 meters per second and ends up 60 seconds later with a speed of 400 meters per second, what was the angular acceleration?

A. The correct answer is 1.6 radians per second^2.

1. Use this equation:

$$a = \frac{\Delta v}{\Delta t} = r\alpha$$

2. Plug in the numbers:

$$a = \frac{\Delta v}{\Delta t} = \frac{(400 \text{ m/s} - 0 \text{ m/s})}{(60 \text{ s})} = r\alpha$$

3. Divide both sides by r:

$$\frac{(400 \text{ m/s} - 0 \text{ m/s})}{(4.3 \text{ m})(60 \text{ s})} = \alpha$$

4. Do the math:

$$\frac{(400 \text{ m/s} - 0 \text{ m/s})}{(4.3 \text{ m})(60 \text{ s})} = \alpha = 1.6 \text{ radians/s}^2$$

Practice Questions

1. If a point on the edge of a tire with a radius of 0.50 meters starts at rest and ends up 3.5 minutes later at 88 meters per second (about 197 miles per hour), what was the magnitude of its average angular acceleration?

2. You're flying a toy plane on a string, and it's going around at 20.0 meters per second, 10.0 meters from you. If it accelerates to a final velocity of 30.0 meters per second in 80.0 seconds, what is its angular acceleration?

Practice Answers

1. **0.84 radians/s^2.** Use this equation:

$$a = \frac{\Delta v}{\Delta t} = r\alpha$$

Plug in the numbers:

$$a = \frac{\Delta v}{\Delta t} = \frac{88 \text{ m/s} - 0 \text{ m/s}}{(3.5 \text{ min})(60 \text{ s})} = r\alpha$$

Divide both sides by *r*:

$$\frac{88 \text{ m/s} - 0 \text{ m/s}}{(0.5 \text{ m})(3.5 \text{ min})(60 \text{ s})} = \alpha$$

Do the math:

$$\frac{88 \text{ m/s} - 0 \text{ m/s}}{(0.5 \text{ m})(3.5 \text{ min})(60 \text{ s})} = \alpha = 0.84 \text{ radians/s}^2$$

2. **1.25 × 10^{-2} radians/s^2.** Use this equation:

$$a = \frac{\Delta v}{\Delta t} = r\alpha$$

Plug in the numbers:

$$a = \frac{\Delta v}{\Delta t} = \frac{(30.0 \text{ m/s} - 20.0 \text{ m/s})}{(80.0 \text{ s})} = r\alpha$$

Divide both sides by *r*:

$$\frac{(30.0 \text{ m/s} - 20.0 \text{ m/s})}{(10.0 \text{ m})(80.0 \text{ s})} = \alpha$$

Do the math:

$$\frac{(30.0 \text{ m/s} - 20.0 \text{ m/s})}{(10.0 \text{ m})(80.0 \text{ s})} = \alpha = 1.25 \times 10^{-2} \text{ radians/s}^2$$

Finding centripetal acceleration

Newton's first law says that when there are no net forces, an object in motion will continue to move uniformly in a straight line (see Chapter 5). For an object to move in a circle, a force has to cause the change in direction — this force is called the *centripetal force*. Centripetal force is always directed toward the center of the circle.

The *centripetal acceleration* is proportional to the centripetal force (obeying Newton's second law; see Chapter 5). This is the component of the object's acceleration in the radial direction (directed toward the center of the circle), and it's the rate of change in the object's velocity that keeps the object moving in a circle; this force doesn't change the magnitude of the velocity, only the direction.

You can connect angular quantities, such as angular velocity, to centripetal acceleration. Centripetal acceleration is given by the following equation (for more on the equation, see Chapter 7):

$$a_c = \frac{v^2}{r}$$

where v is the velocity and r is the radius. Linear velocity is easy enough to tie to angular velocity because $v = r\omega$ (see the earlier section "Finding tangential velocity"). Therefore, you can rewrite the acceleration formula as

$$a_c = \frac{(r\omega)^2}{r}$$

Remember: The centripetal-acceleration equation simplifies to

$$a_c = r\omega^2$$

Nothing to it. The equation for centripetal acceleration means that you can find the centripetal acceleration needed to keep an object moving in a circle given the circle's radius and the object's angular velocity.

Example

Q. What is the centripetal acceleration of the moon around the Earth? The average radius of the moon's orbit is 3.85×10^8 meters, and the moon orbits the Earth about once every 28 days.

A. The correct answer is 2.6×10^{-3} meters per second^2.

1. Start with the new equation:

$$a_c = r\omega^2$$

2. Because the moon makes a complete orbit around the Earth in about 28 days, it travels 2π radians around the Earth in that period, so its angular velocity is

$$\omega = \frac{\Delta\theta}{\Delta t} = \frac{2\pi \text{ radians}}{28 \text{ days}}$$

3. Convert 28 days to seconds:

$$\frac{28 \text{ days}}{1} \times \frac{24 \text{ hr}}{1 \text{ day}} \times \frac{60 \text{ min}}{1 \text{ hr}} \times \frac{60 \text{ s}}{1 \text{ min}} \approx 2.42 \times 10^6 \text{ s}$$

4. Find the angular velocity:

$$\omega = \frac{\Delta\theta}{\Delta t}$$
$$= \frac{2\pi \text{ radians}}{2.42 \times 10^6 \text{ s}}$$
$$= 2.60 \times 10^{-6} \text{ radians/s}$$

5. Plug the numbers into the centripetal-acceleration formula:

$$a_c = r\omega^2$$
$$= \left(3.85 \times 10^8 \text{ m}\right)\left(2.60 \times 10^{-6} \text{ s}^{-1}\right)^2$$
$$\approx 2.60 \times 10^{-3} \text{ m/s}^2$$

Practice Questions

1. An ant hitches a ride on the second hand of a clock, 7.4 centimeters away from the center of the clock. What is the centripetal acceleration of the ant?

2. What is the gravitational force that the sun exerts on the Earth? The mass of the Earth is 5.97×10^{24} kilograms, and the distance between the Earth and the sun is about 150,000,000 kilometers.

Practice Answers

1. $\mathbf{8.1 \times 10^{-4}}$ **m/s²**. Use the centripetal-acceleration equation:

$$a_c = r\omega^2$$

Plug in the numbers:

$$a_c = r\omega^2$$
$$= (0.074 \text{ m})\left(\frac{2\pi}{60 \text{ s}}\right)^2$$
$$= 8.1 \times 10^{-4} \text{ m/s}^2$$

2. $\mathbf{5.3 \times 10^{24}}$ **N.** Convert 1 year to seconds:

$$1 \text{ yr} \times \frac{365.24 \text{ days}}{1 \text{ yr}} \times \frac{24 \text{ hr}}{1 \text{ day}} \times \frac{60 \text{ min}}{1 \text{ hr}} \times \frac{60 \text{ s}}{1 \text{ min}} = 3.1557 \times 10^7 \text{ s} \approx \pi \times 10^7 \text{ s}$$

Use Newton's second law with the centripetal-acceleration equation:

$$F = ma_c$$
$$F = mr\omega^2$$

Plug in the numbers:

$$F = mr\omega^2$$
$$= (5.97 \times 10^{24} \text{ kg})(1.5 \times 10^{14} \text{ m}) \left(\frac{2\pi}{3.1557 \times 10^7 \text{ s}} \right)^2$$
$$\approx 5.3 \times 10^{24} \text{ N}$$

APPLYING VECTORS TO ROTATION

Angular displacement, angular velocity, and angular acceleration are each vector quantities. When you consider circular motion in a plane, these vectors have only one component, which is a scalar number; in that case, you don't have to consider the direction very much. However, when you have circular motion in more than one plane (as with the motions of the planets, which orbit on very slightly different planes) or when the plane of rotation changes (like in a wobbling spinning top, for example), then the direction of these vectors becomes significant.

Remember: Angular velocity and angular acceleration are vectors that are directed along the axis of the rotation.

In this section, you hear more about the directions of the angular vectors. For the rest of this section, the quantities $\Delta\boldsymbol{\theta}$, $\boldsymbol{\omega}$, and $\boldsymbol{\alpha}$ appear in bold type because you're explicitly dealing with vectors.

Calculating angular velocity

When a wheel is spinning, it has not only an angular speed but also a direction. Here's what the angular velocity vector tells you:

- The size of the angular velocity vector tells you the angular speed.

- The direction of the vector tells you the axis of the rotation as well as whether the rotation is clockwise or counterclockwise.

Say that a wheel has a constant angular speed, ω; which direction does its angular velocity, $\boldsymbol{\omega}$, point? It can't point along the rim of the wheel, as tangential velocity does, because its direction would then change every second. In fact, the only real choice for its direction is perpendicular to the wheel.

The direction of the angular velocity always takes people by surprise: Angular velocity, $\boldsymbol{\omega}$, points along the axle of a wheel (see Figure 11-2). Because the angular velocity vector points the way it does, it has no component along the wheel. The wheel is spinning, so the tangential (linear) velocity at any point on the wheel is constantly changing direction — except for at the very center point of the wheel, where the base of the angular velocity vector sits. If the wheel is lying flat on the ground, the vector's head points up or down, away from the wheel, depending on which direction the wheel is rotating.

Tip: You can use the right-hand rule to determine the direction of the angular velocity vector. Wrap your right hand around the wheel so that your fingers point in the direction of the tangential motion at any point — the fingers on your right hand should go in the same direction as the wheel's rotation. When you wrap your right hand around the wheel, your thumb points in the direction of the angular velocity vector, ω.

Figure 11-2 shows a wheel lying flat, turning counterclockwise when viewed from above. Wrap your fingers in the direction of rotation. Your thumb, which represents the angular velocity vector, points up; it runs along the wheel's axle. If the wheel were to turn clockwise instead, your thumb — and the vector — would have to point down, in the opposite direction.

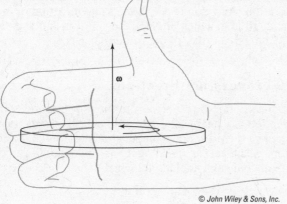

Figure 11-2:
Angular velocity points in a direction perpendicular to the wheel.

© John Wiley & Sons, Inc.

Example

Q. A helicopter's blades are rotating in a horizontal plane, and they're going counterclockwise when viewed from above. Which way does ω point?

A. The correct answer is upward.

1. Curl your right hand in the direction of rotational motion — counterclockwise.

2. Your right thumb points upward, indicating the direction of the ω vector.

Practice Questions

1. Suppose that you're flying a toy plane on a string, and it's going around clockwise as viewed from above. Which way does ω point?

2. Suppose that you're driving forward. Which way does ω point for the left front tire?

Practice Answers

1. **Downward.** Curl your right hand in the direction of rotational motion — clockwise. Your right thumb points downward, indicating the direction of the ω vector.

2. **To the left.** Curl your right hand in the direction of rotational motion. Your right thumb points to the left, indicating the direction of the ω vector.

Figuring angular acceleration

In this section, you find out how the angular acceleration and angular velocity relate to each other in terms of their magnitude and direction. You first see what happens in the simplest case, where the angular acceleration and velocity are in the same direction or in opposite directions. Then you look at a situation in which angular acceleration and angular velocity are at an angle to each other, leading to a tilting of the rotational axis.

Changing the speed and reversing direction

If the angular velocity vector points out of the plane of rotation (see the preceding section), what happens when the angular velocity changes — when the wheel speeds up or slows down? A change in velocity signifies the presence of angular acceleration. Like angular velocity, ω, angular acceleration, α, is a vector, meaning it has a magnitude and a direction. Angular acceleration is the rate of change of angular velocity:

$$\alpha = \frac{\Delta\omega}{\Delta t}$$

For example, look at Figure 11-3, which shows what happens when angular acceleration affects angular velocity. In this case, α points in the same direction as ω in Figure 11-3a. When the angular acceleration vector, α, points along the angular velocity, ω, the magnitude of ω will increase as time goes on, as Figure 11-3b shows.

Remember: Just as an object's linear velocity and linear acceleration may be in opposite directions, the angular acceleration also doesn't have to be in the same direction as the angular velocity vector (as Figure 11-4a shows). If the angular acceleration is directed in the opposite direction of the angular velocity, then the magnitude of the angular velocity decreases at a rate given by the magnitude of the angular acceleration.

Figure 11-3:
Angular acceleration in the same direction as the angular velocity.

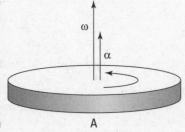

A

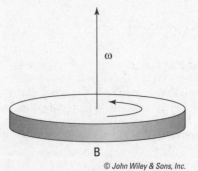

B

Just as in the case of linear velocity and acceleration, the angular acceleration gives the rate of change of angular velocity: The magnitude of the angular acceleration gives the rate at which the angular velocity changes, and the direction gives the direction of the change. You can see a decreased angular velocity in Figure 11-4b.

Figure 11-4:
Angular
accelera-
tion in the
direction
opposite
the angular
velocity
reduces
the angular
speed.

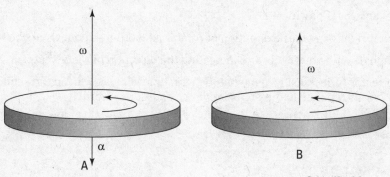

© John Wiley & Sons, Inc.

Tilting the axle

The *angular acceleration* is the rate of change of angular velocity — the change can be to the direction instead of the magnitude. For example, suppose you take hold of the axle of the spinning wheel in Figure 11-3 and tilt it. You'd change the angular velocity of the wheel but not by changing its magnitude (the angular speed of the wheel would remain constant); rather, you'd change the direction of the angular velocity by changing the axis of rotation — this is an angular acceleration that's directed perpendicular to the angular velocity, as in Figure 11-5.

Figure 11-5:
Angular
acceleration
perpen-
dicular to
the angular
velocity tilts
the axis of
rotation.

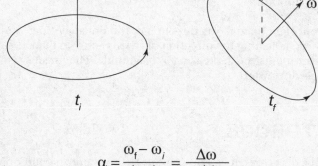

$$\alpha = \frac{\omega_f - \omega_i}{t_f - t_i} = \frac{\Delta \omega}{\Delta t}$$

© John Wiley & Sons, Inc.

Example

Q. A helicopter's blades are rotating in a horizontal plane, and they're going counterclockwise when viewed from above. As they slow down, which way does α point?

A. The correct answer is downward.

1. Curl your right hand in the direction of rotational motion — counterclockwise.

2. Your right thumb points upward, indicating the direction of the ω vector.

3. The ω vector is decreasing in magnitude with time while staying in the same direction, which means α points in the opposite direction — downward.

Practice Questions

1. Suppose that you're flying a toy plane on a string, and it's going around clockwise as viewed from above. In time, the plane is slowing down. Which way does α point?

2. A flying disc tossed from one player to another spins as it flies and slows down. If it's spinning counterclockwise when viewed from above, which way does α point?

Practice Answers

1. **Upward.** Curl your right hand in the direction of rotational motion. Your right thumb points downward, indicating the direction of the ω vector. In time, the rotation is slowing, so the ω vector must be decreasing in magnitude. That means α points in the opposite direction — upward.

2. **Downward.** Curl your right hand in the direction of rotational motion. Your right thumb points upward, indicating the direction of the ω vector. In time, the rotation is slowing, so the ω vector must be decreasing in magnitude. That means α points in the opposite direction — downward.

DOING THE TWIST: TORQUE

For extended objects (rods, disks, or cubes, for example), which, unlike point objects, have their mass distributed through space, you have to take into account where the force is applied. Enter torque. *Torque* is a measure of the ability of a force to cause rotation. In physics terms, the torque exerted on an object depends on the force itself (its magnitude and direction) and where you exert the force. You go from the strictly linear idea of force as something that acts in a straight line (such as when you push a refrigerator up a ramp) to its angular counterpart, torque.

Tip: Just as force causes acceleration, torque causes angular acceleration, so you can think of torque as the angular equivalent of force (see Chapter 12 for more info on that aspect of torque).

Torque brings forces into the rotational world. Most objects aren't just points or rigid masses, so if you push them, they not only move but also turn. For example, if you apply a force tangentially to a merry-go-round, you don't move the merry-go-round away from its current location — you cause it to start spinning. Rotational motions and the forces behind them are the focus of this chapter and Chapter 12.

Look at Figure 11-6, which shows a seesaw with a mass m on it. If you want to balance the seesaw, you can't have a larger mass, M, placed on a similar spot on the other side of the seesaw. Where you put the larger mass M determines whether the seesaw balances. As you can see in Figure 11-6a, if you put the mass M on the pivot point — also called the *fulcrum* — of the seesaw, you don't have balance. The larger mass exerts a force on the seesaw, but the force doesn't balance it.

As you can see in Figure 11-6b, as you increase the distance you put the mass M away from the fulcrum, the balance improves. In fact, if $M = 2m$, you need to put the mass M exactly half as far from the fulcrum as the mass m is.

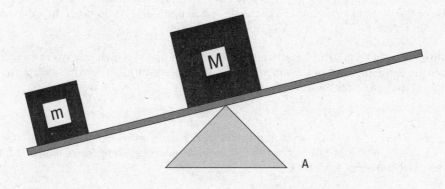

A

Figure 11-6:
A seesaw demon-
strates
torque in
action.

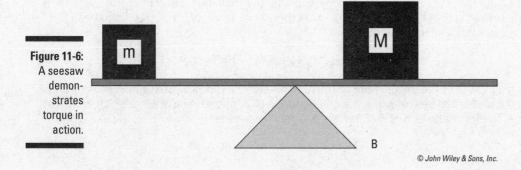

B

The torque is a vector. The magnitude of the torque tells you the ability of the torque to generate rotation; more specifically, the magnitude of the torque is proportional to the angular acceleration it generates. The direction of the torque is along the axis of this angular acceleration. This section starts by considering torques and forces that are in a plane, so you only need to think about the magnitude of the torque and not the full vector. Later, we explain a little more about the direction of the torque vector.

Mapping out the torque equation

Remember: How much torque you exert on an object depends on the following:

- The force you exert, F.

- Where you apply the force; the *lever arm* — also called the *moment arm* — is the perpendicular distance from the pivot point to the point at which you exert your force and is related to the distance from the axis, r, by $l = r \sin \theta$, where θ is the angle between the force and a line from the axis to the point where the force is applied.

Assume that you're trying to open a door, as in the various scenarios in Figure 11-7. You know that if you push on the hinge, as in Figure 11-7a, the door won't open; if you push the middle of the door, as in Figure 11-7b, the door will open; but if you push the edge of the door, as in Figure 11-7c, the door will open more easily.

In Figure 11-7, the lever arm, l, is distance r from the hinge at which you exert your force. The torque is the product of the magnitude of the force multiplied by the lever arm. It has a special symbol, the Greek letter τ (tau):

$$\tau = Fl$$

The units of torque are force units multiplied by distance units, which is newton-meters in the MKS system and foot-pounds in the foot-pound-second system (see Chapter 2 for more on these measurement systems).

For example, the lever arm in Figure 11-7 is distance r (because this is the distance perpendicular to the force), so $\tau = Fr$. If you push with a force of 200 newtons and r is 0.5 meters, what's the torque you see in the figure? In Figure 11-7a, you push on the hinge, so your distance from the pivot point is 0, which means the lever arm is 0. Therefore, the magnitude of the torque is 0. In Figure 11-7b, you exert the 200 newtons of force at a distance of 0.5 meters perpendicular to the hinge, so

$$\tau = Fl = (200 \text{ N})(0.5 \text{ m}) = 100 \text{ N} \cdot \text{m}$$

The magnitude of the torque here is 100 newton-meters. But now take a look at Figure 11-7c. You push with 200 newtons of force at a distance of $2r$ perpendicular to the hinge, which makes the lever arm $2r$ or 1.0 meter, so you get this torque:

$$\tau = Fl = (200 \text{ N})(1.0 \text{ m}) = 200 \text{ N} \cdot \text{m}$$

Now you have 200 newton-meters of torque, because you push at a point twice as far away from the pivot point. In other words, you double the magnitude of your torque. But what would happen if, say, the door were partially open when you exerted your force? Well, you would calculate the torque easily, if you have lever-arm mastery.

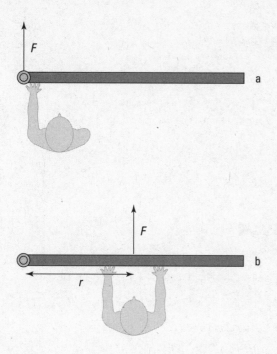

Figure 11-7:
The torque
you exert
on a door
depends on
where you
push it.

Understanding lever arms

If you push a partially open door in the same direction as you push a closed door, you
create a different torque because of the non-right angle between your force and the door.

Take a look at Figure 11-8a to see a person obstinately trying to open a door by pushing
along the door toward the hinge. You know this method won't produce any turning motion,
because the person's force has no lever arm to produce the needed turning force. In this
case, the lever arm is 0, so it's clear that even if you apply a force at a given distance away
from a pivot point, you don't always produce a torque. The direction you apply the force
also counts, as you know from your door-opening expertise.

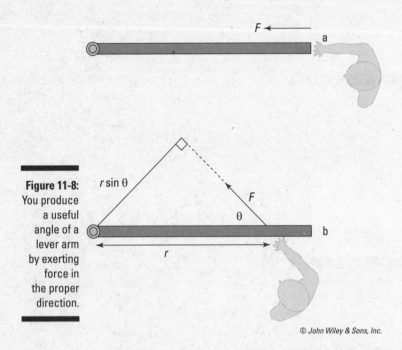

Figure 11-8: You produce a useful angle of a lever arm by exerting force in the proper direction.

Figuring out the torque generated

If you apply a force F at a displacement r from a pivot point where the angle between that displacement and F is θ, the torque you produce is

$$\tau = rF \sin\theta$$

Only the component of the distance perpendicular to the force contributes to the torque, which gives you the factor of $r \sin\theta$ in the torque formula. (Alternatively, you can think of the $\sin\theta$ as picking out the component of the force perpendicular to the distance.)

Example

Q. You push a merry-go-round at its edge, perpendicular to the radius. If the merry-go-round has a diameter of 3.0 meters and you push with a force of 200 newtons, what torque are you applying?

A. The correct answer is 300 newton-meters.

1. Use the equation $\tau = rF \sin\theta$.

2. Plug in the numbers:

$$\tau = rF \sin\theta = (1.5\,\mathrm{m})(200\,\mathrm{N}) \sin 90° = 300\,\mathrm{N \cdot m}$$

Practice Questions

1.   You're opening a door by pushing on its outer edge with a force of 100.0 newtons. If the door is 1.3 meters wide, what torque are you applying if you push perpendicular to the door?

2.   The hot water tap in your shower is stuck. In anger, you apply 200.0 newtons of force to the outer edge of the handle, which has a turning radius of 10.0 centimeters. What torque are you applying?

3.   You're shoveling snow, holding the shovel handle in your right hand. Your left hand is placed halfway down the shovel's shaft and is providing the lifting motion. If you apply 200.0 newtons of force at an angle of 78° to the shovel and the shovel is 1.5 meters long, how much torque are you applying?

4.   You apply force on a wrench to loosen a pipe. If the wrench is 25 centimeters long and you apply 150 newtons at an angle of 67° with respect to the wrench, what torque are you applying?

Practice Answers

1.   **130 N · m.** Use the equation $\tau = rF \sin \theta$.

     Plug in the numbers:

     $$\tau = rF \sin \theta = (100.0 \text{ N})(1.3 \text{ m}) \sin 90° = 130 \text{ N·m}$$

2.   **20 N · m.** Use the equation $\tau = rF \sin \theta$.

     Plug in the numbers:

     $$\tau = rF \sin \theta = (200.0 \text{ N})(0.10 \text{ m}) \sin 90° = 20 \text{ N·m}$$

3.   **150 N · m.** Use the equation $\tau = rF \sin \theta$.

     Plug in the numbers:

     $$\tau = rF \sin \theta = (200.0 \text{ N})(0.75 \text{ m}) \sin 78° = 147 \text{ N·m}$$

     which rounds to 150 newton-meters with significant digits.

4.   **35 N · m.** Use the equation $\tau = rF \sin \theta$.

     Plug in the numbers:

     $$\tau = rF \sin \theta = (150 \text{ N})(0.25 \text{ m}) \sin 67° = 35 \text{ N·m}$$

Recognizing that torque is a vector

Remember: Torque is a vector, so it has not only magnitude but also direction. The direction of the torque is the same as the angular acceleration that it causes. It's perpendicular to the force and the lever arm in a right-hand fashion.

To get a little more technical, torque is given by the *cross-product* of the vector that points from the axis of rotation to the point at which the force is applied, **r,** and the force vector, **F.** The cross-product is written as an ×, so mathematically, the torque vector is the following:

$$\tau = \mathbf{r} \times \mathbf{F}$$

This equation is really a fancy mathematical way of saying that the torque vector has a magnitude of $rF \sin\theta$ and that the direction of the torque vector is as Figure 11-9 shows.

The right-hand rule is a useful way of remembering the direction of torque. If you point the thumb of your right hand in the direction of the radius vector **r** (which points *away* from the axis of rotation) and your fingers in the direction of the force vector **F,** then your palm faces the direction of the torque vector τ (the torque vector is pointing *out* of your palm).

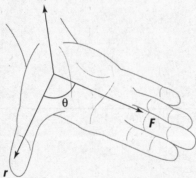

Figure 11-9:
A turning motion toward larger positive angles indicates a positive vector.

© John Wiley & Sons, Inc.

Example

Q. You give your little sister a northward push on a swing. What direction is the torque you produced?

A. The correct answer is east.

1. Draw a diagram of the swing set and compass points to orient yourself:

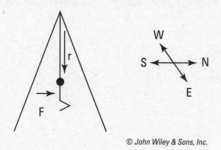

© John Wiley & Sons, Inc.

2. Point your right thumb downward in the direction of the radius.

3. Point your fingers northward in the direction of the force vector.

4. Your palm is facing west in the direction of the torque vector.

Practice Questions

1. You push a merry-go-round westward at its edge, perpendicular to the radius. If your push causes the merry-go-round to spin clockwise as viewed from above, what direction is the torque?

2. You pull south on the doorknob to open the door. If the doorknob is on the left side of the door, what direction is the torque you exerted on the door?

Practice Answers

1. **Down.** Draw a diagram of the merry-go-round and compass points to orient yourself. Point your right thumb southward in the direction of the radius. Point your fingers westward in the direction of the force vector. Your palm is facing upward in the direction of the torque vector.

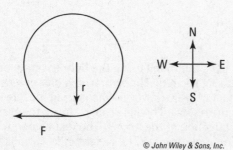

© John Wiley & Sons, Inc.

2. **Up.** Draw a diagram of the door and compass points to orient yourself. Point your right thumb westward in the direction of the radius. Point your fingers southward in the direction of the force vector. Your palm is facing downward in the direction of the torque vector.

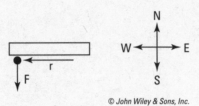

© John Wiley & Sons, Inc.

SPINNING AT CONSTANT VELOCITY: ROTATIONAL EQUILIBRIUM

You may know equilibrium as a state of balance, but what's equilibrium in physics terms? When you say an object has *equilibrium,* you mean that the motion of the object isn't changing; in other words, the object has no acceleration (it can have motion, however, as in constant velocity and/or constant angular velocity). As far as linear motion goes, the vector sum of all forces acting on the object must be 0 for the object to be in equilibrium. The net force acting on the object is 0: $\Sigma\mathbf{F} = 0$.

Remember: Equilibrium occurs in rotational motion in the form of rotational equilibrium. When an object is in *rotational equilibrium,* it has no angular acceleration — the object may be rotating, but it isn't speeding up or slowing down or changing directions (its tilt angle), which means its angular velocity is constant. When an object has rotational equilibrium, you see no net turning force on the object, which means that the net torque on the object must be 0:

$$\sum \tau = 0$$

This equation represents the rotational equivalent of linear equilibrium. Rotational equilibrium is a useful idea because given a set of torques operating on an extended object, you can determine what torque is necessary to stop the object from rotating. In this section, you try out some problems that involve objects in rotational equilibrium.

Examples

Q. Say that Hercules wants to lift a massive dumbbell using the deltoid (shoulder) muscle in his right arm and hold the weight at arm's length. His arm, which has a weight of magnitude $F_a = 56.0$ newtons, can exert a force F of 18,400 newtons. His deltoid muscle is attached to the arm at 13.0°, as the following figure shows. The figure also shows the distances between the pivot point and the points of application of the forces: The distance to the muscle is 0.150 meters, to the effective point of application of the weight of the arm is 0.310 meters (half the length of the arm), and to the dumbbell is 0.620 meters. The weight of the dumbbell has magnitude F_d.

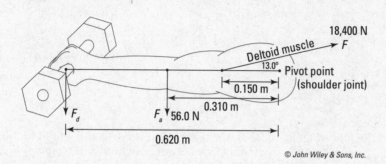

© John Wiley & Sons, Inc.

What is the maximum weight of the dumbbell Hercules can hold at arm's length?

A.   The correct answer is 973 newtons.

1.   Hercules's arm isn't moving, so $\Sigma \mathbf{F} = 0$ and $\Sigma \tau = 0$.

2.   Three forces are acting on the arm to cause torques around the arm joint: the y component of F (the pull of Hercules's deltoid muscle), F_a (the weight of his arm), and F_d (the weight of the dumbbell).

3.   The component of F in the y direction is $F_y = (18,400 \text{ N}) \sin 13.0°$. The magnitude of the weight of Hercules's arm is $F_a = 56.0$ newtons.

4.   The direction of the torque is in the direction perpendicular to the plane of the figure. If positive values correspond to counterclockwise-acting torques and negative values correspond to clockwise-acting torques, the torque from the y component of the muscle-pull F is the following:

$$\tau_M = F_y (0.150 \text{ m})$$
$$= (18,400 \text{ N}) \sin 13.0° (0.150 \text{ m})$$

This torque is positive because it leads to a turning force in the counterclockwise direction, as the figure shows.

5.   The torque from the weight of Hercules's arm is

$$\tau_a = (56.0 \text{ N})(-0.310 \text{ m})$$

This torque is negative because the lever arm is negative, so the force causes a clockwise torque, as the figure shows.

6.   The torque due to the weight of the dumbbell is

$$\tau_d = -F_d (0.620 \text{ m})$$

This is negative for the same reason that τ_a is negative.

7. Use $\Sigma\tau = 0$ to solve for the weight of the dumbbell:

$$\tau_M + \tau_a + \tau_d = \sum\tau$$

$$(18,400 \text{ N})(0.150 \text{ m})\sin 13.0° + (56.0 \text{ N})(-0.310 \text{ m}) + (-F_d)(0.620 \text{ m}) = 0$$

$$(604 \text{ N}\cdot\text{m}) + (-F_d)(0.620 \text{ m}) = 0$$

$$F_d(0.620 \text{ m}) = 604 \text{ N}\cdot\text{m}$$

$$F_d = \frac{604 \text{ N}\cdot\text{m}}{0.620 \text{ m}}$$

$$F_d \approx 973 \text{ N}$$

Q. Will the ladder in the following figure fall if θ is 45° and the static coefficient of friction with the floor is 0.70? (Assume that the frictional force between the wall and the ladder is insignificant.) Here are the forces involved:

F_W = Force exerted by the wall on the ladder

W_P = Weight of the person = 450 N

W_L = Weight of the ladder = 200 N (you can assume it's concentrated at the ladder's center)

F_F = Force of friction holding the ladder in place

F_N = Normal force

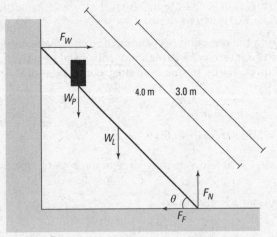

© John Wiley & Sons, Inc.

A. The correct answer is that the ladder won't slip.

1. You want the ladder to be in both linear ($\Sigma F = 0$) and rotational ($\Sigma\tau = 0$) equilibrium. To be in linear equilibrium, the force exerted by the wall on the ladder, F_W, must be the same as the force of friction in magnitude but opposite in direction because those are the only two horizontal forces. So if you can find F_W, you know what the force of friction, F_F, needs to be.

2. To find F_W, take a look at the torques around the bottom of the ladder, using that point as the pivot point. All the torques around that point have to add up to 0. The torque due to the force from the wall against the ladder is

$$-F_W(4.0\text{ m})\sin 45° = -2.83F_W$$

This torque is negative because it tends to produce a clockwise motion toward smaller angles.

3. The torque due to the person's weight is

$$W_P(3.0\text{ m})\cos 45° = (450\text{ N})(3.0\text{ m})(0.707) = 954\text{ N}\cdot\text{m}$$

4. The torque due to the ladder's weight is

$$W_L(2.0\text{ m})\cos 45° = (200\text{ N})(2.0\text{ m})(0.707) = 283\text{ N}\cdot\text{m}$$

5. Both torques are positive, so because $\Sigma\tau = 0$, you get this result when you add all the torques together:

$$954\text{ N}\cdot\text{m} + 283\text{ N}\cdot\text{m} - 2.83F_W = 0$$

6. Solve for F_W:

$$F_W = \frac{954\text{ N}\cdot\text{m} + 283\text{ N}\cdot\text{m}}{2.83\text{ m}} = 437\text{ N}$$

That means the force the wall exerts on the ladder is 437 newtons. Note that this force is also equal to the frictional force of the bottom of the ladder on the floor because F_W and the frictional force are the only two horizontal forces in the whole system. So the force of friction needed is

$$F_F = 437\text{ N}$$

7. What is the maximal force of friction possible given the static coefficient of friction between the ladder and the floor? Use this equation:

$$F_{F\text{ maximal}} = \mu_s F_N$$

8. F_N is the normal force of the floor pushing up on the ladder, and it must balance all the downward-pointing forces in this problem because of linear equilibrium. That means you have the following:

$$F_N = W_P + W_L = 650\text{ N}$$

9. Plug in the numbers, using the value of μ_s 0.70:

$$F_{F\text{ maximal}} = \mu_s F_N = (0.70)(650\text{ N}) = 455\text{ N}$$

That's your answer — you need 437 newtons, and you actually have up to 455 newtons available, so the ladder isn't going to slip.

Practice Questions

1. You're opening a door by pushing on its outer edge with a force of 100 newtons at 90°, and someone is trying to keep the door shut by pulling on it one-third of the door's width from the hinge. What force does the other person need to supply perpendicular to the door to keep the door in rotational equilibrium?

2. You're on a teeter-totter with a total length of 2L with a person twice your weight. The other person's sitting only one-third of the distance from the pivot point, however. Where must you sit to balance the person out?

3. A motor can provide 5,000 newton-meters of torque. If you can provide 33 newtons of force, how far away from the motor's axle must you be to produce rotational equilibrium?

4. The cap on top of the oil well is frozen. If it takes 450 newton-meters of torque to free it and you have a wrench that's 0.40 meters long, how much force must you apply?

Practice Answers

1. **300 N.** Use the equation $\Sigma \tau = 0$.

   Find the torque you apply by using this equation, where L is the width of the door:

   $$\tau = rF_{you} \sin \theta = (100 \text{ N})(L)\sin 90° = 100 L \text{ N} \cdot \text{m}$$

   Find the torque the other person applies by using this equation, where F is the force that person applies:

   $$\tau = rF_{other} \sin \theta = (F)(L/3)\sin 90° = F(L/3) \text{ N} \cdot \text{m}$$

   To have rotational equilibrium, set these two torques equal and solve for F:

   $$(100 \text{ N})L = F(L/3)$$
   $$(3)(100 \text{ N}) = F = 300 \text{ N}$$

2. **2L/3.** Use the equation $\Sigma \tau = 0$.

   Find the torque the other person applies by using this formula, where F is the force that person applies and m is your mass:

   $$\tau = rF_{other} \sin \theta = (2mg)(L/3)\sin 90° = 2mg(L/3) \text{ N} \cdot \text{m}$$

   Find the torque you apply by using this equation, where x is your distance to the pivot:

   $$\tau = rF_{you} \sin \theta = (mg)x \sin 90° = mgx \text{ N} \cdot \text{m}$$

To have rotational equilibrium, set these two torques equal and solve for x:

$$2mg(L/3) \text{ N} \cdot \text{m} = mgx \text{ N} \cdot \text{m}$$
$$x = 2L/3$$

3.   **150 m.** Use the equation $\Sigma\tau = 0$.

Find the torque you apply:

$$\tau = rF\sin\theta = (33 \text{ N})x\sin 90° = 33x \text{ N} \cdot \text{m}$$

Set that equal to the torque of the motor and solve for x:

$$(33 \text{ N})x = 5{,}000 \text{ N} \cdot \text{m}$$
$$x = \frac{5{,}000 \text{ N} \cdot \text{m}}{33 \text{ N}} = 151 \text{ m}$$

which rounds to 150 meters with significant figures.

4.   **1,100 N.** Use the equation $\Sigma\tau = 0$.

Find the torque you apply:

$$\tau = rF\sin\theta = (F)(0.4 \text{ m})\sin 90° = F(0.4) \text{ N} \cdot \text{m}$$

Set that equal to the torque needed and solve for F:

$$F(0.4) \text{ N} \cdot \text{m} = 450 \text{ N} \cdot \text{m}$$
$$F = \frac{450 \text{ N} \cdot \text{m}}{0.4 \text{ N} \cdot \text{m}} = 1{,}125 \text{ N}$$

which rounds to 1,100 newtons with significant figures.

12

ROUND AND ROUND WITH ROTATIONAL DYNAMICS

This chapter is all about applying forces and seeing what happens in the rotational world. You find out what Newton's second law (force equals mass times acceleration) becomes for rotational motion, you see how inertia comes into play in rotational motion, and you get the story on rotational kinetic energy, rotational work, and angular momentum.

ROLLING UP NEWTON'S SECOND LAW INTO ANGULAR MOTION

Newton's second law, net force equals mass times acceleration ($\mathbf{F} = m\mathbf{a}$; see Chapter 5), is a physics favorite in the linear world because it ties together the vectors' force and acceleration. But if you have to talk in terms of angular kinetics rather than linear motion, what happens? Can you get Newton spinning?

Angular kinetics has equivalents (or *analogs*) for linear equations (see Chapter 11). So what's the angular analog for $\mathbf{F} = m\mathbf{a}$? You may guess that $\mathbf{F}$, the linear force, becomes τ. And you may also guess that $\mathbf{a}$, linear acceleration, becomes α, angular acceleration. But what the heck is the angular analog of m, mass? The answer is *rotational inertia, I,* and you come to this answer by converting tangential acceleration to angular acceleration. As we show you in this section, your final formula is $\Sigma\tau = I\alpha$, the angular form of Newton's second law.

Switching force to torque

You can start the linear-to-angular conversion process with a simple example. Say that you're whirling a ball in a circle on the end of a string, as in Figure 12-1. You apply a tangential force (along the circle) to the ball, making it speed up (keep in mind that this force isn't directed toward the center of the circle, as when you have a centripetal force; see Chapter 11). You want to write Newton's second law in terms of torque rather than force.

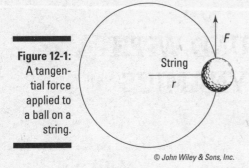

Figure 12-1: A tangential force applied to a ball on a string.

Start by working with only the magnitudes of the vector quantities, saying that

$$F = ma$$

To put this equation in terms of angular quantities, such as torque, multiply by the radius of the circle, r (see Chapter 11 for details on the relationship between angular and linear quantities):

$$Fr = mra$$

Because you're applying tangential force to the ball, the force and the circle's radius are at right angles (refer to Figure 12-1), so you can replace Fr with torque:

$$\tau = mra$$

You're now partly done making the transition to rotational motion. Instead of working with linear force, you're working with torque, which is linear force's rotational analog.

Converting tangential acceleration to angular acceleration

To move from linear motion to angular motion, you have to convert a, tangential acceleration, to α, angular acceleration. Great, but how do you make the conversion? You can multiply angular acceleration by the radius to get the linear equivalent, which is the magnitude of the tangential acceleration (see Chapter 11): $a = r\alpha$. Substitute $r\alpha$ for a in the equation for the angular equivalent of Newton's second law, $\tau = mra$:

$$\tau = mr(r\alpha) = mr^2\alpha$$

Now you've related the magnitude of the torque to magnitude of the angular acceleration. The direction of the angular acceleration and the torque turns out to be the same, so this equation is true for vectors, too:

$$\tau = mr^2\alpha$$

Factoring in the moment of inertia

To go from linear force, $\mathbf{F} = m\mathbf{a}$, to torque (linear force's angular equivalent), you have to find the angular equivalent of acceleration and mass. In the preceding section, you find angular acceleration, giving you the equation $\tau = mr^2\alpha$.

In this equation, mr^2 is the rotational analog of mass, officially called the *moment of inertia* (sometimes referred to as the *rotational inertia*). The *moment of inertia* is a measure of how resistant an object is to changes in its rotational motion.

In physics, the symbol for inertia is *I,* so you can write the equation for torque as follows:

$$\sum\tau = I\alpha$$

The symbol Σ means *sum of,* so $\Sigma\tau$ means *net torque.* The units of moment of inertia are kilogram-square meters ($kg\cdot m^2$). Note how close the torque equation is to the equation for net force, which follows:

$$\sum\mathbf{F} = m\mathbf{a}$$

Remember: $\sum\tau = I\alpha$ is the angular form of Newton's second law for rotating bodies: net torque equals moment of inertia multiplied by angular acceleration.

Example

Q. You're whipping a cannonball around on the end of a 0.50-meter steel rod. If the cannonball has a mass of 10.0 kilograms, what torque do you need to apply to get an angular acceleration of 0.50 radians per second^2?

A. The correct answer is 1.25 newton-meters.

1. Use the equation $\tau = I\alpha = mr^2\alpha$.

2. Plug in the numbers:

$$\tau = mr^2\alpha = (10.0 \text{ kg})(0.50 \text{ m})^2(0.50 \text{ radians/s}^2) = 1.25 \text{ N}\cdot\text{m}$$

The cannonball has so much more mass than the rod that you can ignore the rod's mass entirely.

Practice Questions

1. You're rotating a 10.0-kilogram cannonball at the end of a 1.0-meter rod in a circle and want an angular acceleration of 1.0 radian per second^2. What torque do you need to supply?

2. A 100.0-gram clock pendulum on the end of a 1.0-meter rod has an angular acceleration of 2.0 radians per second^2. What torque is being applied?

Practice Answers

1. **10 N · m.** Use the equation $\tau = mr^2\alpha$.

   Plug in the numbers:

   $$\tau = mr^2\alpha = (10.0\,\text{kg})(1.0\,\text{m})^2(1.0\,\text{radian/s}^2) = 10\,\text{N}\cdot\text{m}$$

2. **0.20 N · m.** Use the equation $\tau = mr^2\alpha$.

   Plug in the numbers:

   $$\tau = mr^2\alpha = (0.1\,\text{kg})(1.0\,\text{m})^2(2.0\,\text{radians/s}^2) = 0.20\,\text{N}\cdot\text{m}$$

MOMENTS OF INERTIA: LOOKING INTO MASS DISTRIBUTION

Remember: The moment of inertia depends not only on the mass of the object but also on how the mass is distributed. For example, if two disks have the same mass but one has all the mass around the rim and one is solid, then the disks would have different moments of inertia.

Calculating moments of inertia is fairly simple if you only have to examine the orbital motion of small point-like objects, where all the mass is concentrated at one particular point at a given radius r. For instance, for a golf ball you're whirling around on a string, the moment of inertia depends on the radius of the circle the ball is spinning in:

$$I = mr^2$$

Here, r is the radius of the circle, from the center of rotation to the point at which all the mass of the golf ball is concentrated.

Crunching the numbers can get a little sticky when you enter the non–golf ball world, however, because you may not be sure of which radius to use. What if you're spinning a rod around? All the mass of the rod isn't concentrated at a single radius. When you have an extended object, such as a rod, each bit of mass is at a different radius. You don't have an easy way to deal with this, so you have to sum up the contribution of each particle of mass at each different radius like this:

$$I = \sum mr^2$$

You can use this concept of adding up the moments of inertia of all the elements to get the total to work out the moment of inertia of any distribution of mass. Here's an example using two point masses, which is a bit more complex than a single point mass. Say you have two golf balls, and you want to know what their combined moment of inertia is. If you have a golf ball at radius r_1 and another at r_2, the total moment of inertia is

$$I = \sum mr^2 = m\left(r_1^2 + r_2^2\right)$$

So how do you find the moment of inertia of, say, a disk rotating around an axis stuck through its center? You have to break the disk up into tiny balls and add them all up. Trusty physicists have already completed this task for many standard shapes; we provide a list of objects you're likely to encounter, and their moments of inertia, in Table 12-1. Figure 12-2 depicts the shapes that these moments of inertia correspond to.

Check out the following examples to see advanced moments of inertia in action.

Table 12-1   Moments of Inertia for Various Shapes and Solids

Shape	Moment of Inertia
(a) Solid cylinder or disk of radius r	$I = \frac{1}{2}mr^2$
(b) Hollow cylinder of radius r	$I = mr^2$
(c) Solid sphere of radius r	$I = \frac{2}{5}mr^2$
(d) Hollow sphere of radius r	$I = \frac{2}{3}mr^2$
(e) Rectangle rotating around an axis along one edge, where the other edge has length r	$I = \frac{1}{3}mr^2$
(f) Rectangle with sides r_1 and r_2 rotating around a perpendicular axis through the center	$I = \left(\frac{1}{12}\right)m\left(r_1^2 + r_2^2\right)$
(g) Thin rod of length r rotating about its middle	$I = \frac{1}{12}mr^2$
(h) Thin rod of length r rotating about one end	$I = \frac{1}{3}mr^2$

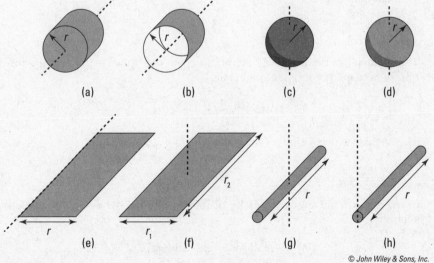

Figure 12-2: The shapes corresponding to the moments of inertia in Table 12-1.

(a)   (b)   (c)   (d)

(e)   (f)   (g)   (h)

© John Wiley & Sons, Inc.

Examples

Q.   A solid cylinder with a mass of 5.0 kilograms is rolling down a ramp. If it has a radius of 10 centimeters and an angular acceleration of 3.0 radians per second², what torque is operating on it?

A.   The correct answer is 0.075 newton-meters.

    1.   Use the equation $\tau = I\alpha$.

    2.   In this case, $I = (1/2)mr^2$

    3.   Plug in the numbers:

$$\tau = \frac{1}{2}mr^2\alpha$$
$$= \frac{1}{2}(5.0 \text{ kg})(0.10 \text{ m})^2(3.0 \text{ radians/s}^2)$$
$$= 0.075 \text{ N} \cdot \text{m}$$

Q.   DVD players change the angular speed of the DVD to keep the section of the DVD under the laser head moving at constant linear speed. Say that a DVD has a mass of 30.0 grams and a diameter of 12.0 centimeters. It starts at 700.0 revolutions per second when you first hit play and winds down to about 200.0 revolutions per second at the end of the DVD 50.0 minutes later. What's the average torque needed to create this acceleration?

A.   The correct answer is -5.65×10^{-5} Newton-meters.

    1.   Start with the torque equation:

$$\tau = I\alpha$$

    2.   A DVD is a disk shape rotating around its center, so from Table 12-1, you know that its moment of inertia is

$$I = \frac{1}{2}mr^2$$

    3.   The diameter of the DVD is 12 centimeters, so the radius is 6.0 centimeters. Put in the numbers to get the moment of inertia:

$$I = \frac{1}{2}mr^2 = \frac{1}{2}(0.030 \text{ kg})(0.060 \text{ m})^2 = 5.4 \times 10^{-5} \text{ kg} \cdot \text{m}^2$$

    4.   The angular acceleration α (see Chapter 11 for details) is

$$\alpha = \frac{\Delta\omega}{\Delta t}$$

    5.   Convert the initial angular velocity of 700 revolutions per second to radians per second:

$$\omega_i = \frac{700 \text{ revolutions}}{1 \text{ s}} \times \frac{2\pi \text{ radians}}{1 \text{ revolution}} = 1,400\pi \text{ radians/s}$$

6. Similarly, convert the units on the final angular velocity:

$$\omega_f = \frac{200 \text{ revolutions}}{1 \text{ s}} \times \frac{2\pi \text{ radians}}{1 \text{ revolution}} = 400\pi \text{ radians/s}$$

7. Convert the time of 50 minutes to seconds:

$$\Delta t = 50 \text{ min} \times \frac{60 \text{ s}}{1 \text{ min}} = 3,000 \text{ s}$$

8. Plug in the angular velocities and time into the angular acceleration formula:

$$\alpha = \frac{\Delta \omega}{\Delta t} = \frac{\omega_f - \omega_i}{\Delta t}$$
$$= \frac{(400\pi - 1,400\pi) \text{ radians/s}}{3,000 \text{ s}}$$
$$= \frac{-1,000\pi \text{ radians}}{3,000 \text{ s}^2}$$
$$= -1.047 \text{ radians/s}^2$$

As previously defined, the component of the angular velocity along the axis of rotation is positive. The negative acceleration then leads to a reduction in this angular velocity.

9. Plug the moment of inertia and angular acceleration into the torque equation:

$$\tau = I\alpha = \left(5.4 \times 10^{-5} \text{ kg} \cdot \text{m}^2\right)\left(-1.047 \text{ s}^{-2}\right) = -5.65 \times 10^{-5} \text{ N} \cdot \text{m}$$

Practice Questions

1. You're spinning a 5.0-kilogram ball with a radius of 0.5 meters. If it's accelerating at 4.0 radians per second[2], what torque are you applying?

2. A tire with a radius of 0.50 meters and mass of 1.0 kilogram is rolling down a street. If it's accelerating with an angular acceleration of 10.0 radians per second[2], what torque is operating on it?

3. You're spinning a hollow sphere with a mass of 10.0 kilograms and radius of 1.0 meter. If it has an angular acceleration of 15 radians per second[2], what torque are you applying?

4. You're throwing a 300.0-gram flying disc with a radius of 10 centimeters, accelerating it with an angular acceleration of 20.0 radians per second[2]. What torque are you applying?

Practice Answers

1. **2 N · m.** Use the equation $\tau = I\alpha$.

    In this case, $I = (2/5)mr^2$.

    Plug in the numbers:

    $$\tau = \frac{2}{5}mr^2\alpha = \frac{2}{5}(5.0 \text{ kg})(0.5 \text{ m})^2 (4.0 \text{ radians/s}^2) = 2 \text{ N·m}$$

2. **2.5 N · m.** Use the equation $\tau = I\alpha$.

    In this case, $I = mr^2$.

    Plug in the numbers:

    $$\tau = mr^2\alpha = (1.0 \text{ kg})(0.50 \text{ m})^2 (10.0 \text{ radians/s}^2) = 2.5 \text{ N·m}$$

3. **100 N · m.** Use the equation $\tau = I\alpha$.

    In this case, $I = (2/3)mr^2$.

    Plug in the numbers:

    $$\tau = \frac{2}{3}mr^2\alpha = \frac{2}{3}(10.0 \text{ kg})(1.0 \text{ m})^2 (15 \text{ radians/s}^2) = 100 \text{ N·m}$$

4. **0.03 N · m.** Use the equation $\tau = I\alpha$.

    In this case, $I = (1/2)mr^2$.

    Plug in the numbers:

    $$\tau = \frac{1}{2}mr^2\alpha = \frac{1}{2}(0.3 \text{ kg})(0.10 \text{ m})^2 (20.0 \text{ radians/s}^2) = 0.03 \text{ N·m}$$

WRAPPING YOUR HEAD AROUND ROTATIONAL WORK AND KINETIC ENERGY

One major player in the linear-force game is *work* (see Chapter 9); when force and distance are in the same direction, the equation for work is work equals force times distance, or $W = Fs$. Work has a rotational analog. To relate a linear force acting for a certain distance with the idea of rotational work, you convert force to torque (its angular equivalent) and distance to angle. We show you how to derive the rotational-work equation in this section. We also show you what happens when you do work by turning an object, creating rotational motion — your work goes to increasing the kinetic energy.

Putting a new spin on work

When force moves an object through a distance, work is done on the object (refer to Chapter 9). Similarly, when a torque rotates an object through an angle, work is done. In this section, you work out how much work is done when you rotate a wheel by pulling a string attached to the wheel's outside edge (see Figure 12-3).

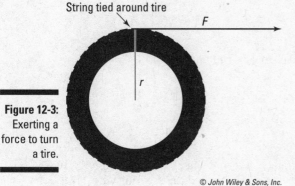

String tied around tire

F

r

Figure 12-3:
Exerting a
force to turn
a tire.

© John Wiley & Sons, Inc.

Work is the amount of force applied to an object multiplied by the distance it's applied. In this case, a force F is applied with the string. Bingo! The string lets you make the handy transition between linear and rotational work. So how much work is done? Use the following equation:

$$W = Fs$$

where s is the distance the person pulling the string applies the force over. In this case, the distance s equals the radius multiplied by the angle through which the wheel turns, $s = r\theta$, so you get

$$W = Fr\theta$$

However, the torque, τ, equals Fr in this case, because the string is acting at right angles to the radius (see Chapter 11). So you're left with

$$W = \tau\theta$$

When the string is pulled, applying a constant torque that turns the wheel, the work done equals $\tau\theta$. This makes sense, because linear work is Fs, and to convert to rotational work, you convert from force to torque and from distance to angle. The units here are the standard units for work — joules in the MKS system.

Warning: You have to give the angle in radians for the conversion between linear work and rotational work to come out right.

Example

Q. If you apply a torque of 500.0 newton-meters to a tire and turn it through an angle of 2π radians, what work have you done?

A. The correct answer is 3,140 joules.

1. Use the equation $W = \tau\theta$.

2. Plug in the numbers:

$$W = \tau\theta = (500.0\ \text{N}\cdot\text{m})(2\pi\ \text{radians}) = 3{,}140\ \text{J}$$

Practice Questions

1. You've done 20.0 joules of work turning a steering wheel. If you're applying 10.0 newton-meters of torque, what angle have you turned the steering wheel through?

2. You've done 350 joules of work turning a bicycle tire. If you're applying 150 newton-meters of torque, what angle have you turned the wheel through?

Practice Answers

1. **2 radians.** Use the equation $W = \tau\theta$.

   Solve for θ and plug in the numbers:

   $$\theta = \frac{W}{\tau} = \frac{20.0\ \text{J}}{10.0\ \text{N}\cdot\text{m}} = 2\ \text{radians}$$

2. **2.3 radians.** Use the equation $W = \tau\theta$.

   Solve for θ and plug in the numbers:

   $$\theta = \frac{W}{\tau} = \frac{350\ \text{J}}{150\ \text{N}\cdot\text{m}} = 2.3\ \text{radians}$$

Moving along with rotational kinetic energy

If you put a lot of work into turning an object, the object starts spinning. And when an object is spinning, all its pieces are moving, which means that it has kinetic energy. For spinning objects, you have to convert from the linear concept of kinetic energy to the rotational concept of kinetic energy.

You can calculate the kinetic energy of a body in linear motion with the following equation (see Chapter 9):

$$KE = \frac{1}{2}mv^2$$

where m is the mass of the object and v is the speed. This formula applies to every bit of the object that's rotating — each bit of mass has this kinetic energy.

To go from the linear version to the rotational version, you have to go from mass to moment of inertia, I, and from velocity to angular velocity, ω. You can tie an object's tangential speed to its angular speed like this (see Chapter 11):

$$v = r\omega$$

where r is the radius and ω is its angular speed. Plugging v's equivalent into the kinetic-energy equation gives you the following:

$$KE = \frac{1}{2}mv^2 = \frac{1}{2}m\left(r^2\omega^2\right)$$

The equation looks okay so far, but it holds true only for the one single bit of mass under discussion — each other bit of mass may have a different radius, so you're not finished. You have to sum up the kinetic energy of every bit of mass like this:

$$KE = \frac{1}{2}\sum\left(mr^2\omega^2\right)$$

You can simplify this equation. Start by noticing that even though each bit of mass may be different and be at a different radius, each bit has the same angular speed (they all turn through the same angle in the same time). Therefore, you can take the ω out of the summation:

$$KE = \frac{1}{2}\sum\left(mr^2\right)\omega^2$$

This makes the equation much simpler, because $\sum\left(mr^2\right)$ equals the moment of inertia, I (see the section "Rolling Up Newton's Second Law into Angular Motion," earlier in this chapter). Making this substitution takes all the dependencies on the individual radius of each bit of mass out of the equation, giving you

$$KE = \frac{1}{2}I\omega^2$$

Now you have a simplified equation for rotational kinetic energy. The equation proves useful because rotational kinetic energy is everywhere. A satellite spinning around in space has rotational kinetic energy. A barrel of beer rolling down a ramp from a truck has rotational kinetic energy. The latter example (not always with beer trucks, of course) is a common thread in physics problems.

Example

Q. You have a 100-kilogram solid sphere with a radius equal to 2.0 meters. If it's rotating at ω = 10.0 radians per second, what is its rotational kinetic energy?

A. The correct answer is 800 joules.

1. Use this equation:

$$KE = \frac{1}{2}I\omega^2$$

2. For a solid sphere, $I = (2/5)mr^2$

3. Plug in the numbers:

$$KE = \frac{1}{2}I\omega^2$$
$$= \frac{1}{2}\left(\frac{2}{5}mr^2\right)\omega^2$$
$$= \left(\frac{1}{2}\right)\left(\frac{2}{5}\right)(100 \text{ kg})(2.0 \text{ m})^2(10.0 \text{ radians/s})^2$$
$$= 800 \text{ J}$$

Practice Questions

1. How much rotational kinetic energy does a spinning tire of mass 10.0 kilograms and radius 0.50 meters have if it's spinning at 40.0 rotations per second?

2. How much work do you do to spin a hollow sphere, which has a mass of 10.0 kilograms and a radius of 0.50 meters, from 0.0 radians per second to 200.0 radians per second?

Practice Answers

1. **7.9×10^4 J.** Use the equation for kinetic energy:

$$KE = \frac{1}{2}I\omega^2$$

For a spinning tire, $I = mr^2$.

Convert 40 rotations per second to radians per second. One rotation is 2π radians, so 40 rotations per second is $40(2\pi) = 80\pi$ radians/s.

Plug in the numbers:

$$KE = \frac{1}{2}I\omega^2 = \frac{1}{2}(10.0 \text{ kg})(0.50 \text{ m})^2(80\pi)^2 = 7.9 \times 10^4 \text{ J}$$

2. **33,000 J.** The work you do goes into the sphere's final kinetic energy, so use the equation for kinetic energy:

$$KE = \frac{1}{2}I\omega^2$$

For a hollow sphere, $I = (2/3)mr^2$.

Plug in the numbers:

$$KE = \frac{1}{2}I\omega^2 = \frac{1}{2}\left(\frac{2}{3}\right)(10.0 \text{ kg})(0.50 \text{ m})^2(200.0 \text{ radians/s})^2 = 33,300 \text{ J}$$

which rounds to 33,000 joules with significant figures.

Let's roll! Finding rotational kinetic energy on a ramp

Objects can have both linear and rotational kinetic energy. This fact is an important one, if you think about it, because when objects start rolling down ramps, any previous ramp expertise you have goes out the window. Why? Because when an object rolls down a ramp instead of sliding, some of its gravitational potential energy (see Chapter 9) goes into its linear kinetic energy, and some of it goes into its rotational kinetic energy.

Look at Figure 12-4, where you're pitting a solid cylinder against a hollow cylinder in a race down the ramp. Each object has the same mass. Which cylinder is going to win? In other words, which cylinder will have the higher speed at the bottom of the ramp? When looking only at linear motion, you can handle a problem like this by setting the potential energy equal to the final kinetic energy (assuming no friction!) like this:

$$PE = KE$$
$$mgh = \frac{1}{2}mv^2$$

where m is the mass of the object, g is the acceleration due to gravity, and h is the height at the top of the ramp. This equation would let you solve for the final speed.

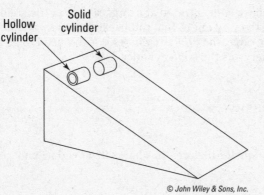

Figure 12-4:
A solid cylinder and a hollow cylinder ready to race down a ramp.

Hollow cylinder

Solid cylinder

© John Wiley & Sons, Inc.

But the cylinders are rolling in this case, which means that the initial gravitational potential energy becomes *both* linear kinetic energy and rotational kinetic energy. You can now write the equation as

$$mgh = \frac{1}{2}mv^2 + \frac{1}{2}I\omega^2$$

You can relate v and ω together with the equation $v = r\omega$, which means that $\omega = v/r$, so

$$mgh = \frac{1}{2}mv^2 + \frac{1}{2}I\left(\frac{v^2}{r^2}\right)$$

You want to solve for v, so try grouping things together. You can factor $(1/2)v^2$ out of the two terms on the right:

$$mgh = \frac{1}{2}mv^2 + \frac{1}{2}I\left(\frac{v^2}{r^2}\right)$$
$$mgh = \frac{1}{2}\left(m + \frac{I}{r^2}\right)v^2$$

Isolating v, you get the following:

$$v = \sqrt{\frac{2mgh}{m + I/r^2}}$$

For the hollow cylinder, the moment of inertia equals mr^2, as you can see in Table 12-1. For a solid cylinder, on the other hand, the moment of inertia equals $(1/2)mr^2$. Substituting for I for the hollow cylinder gives you the hollow cylinder's velocity:

$$v = \sqrt{gh}$$

Substituting for I for the solid cylinder gives you the solid cylinder's velocity:

$$v = \sqrt{\frac{4gh}{3}}$$

Now the answer becomes clear. The solid cylinder will be going $\sqrt{4/3}$ times as fast as the hollow cylinder, or about 1.15 times as fast, so the solid cylinder will win.

The hollow cylinder has as much mass concentrated at a large radius as the solid cylinder has distributed from the center all the way out to that radius, so this answer makes sense. With that large mass way out at the edge, the hollow cylinder doesn't need to go as fast to have as much rotational kinetic energy as the solid cylinder.

Example

Q. If a solid sphere is at the top of a 3.0-meter high ramp, what is its speed when it reaches the bottom of the ramp?

A. The correct answer is 6.5 meters per second.

1. Use this equation:

$$v = \sqrt{\frac{2mgh}{m + I/r^2}}$$

2. For a solid sphere, $I = (2/5)mr^2$. That means v is equal to this:

$$v = \sqrt{\frac{2mgh}{m + 2/5\,m}}$$

3. That breaks down to:

$$v = \sqrt{\frac{2gh}{1 + 2/5}}$$

4. Plug in the numbers:

$$v = \sqrt{\frac{2\left(9.8 \text{ m/s}^2\right)\left(3.0 \text{ m}\right)}{1 + 2/5}} = 6.5 \text{ m/s}$$

Practice Questions

1. If a hollow cylinder is at the top of a 4.0-meter high ramp, what is its speed when it reaches the bottom of the ramp?

2. If a solid cylinder is at the top of a 2.0-meter high ramp, what is its speed when it reaches the bottom of the ramp?

3. A tire is rolling down a ramp, starting at a height of 3.5 meters. What is its speed when it reaches the bottom of the ramp?

4. A basketball (that is, a hollow sphere) is rolling down a ramp, starting at a height of 4.8 meters. What is its speed when it reaches the bottom of the ramp?

Practice Answers

1. **6.3 m/s.** Use this equation:

$$v = \sqrt{\frac{2mgh}{m + I/r^2}}$$

For a hollow cylinder, $I = mr^2$. That means v is equal to this:

$$v = \sqrt{\frac{2mgh}{m + m}}$$

This equation breaks down to

$$v = \sqrt{gh}$$

Plug in the numbers:

$$v = \sqrt{\left(9.8 \text{ m/s}^2\right)\left(4.0 \text{ m}\right)} = 6.3 \text{ m/s}$$

2.   **5.1 m/s.** Use this equation:

$$v = \sqrt{\frac{2mgh}{m + I/r^2}}$$

For a solid cylinder, $I = (1/2)mr^2$. That means v is equal to this:

$$v = \sqrt{\frac{2mgh}{m + 1/2\,m}}$$

This equation breaks down to

$$v = \sqrt{\frac{2gh}{1 + 1/2}}$$

Plug in the numbers:

$$v = \sqrt{\frac{2\left(9.8 \text{ m/s}^2\right)\left(2.0 \text{ m}\right)}{1 + 1/2}} = 5.1 \text{ m/s}$$

3.   **5.9 m/s.** Use this equation:

$$v = \sqrt{\frac{2mgh}{m + I/r^2}}$$

For a tire, $I = mr^2$. That means v is equal to this:

$$v = \sqrt{\frac{2mgh}{m + m}}$$

This equation breaks down to

$$v = \sqrt{gh}$$

Plug in the numbers:

$$v = \sqrt{\left(9.8 \text{ m/s}^2\right)\left(3.5 \text{ m}\right)} = 5.9 \text{ m/s}$$

4.   7.5 m/s. Use this equation:

$$v = \sqrt{\frac{2mgh}{m + I/r^2}}$$

For a hollow sphere, $I = (2/3)mr^2$. That means v is equal to this:

$$v = \sqrt{\frac{2mgh}{m + 2/3\,m}}$$

This equation breaks down to

$$v = \sqrt{\frac{2gh}{1 + 2/3}}$$

Plug in the numbers:

$$v = \sqrt{\frac{2(9.8 \text{ m/s}^2)(4.8 \text{ m})}{1 + 2/3}} = 7.5 \text{ m/s}$$

CAN'T STOP THIS: ANGULAR MOMENTUM

Picture a small child on a spinning playground ride, such as a merry-go-round, and she's yelling that she wants to get off. You have to stop the spinning ride, but it's going to take some effort. Why? Because it has *angular momentum*.

Linear momentum, **p,** is defined as the product of mass and velocity:

$$\mathbf{p} = m\mathbf{v}$$

This is a quantity that is conserved when there are no external forces acting. The more massive and faster moving an object, the greater the magnitude of momentum.

Physics also features angular momentum, **L.** The equation for angular momentum looks like this:

$$\mathbf{L} = I\omega$$

where I is the moment of inertia and ω is the angular velocity.

Remember: Angular momentum is a vector quantity, meaning it has a magnitude and a direction. The vector points in the same direction as the ω vector (that is, in the direction the thumb of your right hand points when you wrap your fingers around in the direction the object is turning).

The units of angular momentum are I multiplied by the units of ω, or kg·m^2/s in the MKS system.

The important idea about angular momentum, much as with linear momentum, is that it's conserved.

Remember: The *principle of conservation of angular momentum* states that angular momentum is conserved if no net torques are involved.

This principle comes in handy in all sorts of problems, such as when two ice skaters start off holding each other close while spinning but then end up at arm's length. Given their initial angular velocity, you can find their final angular velocity, because angular momentum is conserved:

$$I_1\omega_1 = I_2\omega_2$$

If you can find the initial moment of inertia and the final moment of inertia, you're set. But you also come across less obvious cases where the principle of conservation of angular momentum helps out. For example, satellites don't have to travel in circular orbits; they can travel in ellipses. And when they do, the math can get a lot more complicated. Lucky for you, the principle of conservation of angular momentum can make the problems simple.

Example

Q. Say that NASA planned to put a satellite into a circular orbit around Pluto for studies, but the situation got a little out of hand and the satellite ended up with an elliptical orbit. At its nearest point to Pluto, 6.0×10^6 meters, the satellite zips along at 900.0 meters per second.

The satellite's farthest point from Pluto is 2.0×10^7 meters. What is its speed at that point?

A. The correct answer is 270 meters per second.

1. Angular momentum is conserved because there are no external torques the satellite must deal with (gravity always acts perpendicular to the orbital radius). Because angular momentum is conserved, you can say that

$$I_1\omega_1 = I_2\omega_2$$

2. Because the satellite is so small compared to the radius of its orbit at any location, treat the satellite as a point mass with a moment of inertia $I = mr^2$ (refer to the earlier section "Factoring in the moment of inertia").

3. The magnitude of the angular velocity equals v/r.

4. Plug into the angular momentum conservation equation:

$$I_1\omega_1 = I_2\omega_2$$
$$mr_1v_1 = mr_2v_2$$

5. Put v_2 on one side of the equation by dividing by mr_2:

$$v_2 = \frac{r_1 v_1}{r_2}$$

6. Plug in the numbers:

$$v_2 = \frac{\left(6.0\times10^6 \text{ m}\right)\left(900 \text{ m/s}\right)}{\left(2.0\times10^7 \text{ m}\right)} = 270 \text{ m/s}$$

At its farthest point, the satellite will be moving at a leisurely 270 meters per second.

Practice Questions

1. A merry-go-round with a mass of 500.0 kilograms and radius of 2.0 meters is rotating at 3.0 radians per second when two children with a combined mass of 70.0 kilograms jump on the outer rim. What is the new angular speed of the merry-go-round?

2. A 2,000.0-kilogram space station, which is a hollow cylinder with a radius of 2.0 meters, is rotating at 1.0 radians per second when an astronaut with a mass of 80.0 kilograms lands on the outside of the station. What is the station's new angular speed?

Practice Answers

1. **2.3 radians/s.** Use this equation:

$$I_1\omega_1 = I_2\omega_2$$

For a solid disc like the merry-go-round, $I = (1/2)mr^2$.

When the children jump on, that adds $m_c r^2$ to I, where m_c is the mass of the children. In other words:

$$\frac{1}{2}mr^2\omega_1 = \left(\frac{1}{2}mr^2 + m_c r^2\right)\omega_2$$

Solve for ω_2 to get

$$\omega_2 = \frac{1/2 \; mr^2\omega_1}{1/2 \; mr^2 + m_c r^2}$$

Plug in the numbers:

$$\omega_2 = \frac{3,000 \text{ kg}\cdot\text{m}^2\cdot\text{radians/s}}{1,000 \text{ kg}\cdot\text{m}^2 + 280 \text{ kg}\cdot\text{m}^2} = 2.3 \text{ radians/s}$$

2. **0.96 radians/s.** Use this equation:

$$I_1\omega_1 = I_2\omega_2$$

For a hollow cylinder like the space station, $I = mr^2$.

When the astronaut lands, that adds $m_a r^2$ to I, where m_a is the mass of the astronaut. That change means the following equation applies:

$$mr^2\omega_1 = \left(mr^2 + m_a r^2\right)\omega_2$$

Solve for ω_2:

$$\omega_2 = \frac{mr^2\omega_1}{mr^2 + m_a r^2}$$

Plug in the numbers:

$$\omega_2 = \frac{8,000 \text{ kg}\cdot\text{m}^2\cdot\text{radians/s}}{8,000 \text{ kg}\cdot\text{m}^2 + 320 \text{ kg}\cdot\text{m}^2} = 0.96 \text{ radians/s}$$

13

SPRINGS 'N' THINGS: SIMPLE HARMONIC MOTION

IN THIS CHAPTER

- Understanding force when you stretch or compress springs
- Going over the basics of simple harmonic motion
- Mustering up the energy for simple harmonic motion
- Predicting a pendulum's motion and period

I n this chapter, we shake things up with a new kind of motion: periodic motion, which occurs when objects are bouncing around on springs or bungee cords or are swooping on the end of a pendulum. This chapter is all about describing their motion. Not only can you describe their motion in detail, but you can also predict how much energy bunched-up springs have, how long a pendulum will take to swing back and forth, and more.

BOUNCING BACK WITH HOOKE'S LAW

Objects that can stretch but return to their original shapes are called *elastic*. Elasticity is a valuable property, because it allows you to use objects such as springs for all kinds of applications: as shock absorbers in lunar landing modules, as timekeepers in clocks and watches, and even as hammers of justice in mousetraps.

In this section, we introduce Hooke's law, which relates forces to how much a spring is stretched or compressed.

Stretching and compressing springs

Robert Hooke, a physicist from England, undertook the study of elastic materials in the 1600s. He discovered a new law, not surprisingly called Hooke's law, which states that stretching or compressing an elastic material requires a force that's directly proportional to the amount of stretching or compressing you do. Hooke's law is represented by this equation:

$$F = -k\Delta x$$

Remember: This deceptively simple equation is at the heart of explaining the motion of objects on springs. It says that the force on an object in simple harmonic motion is proportional to the displacement (that's Δx) from the equilibrium position.

Pushing or pulling back: The spring's restoring force

Hooke's law gives the force a spring exerts on an object attached to it with the following equation:

$$F = -k\Delta x$$

where the minus sign shows that this force is in the opposite direction of the force that's stretching or compressing the spring. The constant of proportionality is k (Hooke's constant, also called a spring constant), which must be measured for every situation (because no two springs are exactly identical, for example). The negative sign in the equation indicates that k is a restoring force — that is, that the force points toward the equilibrium position of the object.

What are the units of spring constants? Just check out the equation: Because $F = -k\Delta x$, k must have the unit newtons per meter, or N/m.

You'll commonly see Hooke's law applied to springs. Don't get confused by the minus sign in this equation; it's just there to indicate that the force opposes the displacement, which you know is true about springs.

Figure 13-1 shows a ball attached to a spring. You can see that if the spring isn't stretched or compressed, it exerts no force on the ball. If you push the spring, however, it pushes back, and if you pull the spring, it pulls back.

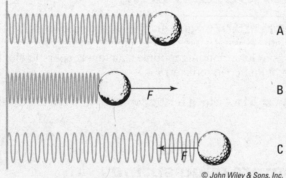

Figure 13-1: The direction of force exerted by a spring.

Remember: Hooke's law is valid as long as the elastic material you're dealing with stays elastic — that is, it stays within its *elastic limit*. If you pull a spring too far, it loses its stretchy ability. As long as a spring stays within its elastic limit, you can say that $F = -k\Delta x$. When a spring stays within its elastic limit and obeys Hooke's law, the spring is called an *ideal spring*.

Example

Q. Suppose that a group of car designers knocks on your door and asks whether you can help design a suspension system. "Sure," you say. They inform you that the car will have a mass of 1,000 kilograms, and you have four shock absorbers to work with. If you want the springs in the shock absorbers to compress by no more than 0.25 meters when the car is empty and at rest, what is the minimum spring constant you should use?

A. The correct answer is 9,800 newtons per meter.

  1. Each spring has to support a mass of at least 250 kilograms, which weighs the following:

  $$F_g = mg = (250 \text{ kg})(9.8 \text{ m/s}^2) = 2,450 \text{ N}$$

  where F_g equals the force of gravity, m equals the mass of the object, and g equals the acceleration due to gravity, 9.8 meters per second^2. The spring in the shock absorber will, at a minimum, have to give you 2,450 newtons of force at the maximum compression of 0.25 meters to oppose the force of gravity.

  2. Solve Hooke's law

  $$F = -k\Delta x$$

  for the spring constant:

  $$k = -\frac{F}{\Delta x}$$

  3. Plug in the numbers:

  $$k = -\frac{F}{\Delta x} = -\frac{2,450 \text{ N}}{(-0.25 \text{ m})} = \frac{2,450 \text{ N}}{0.25 \text{ m}} = 9,800 \text{ N/m}$$

  The force of the spring, F, is positive because it points upward. The displacement, Δx, is negative because the springs are displaced downward from their equilibrium position.

Practice Questions

  1. You have a spring whose spring constant is 200 newtons per meter, and you want to stretch it by 6.0 meters. What force do you need to apply?

  2. You have a spring of spring constant 73 newtons per meter. How much does the spring stretch if you apply a force of 52 newtons?

Practice Answers

1. **1,200 N.** Use the equation $F = k\Delta x$.

   Plug in the numbers:

   $$F = k\Delta x = (200 \text{ N/m})(6.0 \text{ m}) = 1,200 \text{ N}$$

   Note that the sign is positive here because you're pulling on the spring; you're not interested in the force with which the spring pulls back (–1,200 newtons).

2. **0.71 m.** Use the equation $F = k\Delta x$.

   Solve for Δx:

   $$\Delta x = F/k$$

   Plug in the numbers:

   $$\Delta x = F/k = (52 \text{ N})(73 \text{ N/m}) = 0.71 \text{ m}$$

GETTING AROUND TO SIMPLE HARMONIC MOTION

An *oscillatory motion* is one that undergoes repeated cycles. When the net force acting on an object is elastic, the object undergoes a simple oscillatory motion called *simple harmonic motion*. The force that tries to restore the object to its resting position is proportional to the displacement of the object. In other words, it obeys Hooke's law.

Elastic forces suggest that the motion will just keep repeating (that isn't really true, however; even objects on springs quiet down after a while as friction and heat loss in the spring take their toll). This section delves into simple harmonic motion and shows you how it relates to circular motion. Here, you graph motion with the sine wave and explore familiar concepts such as position, velocity, and acceleration.

Around equilibrium: Examining horizontal and vertical springs

Take a look at the golf ball in Figure 13-1. The ball is attached to a spring on a frictionless horizontal surface. Say that you push the ball, compressing the spring, and then you let go; the ball shoots out, stretching the spring. After the stretch, the spring pulls back and once again passes the equilibrium point (where no force acts on the ball), shooting backward past it. This happens because the ball has inertia (see Chapter 5), and when the ball is moving, bringing it to a stop takes some force. Here are the various stages the ball goes through, matching the letters in Figure 13-1 (and assuming no friction):

- **Point A:** The ball is at equilibrium, and no force is acting on it. This point, where the spring isn't stretched or compressed, is called the *equilibrium point*.

- **Point B:** The ball pushes against the spring, and the spring retaliates with force *F* opposing that pushing.

- **Point C:** The spring releases, and the ball springs to an equal distance on the other side of the equilibrium point. At this point, the ball isn't moving, but a force acts on it, *F*, so it starts going back the other direction.

The ball passes through the equilibrium point on its way back to Point B. At the equilibrium point, the spring doesn't exert any force on the ball, but the ball is traveling at its maximum speed. Here's what happens when the golf ball bounces back and forth; you push the ball to Point B, and it goes through Point A, moves to Point C, shoots back to A, moves to B, and so on: B-A-C-A-B-A-C-A, and so on. Point A is the equilibrium point, and both Points B and C are equidistant from Point A.

What if the ball were to hang in the air on the end of a spring, as Figure 13-2 shows? In this case, the ball oscillates up and down. Like the ball on a surface in Figure 13-1, the ball hanging on the end of a spring oscillates around the equilibrium position; this time, however, the equilibrium position isn't the point where the spring isn't stretched.

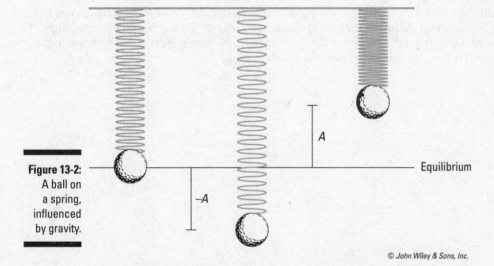

Figure 13-2:
A ball on
a spring,
influenced
by gravity.

Equilibrium

© John Wiley & Sons, Inc.

Remember: The *equilibrium* position is defined as the position at which no net force acts on the ball. In other words, the equilibrium position is the point where the ball can simply sit at rest. When the spring is vertical, the weight of the ball downward matches the pull of the spring upward. If the *y* position of the ball corresponds to the equilibrium point, y_i, the weight of the ball, *mg*, must match the force exerted by the spring. Because $F = ky_i$, you can write the following:

$$mg = ky_i$$

Solving for y_i gives you the distance the spring stretches because of the ball's weight:

$$y_i = \frac{mg}{k}$$

When you pull the ball down or lift it up and then let go, it oscillates around the equilibrium position, as Figure 13-2 shows. If the spring is elastic, the ball undergoes simple harmonic motion vertically around the equilibrium position; the ball goes up a distance A and down a distance $-A$ around that position (in real life, the ball would eventually come to rest at the equilibrium position, because a frictional force would dampen this motion).

Remember: The distance A, or how high the object springs up, is an important one when describing simple harmonic motion; it's called the *amplitude*. The amplitude is simply the maximum extent of the oscillation, or the size of the oscillation.

Catching the wave: A sine of simple harmonic motion

Calculating simple harmonic motion can require time and patience when you have to figure out how the motion of an object changes over time. Imagine that one day you come up with a brilliant idea for an experimental apparatus. You decide to shine a spotlight on a ball bouncing on a spring, casting a shadow on a moving piece of photographic film. Because the film is moving, you get a record of the ball's motion as time goes on. You turn the apparatus on and let it do its thing. See the results in Figure 13-3.

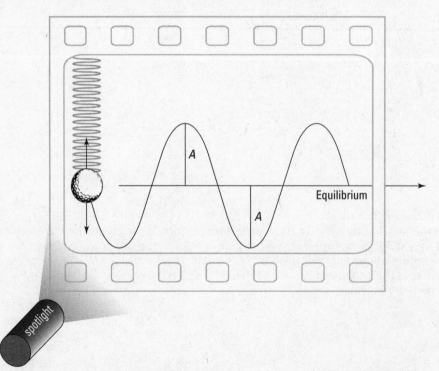

Figure 13-3: Tracking a ball's simple harmonic motion over time.

© John Wiley & Sons, Inc.

The ball oscillates around the equilibrium position, up and down, reaching amplitude *A* at its lowest and highest points. But take a look at the ball's track: You can tell where the ball is moving fastest because that's where the curve has the steepest slope. The ball goes fastest near the equilibrium point because of the acceleration caused by the spring force, which has been applied since the turning point. At the top and bottom, the ball is subject to plenty of force, so it slows down and reverses its motion.

The track of the ball is best modeled with a *sine wave,* which means that its track is a sine wave of amplitude *A.* (**Note:** You can also use a cosine wave, because the shape is the same. The only difference is that when a sine wave is at its peak, the cosine wave is at zero, and vice versa.)

Tip: You can get a clear picture of the sine wave if you plot the sine function on an *xy* graph like this:

$$y = \sin x$$

In the rest of this section, we show you how the sine wave relates circular motion to simple harmonic motion.

Understanding sine waves with a reference circle

Take a look at the sine wave in a circular way. If you attach a ball to a rotating disk (see Figure 13-4) and you shine a spotlight on it, you get the same result as when you have the ball hanging from the spring (Figure 13-3): a sine wave.

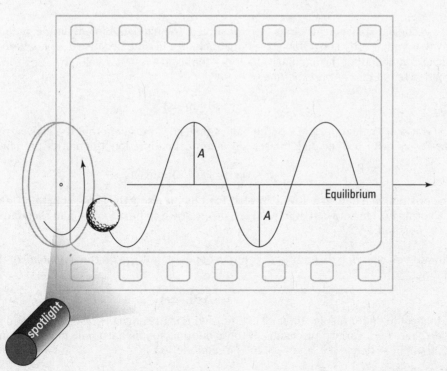

Figure 13-4:
The vertical component of the displacement of an object moving in a circle follows a sine wave.

© John Wiley & Sons, Inc.

The rotating disk, which you can see in Figure 13-5, is often called a *reference circle.* You can see how the vertical component of circular motion relates to the sinusoidal (sine-like) wave of simple harmonic motion. Reference circles can tell you a lot about simple harmonic motion.

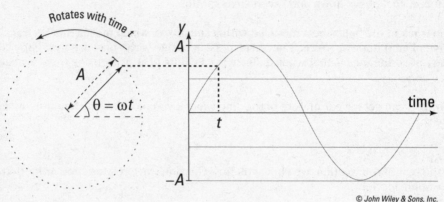

Figure 13-5: A reference circle helps you analyze simple harmonic motion.

As the disk turns, the angle, θ, increases in time. What does the track of the ball look like as the film moves to the right? Using a little trig, you can resolve the ball's motion along the y-axis; all you need is the vertical (y) component of the ball's position. At any one time, the ball's y position is the following:

$$y = A\sin\theta$$

The vertical displacement varies from positive A to negative A in amplitude. In fact, you can say that you already know how θ is going to change in time because $\theta = \omega t$, where ω is the single component of the angular velocity — that is, the angular velocity along the axis of rotation of the disk — and t is time:

$$y = A\sin(\omega t)$$

You can now explain the track of the ball as time goes on, given that the disk is rotating with angular velocity ω. If the disk rotates in the opposite direction, the ball's y-path becomes

$$y = A\sin(-\omega t) = -A\sin(\omega t)$$

The position of a ball on a spring is described by precisely the same equation. Take a look at the spring in Figure 13-6. It starts at rest, drops down to position $-A$, and then moves back up to position A.

How do you describe this motion? In terms of position A, called the *amplitude* of the spring's motion:

$$y = -A\sin(\omega t)$$

In this equation for displacement, t is time, and ω is the *angular speed.* (It's called angular speed here for a variety of reasons; it turns out that simple harmonic motion is only one component — hence the sine — of full circular motion.)

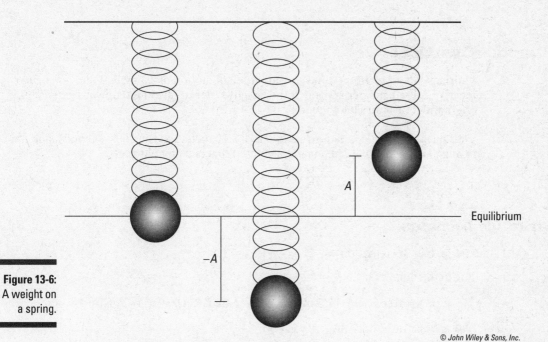

Figure 13-6:
A weight on
a spring.

That equation assumes that you start moving downward from the equilibrium position, at time $t = 0$. If you start at a different time, t_i, you can adjust the equation to match the spring's motion like this:

$$y = -A\sin\left[\omega(t - t_i)\right]$$

If you start at the maximum height at time $t = 0$, the equation becomes

$$y = A\cos(\omega t)$$

Example

Q. A weight on a spring is making that spring oscillate up and down. If the amplitude of the motion is 1.0 meter, the angular speed is 1.0 radians per second, and the weight is initially at its maximum upward position, where will the oscillating weight be after 10.0 seconds?

A. The correct answer is –0.84 meters.

1. Use the equation $y = A\cos(\omega t)$.

2. Plug in the numbers:

$$y = A\cos(\omega t) = (1.0\text{ m})\cos\left[(1.0\text{ radians/s})(10.0\text{ s})\right] = -0.84\text{ m}$$

The negative sign indicates that the spring is extended downward.

Tip: If you plug the numbers in this equation into your calculator, either put it into radian mode because ωt is in radians or convert to degrees by multiplying ωt by $180/\pi$.

Practice Questions

1. A spring with a weight on it has an amplitude of motion of 2.5 meters and an angular speed of 2.0 radians per second. If the weight starts at maximum upward extension, where will it be after 60.0 seconds?

2. A spring starts at maximum extension, is at –3.0 meters at time 60.0 seconds, and has an angular speed of 6.0 radians per second. What is its amplitude?

Practice Answers

1. **2.04 m.** Use the equation $y = A \cos (\omega t)$.

Plug in the numbers:

$$y = A\cos(\omega t)=(2.5\ \text{m})\cos\big[(2.0\ \text{radians/s})(60.0\ \text{s})\big]=2.04\ \text{m}$$

2. **10.6 m.** Use the equation $y = A \cos (\omega t)$.

Solve for A and plug in the numbers:

$$A = y/\cos(\omega t)= -3.0\ \text{m}/\cos\big[(6.0\ \text{radians/s})(60.0\ \text{s})\big]=10.6\ \text{m}$$

Getting periodic

Each time an object moves around a full circle, it completes a *cycle*. The time the object takes to complete the cycle is called the *period,* and it's generally measured in seconds. The letter used for period is *T.*

Look at Figure 13-5 in terms of the *y* motion on the film. During one cycle, the ball moves from $y = A$ to $-A$ and then back to A. When the ball goes from any point on the sine wave and passes through one whole wave (including one peak and one trough) back to the next equivalent point on the sine wave later in time, it completes a cycle. The time the ball takes to move from a certain position back to that same position while moving in the same direction is its period.

Tip: How can you relate the period to something more familiar? When an object moves in a full circle, completing a cycle, the object goes 2π radians. It travels that many radians in *T* seconds, so its angular speed, ω (see Chapter 11), is

$$\omega = \frac{2\pi}{T}$$

Multiplying both sides by *T* and dividing by ω allows you to solve for the period. Now you can relate the period and the angular speed:

$$T = \frac{2\pi}{\omega}$$

Remember: Sometimes you speak in terms of the frequency of periodic motion, not the period. The *frequency* is the number of cycles that are completed per second.

For instance, if the disk from Figure 13-4 rotates at 1,000 full turns per second, the frequency, *f,* would be 1,000 cycles per second. Cycles per second are also called *hertz,* abbreviated Hz, so this frequency would be 1,000 hertz.

Remember: So how do you connect frequency, *f,* to period, *T? T* is the amount of time one cycle takes, so here's how you can define frequency:

$$f = \frac{1}{T}$$

Because $\omega = 2\pi/T$ and $Tf = 1$, you can rewrite the angular-velocity equation in terms of frequency:

$$\omega = \left(\frac{2\pi}{T} \right) Tf = 2\pi f$$

Warning: In simple harmonic motion, the angular velocity, ω, is often referred to as *angular frequency*. Don't confuse the wave's frequency, *f,* with the angular frequency.

Example

Q. A weight on a spring is bouncing up and down with an angular frequency of 12.0 radians per second. How long does it take to complete each cycle?

A. The correct answer is 0.52 seconds.

1. Use this equation:

$$T = \frac{2\pi}{\omega}$$

2. Plug in the numbers:

$$T = \frac{2\pi}{\omega} = \frac{2\pi}{12.0 \text{ radians/s}} = 0.52 \text{ s}$$

Practice Questions

1. A spring has a weight pulling it down, and the angular frequency of oscillation is 1.5 radians per second. What is the weight's period?

2. A spring with a weight on it is bouncing up and down with an angular frequency of 1.3 radians per second. What is its frequency?

Practice Answers

1. **4.2 s.** Use this equation:

$$T = \frac{2\pi}{\omega}$$

Plug in the numbers:

$$T = \frac{2\pi}{\omega} = \frac{2\pi}{1.5 \text{ radians/s}} = 4.2 \text{ s}$$

2. **0.21 cycles/s.** Use this equation:

$$\omega = \frac{2\pi}{T} = 2\pi f$$

Solve for f:

$$f = \frac{\omega}{2\pi}$$

Plug in the numbers:

$$f = \frac{\omega}{2\pi} = \frac{1.3 \text{ radians/s}}{2\pi} = 0.21 \text{ cycles/s}$$

Remembering not to speed away without the velocity

Refer to Figure 13-5, where a ball is rotating on a disk. In the section "Understanding sine waves with a reference circle," earlier in this chapter, you figure out that

$$y = A\cos(\omega t)$$

where y stands for the y coordinate and A stands for the amplitude of the motion if the ball starts at the top. At any point y, the ball also has a certain velocity, which varies in time also. So how can you describe the velocity mathematically? Well, you can relate tangential velocity to angular velocity like this (see Chapter 11):

$$v = r\omega$$

where r represents the radius. Because the radius of the circle is equal to the amplitude of wave it corresponds to, $r = A$. Therefore, you get the following equation:

$$v = A\omega$$

Does this equation get you anywhere? Sure, because the ball's shadow on the film gives you simple harmonic motion. The velocity vector (see Chapter 4) always points tangential to the circle — perpendicular to the radius — so you get the following for the y component of the velocity at any one time:

$$v_y = -A\omega\sin\theta$$

And because the ball is on a rotating disk, you know that $\theta = \omega t$, so

$$v_y = -A\omega \sin(\omega t)$$

Remember: This equation describes the velocity of any object in simple harmonic motion. Note that the velocity changes in time — from 0 to $-A\omega$ and then back to 0 and then to $A\omega$. So the maximum velocity, which happens at the equilibrium point, has a magnitude of $A\omega$. Among other things, this equation says that, for a given angular velocity, the maximum velocity (v) is directly proportional to the amplitude (A) of the motion: Simple harmonic motion of greater amplitude has a larger maximum velocity, and vice versa.

Example

Q. A weight on a spring is bouncing up and down with an angular frequency of 3.4 radians per second and an amplitude of 1.4 meters. If the weight starts at maximum displacement at $t = 0$, what is its speed at $t = 5.0$ seconds?

A. The correct answer is 4.6 meters per second.

1. Use this equation:

$$v_y = -A\omega \sin(\omega t)$$

2. Plug in the numbers:

$$v_y = -A\omega \sin(\omega t)$$
$$= -(1.4 \text{ m})(3.4 \text{ radians/s}) \sin\left[(3.4 \text{ radians/s})(5.0 \text{ s})\right]$$
$$= 4.6 \text{ m/s}$$

Practice Questions

1. If you have a spring with a weight on it, and the weight's angular frequency is 1.7 radians per second and amplitude is 5.6 meters, what is the weight's velocity at 30.0 seconds if it starts at its maximum upward displacement?

2. A spring with a weight on it is bouncing up and down with an angular frequency of 2.7 radians per second and amplitude of 1.7 meters. What is its velocity at 10.0 seconds if it starts at its maximum upward displacement?

Practice Answers

1. **–6.38 m/s.** Use this equation:

$$v_y = -A\omega \sin(\omega t)$$

Plug in the numbers:

$$v_y = -A\omega \sin(\omega t)$$
$$= -(5.6 \text{ m})(1.7 \text{ radians/s}) \sin\left[(1.7 \text{ radians/s})(30.0 \text{ s})\right]$$
$$= -6.38 \text{ m/s}$$

2. **−4.4 m/s.** Use this equation:

$$v_y = -A\omega \sin(\omega t)$$

Plug in the numbers:

$$v_y = -A\omega \sin(\omega t)$$
$$= -(1.7 \text{ m})(2.7 \text{ radians/s}) \sin\left[(2.7 \text{ radians/s})(10.0 \text{ s})\right]$$
$$= -4.4 \text{ m/s}$$

Including the acceleration

You can find the displacement of an object undergoing simple harmonic motion with the equation $y = A\cos(\omega t)$, and you can find the object's velocity with the equation $v = -A\omega \sin(\omega t)$. But you have another factor to account for when describing an object in simple harmonic motion: its acceleration at any particular point. How do you figure it out? No sweat. When an object is going around in a circle, the acceleration is the centripetal acceleration (see Chapter 11), which is

$$a = r\omega^2$$

where r is the radius and ω is the single component of angular velocity (that is, the angular velocity in the direction of the [constant] axis of rotation). And because $r = A$ — the amplitude — you get the following equation:

$$a = A\omega^2$$

This equation represents the relationship between centripetal acceleration, a, and angular velocity, ω. To go from a reference circle (see the earlier section "Understanding sine waves with a reference circle") to simple harmonic motion, you take the component of the acceleration in one dimension — the y direction here — which looks like this:

$$a = -A\omega^2 \cos\theta$$

Remember: The negative sign indicates that the y component of the acceleration is always directed opposite the displacement (the ball always accelerates toward the equilibrium point). And because $\theta = \omega t$, where t represents time, you get the following equation for acceleration:

$$a = -A\omega^2 \cos(\omega t)$$

Now you have the equation to find the acceleration of an object at any point while it's moving in simple harmonic motion.

Example

Q. The diaphragm (a metal disk that acts like an eardrum) in your phone undergoes a motion very similar to simple harmonic motion. The amplitude of the diaphragm's motion is about 1.0×10^{-4} meters. A typical frequency for human speech is about 1.0 kilohertz (1,000 hertz). What is the maximum acceleration of the diaphragm?

A. The correct answer is 3,900 meters per second^2.

1. Use this equation:

$$a_{max} = A\omega^2$$

2. Replace ω with $2\pi f$:

$$a_{max} = A(2\pi f)^2$$

3. Plug in the numbers:

$$a_{max} = A\left(2\pi f\right)^2 = \left(1.0 \times 10^{-4} \text{ m}\right)\left[2\pi\left(1,000 \text{ s}^{-1}\right)\right]^2 \approx 3,900 \text{ m/s}^2$$

Practice Questions

1. You have a spring with a weight pulling it down; the angular frequency is 1.7 radians per second, and the amplitude is 6.0 meters. What is the weight's acceleration at 60 seconds if it starts at its maximum upward displacement?

2. A spring with a weight on it has an angular frequency of 3.7 radians per second and an amplitude of 1.4 meters. What is the weight's acceleration at 15 seconds if it starts at its maximum upward displacement?

Practice Answers

1. **−1.8 m/s^2.** Use this equation:

$$a = -A\omega^2 \cos(\omega t)$$

Plug in the numbers:

$$a = -A\omega^2 \cos(\omega t)$$
$$= -\left(6.0 \text{ m}\right)\left(1.7 \text{ radians/s}\right)^2 \cos\left[\left(1.7 \text{ radians/s}\right)\left(60 \text{ s}\right)\right]$$
$$= -1.8 \text{ m/s}^2$$

2. **−10 m/s².** Use this equation:

$$a = -A\omega^2 \cos(\omega t)$$

Plug in the numbers:

$$a = -A\omega^2 \cos(\omega t)$$
$$= -(1.4 \text{ m})(3.7 \text{ radians/s})^2 \cos\left[(3.7 \text{ radians/s})(15 \text{ s})\right]$$
$$= -10 \text{ m/s}^2$$

Finding the angular frequency of a mass on a spring

If you take the information you know about Hooke's law for springs (see the earlier section "Bouncing Back with Hooke's Law") and apply it to what you know about simple harmonic motion (see the earlier section "Getting Around to Simple Harmonic Motion"), you can find the angular frequencies of masses on springs, along with the frequencies and periods of oscillations. And because you can relate angular frequency and the masses on springs, you can find the displacement, velocity, and acceleration of the masses.

Hooke's law says that

$$F = -k\Delta x$$

where F is the force exerted by the string, k is the spring constant, and Δx is displacement from equilibrium. Because of Isaac Newton (see Chapter 5), you know that force also equals mass times acceleration:

$$F = ma$$

These force equations are in terms of displacement and acceleration, which you see in simple harmonic motion in the following forms (see the preceding section):

- $x = A\cos(\omega t)$
- $a = -A\omega^2 \cos(\omega t)$

Inserting these two equations into the force equations gives you the following:

$$ma = -kx$$
$$m\left[-A\omega^2 \cos(\omega t)\right] = -kA\cos(\omega t)$$

Divide both sides by $-A\cos(\omega t)$, and this equation breaks down to

$$m\omega^2 = k$$

Rearranging to put ω on one side of the equation gives you the formula for angular frequency:

$$\omega = \sqrt{\frac{k}{m}}$$

You may like to check how the units work out. Remember that $1\,N = 1\,kg \cdot m/s^2$, so the units you get from the preceding equation for the angular velocity work out to be

$$\sqrt{\frac{\left(kg \cdot m/s^2\right)/m}{kg}} = \sqrt{\frac{kg/s^2}{kg}} = \sqrt{\frac{1}{s^2}} = s^{-1}$$

You can now find the angular frequency (angular velocity) of a mass on a spring, as it relates to the spring constant and the mass. You can also tie the angular frequency to the frequency and period of oscillation (see the section "Getting periodic," earlier in this chapter) by using the following equation:

$$\omega = \frac{2\pi}{T} = 2\pi f$$

With this equation and the earlier angular-frequency formula, you can write the formulas for frequency and period in terms of k and m:

- $f = \frac{1}{2\pi}\sqrt{\frac{k}{m}}$

- $T = 2\pi\sqrt{\frac{m}{k}}$

Remember: Because you can relate the angular frequency, ω, to the spring constant and the mass on the end of the spring, you can predict the displacement, velocity, and acceleration of the mass, using the following equations for simple harmonic motion (see the section "Catching the wave: A sine of simple harmonic motion," earlier in this chapter):

- $y = A\cos\left(\omega t\right)$

- $v = -A\omega\sin\left(\omega t\right)$

- $a = -A\omega^2\cos\left(\omega t\right)$

Example

Q. A weight on a spring is bouncing up and down. The spring constant is 1.6 newtons per meter, the mass is 1.0 kilograms, and the amplitude is 3.0 meters. What equation describes the spring's motion if it starts at its maximum upward displacement?

A. The correct answer is $y = A\cos(\omega t) = (3.0\ m)\cos[(1.3\ radians/s)t]$.

1. Use this equation:

$$\omega = \sqrt{\frac{k}{m}} = \sqrt{\frac{1.6\ N/m}{1.0\ kg}} = 1.3\ radians/s$$

2. This equation describes the motion:

$$y = A \cos(\omega t) = (3.0 \text{ m})\cos\left[(1.3 \text{ radians/s})t\right]$$

Practice Questions

1. If you have a spring with a mass of 1.0 kilograms on it and a spring constant of 12.0 newtons per meter, what is the spring's period of oscillation?

2. A weight on a spring is moving up and down. The spring constant is 1.9 newtons per meter, the mass is 1.0 kilograms, and the amplitude is 2.4 meters. What equation describes the spring's motion if it starts at its maximum upward displacement?

Practice Answers

1. **1.8 s.** Use this equation and plug in the numbers:

$$\omega = \sqrt{\frac{k}{m}} = \sqrt{\frac{12.0 \text{ N/m}}{1.0 \text{ kg}}} = 3.5 \text{ radians/s}$$

Now use this equation and plug in the numbers:

$$T = \frac{2\pi}{\omega} = \frac{2\pi}{3.5 \text{ radians/s}} = 1.8 \text{ s}$$

2. **$y = A \cos(\omega t) = (2.4 \text{ m})\cos[(1.4 \text{ radians/s})t]$.** Use this equation:

$$\omega = \sqrt{\frac{k}{m}} = \sqrt{\frac{1.9 \text{ N/m}}{1.0 \text{ kg}}} = 1.4 \text{ radians/s}$$

Now use this equation:

$$y = A \cos(\omega t) = (2.4 \text{ m})\cos\left[(1.4 \text{ radians/s})t\right]$$

FACTORING ENERGY INTO SIMPLE HARMONIC MOTION

Along with the actual motion that takes place in simple harmonic motion, you can examine the energy involved. For example, how much energy is stored in a spring when you compress or stretch it? The work you do compressing or stretching the spring must go into the energy stored in the spring. That energy is called *elastic potential energy* and is equal to the force, *F*, times the distance, *s:*

$$W = Fs$$

As you stretch or compress a spring, the force varies, but it varies in a linear way (because in Hooke's law, force is proportional to the displacement). Therefore, you can write the equation in terms of the average force, $\bar{F}$:

$$W = \bar{F}s$$

The distance (or displacement), s, is just the difference in position, $x_f - x_i$, and the average force is $(1/2)(F_f + F_i)$. Therefore, you can rewrite the equation as follows:

$$W = \left[\frac{1}{2}(F_f + F_i)\right](x_f - x_i)$$

Hooke's law says that $F = -kx$. Therefore, you can substitute $-kx_f$ and $-kx_i$ for F_f and F_i:

$$W = \left[-\frac{1}{2}(kx_f + kx_i)\right](x_f - x_i)$$

Distributing and simplifying the equation gives you the equation for work in terms of the spring constant and position:

$$W = \frac{1}{2}kx_i^2 - \frac{1}{2}kx_f^2$$

Remember: The work done on the spring changes the potential energy stored in the spring. Here's how you give that potential energy, or the elastic potential energy:

$$PE = \frac{1}{2}kx^2$$

Remember: You can also note that when you let the spring go with a mass on the end of it, the mechanical energy (the sum of potential and kinetic energy) is conserved:

$$PE_1 + KE_1 = PE_2 + KE_2$$

When you compress the spring, you know that you have potential energy stored up. When the moving mass reaches the equilibrium point and no force from the spring is acting on the mass, you have maximum velocity and therefore maximum kinetic energy — at that point, the kinetic energy is equal to the initial potential energy, by the conservation of mechanical energy (see Chapter 9 for more on this topic).

Example

Q. What is the potential energy in a spring of spring constant 40 newtons per meter that's stretched 5.0 meters from equilibrium?

A. The correct answer is 500 joules.

   1. Use this equation:

$$PE = \frac{1}{2}kx^2$$

2. Plug in the numbers:

$$PE = \frac{1}{2}kx^2$$
$$= \frac{1}{2}(40 \text{ N/m})(5.0 \text{ m})^2$$
$$= 500 \text{ J}$$

Practice Questions

1. If you stretch a spring of spring constant 100 newtons per meter by 5.0 meters, what potential energy is in the spring?

2. If you stretch a spring of spring constant 250 newtons per meter by 6.5 meters, what potential energy is in the spring?

Practice Answers

1. **1,250 J.** Use this equation:

$$PE = \frac{1}{2}kx^2$$

Plug in the numbers:

$$PE = \frac{1}{2}kx^2$$
$$= \frac{1}{2}(100 \text{ N/m})(5.0 \text{ m})^2$$
$$= 1,250 \text{ J}$$

2. **5,300 J.** Use this equation:

$$PE = \frac{1}{2}kx^2$$

Plug in the numbers:

$$PE = \frac{1}{2}kx^2$$
$$= \frac{1}{2}(250 \text{ N/m})(6.5 \text{ m})^2$$
$$= 5,300 \text{ J}$$

SWINGING WITH PENDULUMS

Other objects besides springs, such as pendulums, move in simple harmonic motion. In Figure 13-7, a ball tied to a string swings back and forth.

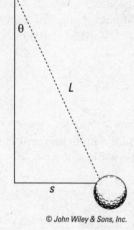

Figure 13-7:
A pendulum moves in simple harmonic motion.

The torque, τ, that comes from gravity is the weight of the ball (which is a force of magnitude mg directed downward — hence the minus sign) multiplied by the lever arm, s (for more on lever arms and torque, see Chapter 11):

$$\tau = -mgs$$

Here's where you make an approximation. For small angles θ, the distance s approximately equals $L\theta$, where L is the length of the pendulum string:

$$\tau = -mgL\theta$$

This equation resembles Hooke's law, $F = -kx$, if you treat mgL as you would a spring constant. In Chapter 12, we show you the relation between torque and angular acceleration, and you see that the angular variables obey the same equations as their linear equivalents. Therefore, the calculation goes in just the same way as for the spring. In the angular variables:

$$\sum \tau = I\alpha$$

Just as in the case of the spring, the pendulum undergoes simple harmonic motion, if the pendulum starts vertical, with

- $\theta = A\sin(\omega t)$
- $\alpha = -A\omega^2 \sin(\omega t)$

Plug in the torque of the pendulum and the values of α and θ to get

$$I\alpha = -mgL\theta$$
$$I\left[-A\omega^2 \sin(\omega t)\right] = -mgLA\sin(\omega t)$$

Then solve for ω to get this:

$$\omega = \sqrt{\frac{mgL}{I}}$$

The moment of inertia equals mr^2 for a point mass (see Chapter 12), which you can use here, assuming that the ball is small compared to the pendulum string. For the pendulum, the radius r is the length of the string, L. This gives you the following equation:

$$\omega = \sqrt{\frac{mgL}{mL^2}}$$
$$= \sqrt{\frac{g}{L}}$$

Now you can plug this angular velocity into the equations for simple harmonic motion. You can also find the period of a pendulum with the following equation:

$$\omega = \frac{2\pi}{T}$$

where T represents period. If you substitute the preceding form for ω, you get this:

$$\sqrt{\frac{g}{L}} = \frac{2\pi}{T}$$

Then rearrange to find the period:

$$T = 2\pi\sqrt{\frac{L}{g}}$$

Note that this period is actually independent of the mass on the pendulum!

Example

Q. What's the angular frequency of a pendulum of length 1.0 meter?

A. The correct answer is 3.1 radians per second.

Use this equation and plug in the numbers:

$$\omega = \sqrt{\frac{g}{L}} = \sqrt{\frac{9.8 \text{ m/s}^2}{1.0 \text{ m}}} = 3.1 \text{ radians/s}$$

Practice Questions

1. What's the period of a pendulum with a length of 1.5 meters?

2. What's the period of a pendulum with a length of 3.0 meters?

3. If you have a pendulum with a length of 1.0 meter with a weight on the end of it, what equation describes the angular position θ of that weight if the amplitude of motion is 0.05 radians and the pendulum starts in its equilibrium position?

4. You have a pendulum with a length of 1.5 meters with a weight on the end of it. What equation describes the angular position θ of that weight if the amplitude of motion is 0.075 radians and the pendulum starts in its equilibrium position?

Practice Answers

1. **2.5 s.** Use this equation:

$$\omega = \sqrt{\frac{g}{L}} = \sqrt{\frac{9.8 \text{ m/s}^2}{1.5 \text{ m}}} = 2.56 \text{ radians/s}$$

Now use this equation:

$$T = \frac{2\pi}{\omega} = \frac{2\pi}{2.56 \text{ radians/s}} = 2.5 \text{ s}$$

2. **3.5 s.** Use this equation:

$$\omega = \sqrt{\frac{g}{L}} = \sqrt{\frac{9.8 \text{ m/s}^2}{3.0 \text{ m}}} = 1.8 \text{ radians/s}$$

Now use this equation:

$$T = \frac{2\pi}{\omega} = \frac{2\pi}{1.8 \text{ radians/s}} = 3.5 \text{ s}$$

3. $\theta = \mathbf{0.05 \sin\left[(3.1 \text{ radians/s})t\right]}$. Use this equation:

$$\omega = \sqrt{\frac{g}{L}} = \sqrt{\frac{9.8 \text{ m/s}^2}{1.0 \text{ m}}} = 3.13 \text{ radians/s}$$

Now use this equation:

$$\theta = A\sin(\omega t) = 0.05 \sin\left[(3.1 \text{ radians/s})t\right]$$

Note that the motion of a pendulum isn't exactly harmonic.

4. $\theta = \mathbf{0.075 \sin\left[(2.6 \text{ radians/s})t\right]}$. Use this equation:

$$\omega = \sqrt{\frac{g}{L}} = \sqrt{\frac{9.8 \text{ m/s}^2}{1.5 \text{ m}}} = 2.6 \text{ radians/s}$$

Use this equation and plug in the numbers:

$$\theta = A\sin(\omega t) = 0.075 \sin\left[(2.6 \text{ radians/s})t\right]$$

PART IV

laying down the
laws of

THERMODYNAMICS

WEB EXTRA *If you want to find out how to get the most work done with the least amount of sweat, check out the article on Carnot engines at www.dummies.com/extras/ucanphysics1.*

IN THIS PART . . .

- Discover how heat makes objects expand and change phase.

- See how heat is transferred through convection, conduction, and radiation.

- Understand the relationship among the pressure, volume, and temperature of a gas.

- Work through the laws of thermodynamics that govern heat processes.

14

TURNING UP THE HEAT WITH THERMODYNAMICS

The concepts of heat and temperature are part of your daily life. Understanding the laws that govern the temperatures of things, how heat flows between them, and how the material and thermal properties depend on each other hasn't only furthered physicists' appreciation of the world and its workings; it has also led to technological and engineering advances. A structurally sound bridge, for example, depends on understanding the thermal expansion of any of the bridge's metal elements. The motor car works because of the thermal energy released from the combustion of gasoline and air. These, and more, are possible only with an understanding of the relationship between materials and their thermal properties.

This chapter explores heat and temperature. Physics gives you plenty of power to predict what goes on when things heat up or cool down. We discuss temperature scales, linear expansion, volume expansion, and how much of liquid at one temperature will change the temperature of another when they're put together.

MEASURING TEMPERATURE

Temperature is a measure of molecular movement — how fast and how much the molecules of whatever substance you're measuring are moving. You always start a calculation or observation in physics by making measurements, and when you're discussing temperature, you have several scales at your disposal: most notably, Fahrenheit, Celsius, and Kelvin.

In the United States, the most common temperature scale is the *Fahrenheit scale,* which measures temperature in degrees. For example, the blood temperature of a healthy human being is 98.6°F — the *F* means you're using the Fahrenheit scale. In Fahrenheit's system, pure water freezes at 32°F and boils at 212°F.

However, the Fahrenheit system wasn't very reproducible in its early days, so scientists developed another system — the *Celsius scale* (formerly called the *centigrade* system). Using this system, pure water freezes at 0°C and boils at 100°C (these measurements are at sea level; they change as you go up in altitude). Here's how you tie the two systems of temperature measurement together:

• **Freezing water:** 32°F = 0°C

• **Boiling water:** 212°F = 100°C

If you do the math, you find 180°F between the points of freezing and boiling in the Fahrenheit system and 100°C in the Celsius system, so the conversion ratio is $180/100 = 18/10 = 9/5$. And don't forget that the measurements are also offset by 32 degrees (the 0-degrees point of the Celsius scale corresponds to the 32-degrees point of the Fahrenheit scale). Putting these ideas together lets you convert from Celsius to Fahrenheit or from Fahrenheit to Celsius pretty easily; just remember these equations:

- $C = \frac{5}{9}(F - 32)$
- $F = \frac{9}{5}C + 32$

For example, the blood temperature of a healthy human being is 98.6°F. What does this equal in Celsius? Just plug the numbers into the equation:

$$C = \frac{5}{9}(F - 32)$$
$$= \frac{5}{9}(98.6 - 32) \approx 37.0°C$$

In the 19th century, William Thompson created a third temperature system, one now in common use in physics — the *Kelvin system* (Thompson later became Lord Kelvin). The Kelvin system has become so central to physics that the Fahrenheit and Celsius systems are defined in terms of the Kelvin system — a system based on the concept of absolute zero.

Molecules move more and more slowly as the temperature lowers. At *absolute zero,* the molecules almost stop, which means you can't cool them anymore. (The molecules only "almost" stop because when you get down to the scale of molecules, you're in the realm of quantum mechanics. When the molecules have as little energy as possible, they still have *zero-point energy.*) No refrigeration system in the world — or in the universe — can go any lower than absolute zero.

The Kelvin system uses absolute zero as its zero point, which makes sense. What's a little odd is that you don't measure temperature in this scale in degrees; you measure it in *kelvins.* A temperature of 100 is 100 kelvins (not 100 degrees Kelvin) in the Kelvin scale. This system has become so widely adopted that the official MKS unit of temperature is the kelvin (in practice, however, you see °C used more often in introductory physics).

Remember: Each kelvin is the same size as a Celsius degree, which makes converting between Celsius degrees and kelvins easy. On the Celsius scale, absolute zero is –273.15°C. This temperature corresponds to 0 kelvins, which you also write as 0 K (not, please note, 0°K).

Tip: To convert between the Celsius and Kelvin scales, use the following formulas:

- **Celsius to Kelvin:** $K = C + 273.15$
- **Kelvin to Celsius:** $C = K - 273.15$

Tip: And to convert from kelvins to Fahrenheit degrees, you can use this formula:

$$F = \frac{9}{5}(K - 273.15) + 32$$
$$= \frac{9}{5}K - 459.67$$

Examples

Q. What is 54° Fahrenheit in Celsius?

A. The correct answer is 12°C.

1. Use this equation:

$$C = \frac{5}{9}\left(F - 32°\right)$$

2. Plug in the numbers:

$$C = \frac{5}{9}\left(F - 32°\right) = \left(0.56\right)\left(54° - 32°\right) = 12°C$$

Q. Helium turns to liquid at 4.2 kelvins; what's that in degrees Celsius?

A. The correct answer is –268.95°C.

1. Use this equation:

$$C = K - 273.15°$$

2. Plug in the numbers:

$$C = K - 273.15°$$
$$= 4.2° - 273.15° = -268.95°C$$

Helium liquefies at –268.95°C. Pretty chilly.

Practice Questions

1. What is 23° Fahrenheit in Celsius?

2. What is 18° Celsius in Fahrenheit?

3. What is 18 kelvins in Celsius?

4. What is 57 kelvins in Fahrenheit?

Practice Answers

1. –5°C. Use this equation:

$$C = \frac{5}{9}\left(F - 32°\right)$$

Plug in the numbers:

$$C = \frac{5}{9}(F - 32°) = (0.56)(23° - 32°) = -5°C$$

2. **64°F.** Use this equation:

$$F = \frac{9}{5}C + 32°$$

Plug in the numbers:

$$F = \frac{9}{5}C + 32° = (1.8)(18°) + 32° = 64°F$$

3. **−255°C.** Use this equation:

$$C = K - 273.15°$$

Plug in the numbers:

$$C = K - 273.15° = 18° - 273.15° = -255°C$$

4. **−357°F.** Convert kelvins to Celsius:

$$C = K - 273.15° = 57° - 273° = -216°C$$

Convert Celsius to Fahrenheit:

$$F = \frac{9}{5}C + 32° = (1.8)(-216°) + 32° = -357°F$$

THE HEAT IS ON: THERMAL EXPANSION

Some screw-top jars can be tough to open, which is maddening when you really want some pickles. Maybe you remember seeing your mom run stubborn jar lids under hot water when you were a kid. She did this because heat makes the lid expand, which usually makes the job of turning it much easier.

On a molecular level, thermal expansion happens because when you heat objects, the molecules bounce around faster, which leads to a physical expansion. (Note that this relationship between heating and expanding isn't true for all materials, however. For example, water becomes denser as you raise its temperature from 0°C to 4°C.)

In this section, we first cover the linear expansion of solids — how solid objects lengthen when temperature rises. We then discuss thermal expansion in 3-D so you can observe volume changes in both solids and liquids. (For info on thermal expansion in gases, flip to Chapter 16.)

Linear expansion: Getting longer

When you talk about the expansion of a solid in any one dimension under the influence of heat, you're talking about *linear expansion*. Figure 14-1 shows an image of this phenomenon.

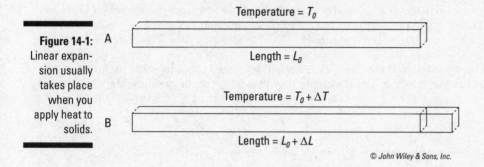

Figure 14-1: Linear expansion usually takes place when you apply heat to solids.

Temperature = T_0

A

Length = L_0

Temperature = $T_0 + \Delta T$

B

Length = $L_0 + \Delta L$

© John Wiley & Sons, Inc.

Under thermal expansion, a solid object's change in length, ΔL, is proportional to the change in temperature, ΔT. You can show this relationship mathematically.

Note: Even though initial values are represented by a subscript i in other chapters (L_i, for example), we use a subscript 0 (L_0, for example), which is what you're more likely to see in other texts for these kinds of equations.

First, suppose you raise the temperature of an object a small amount:

$$T = T_0 + \Delta T$$

where T represents the final temperature, T_0 represents the original temperature, and ΔT represents the change in temperature. The change of temperature results in an expansion in any linear direction of

$$L = L_0 + \Delta L$$

where L represents the final length of the solid, L_0 represents its original length, and ΔL represents the change in length.

When you heat a solid, the solid expands by a few percent, and that percentage is proportional to the change in temperature. In other words, $\Delta L / L_0$ (the fraction by which the solid expands) is proportional to ΔT (the change in temperature).

The constant of proportionality, which helps tell you exactly how much an object will expand, depends on which material you're working with. The constant of proportionality is the *coefficient of linear expansion,* which you give the symbol α. You can write this relationship as an equation this way:

$$\frac{\Delta L}{L_0} = \alpha \Delta T$$

Remember: Here's the linear-expansion equation in standard form, solved for ΔL:

$$\Delta L = \alpha L_0 \Delta T$$

Remember: People usually measure α, the coefficient of linear expansion, in units of $1/°C$ (that is, in $°C^{-1}$). However, because the units of Celsius and Kelvin are the same size, a difference in temperature measured in degrees Celsius is of the same magnitude when measured in kelvins. Therefore, to convert the coefficient of linear expansion from degrees Celsius to kelvins, you only have to swap the symbols.

Physics problems provide these coefficients when you need them to solve the problem. But just in case, here's a useful website that lists many of the coefficients: www.engineeringtoolbox.com/linear-expansion-coefficients-d_95.html.

Example

Q. You're heating a 1.0-meter steel bar, coefficient of linear expansion $1.2 \times 10^{-5}°C^{-1}$, raising its temperature by 5°C. What is the final length of the bar?

A. The correct answer is 1.00006 meters.

   1. Use this equation:

$$\Delta L = \alpha L_0 \Delta T$$

   2. Plug in the numbers:

$$\Delta L = \alpha L_0 \Delta T = \left(1.2 \times 10^{-5}°C^{-1}\right)\left(1.0 \text{ m}\right)\left(5°C\right) = 6.0 \times 10^{-5} \text{ m} = 0.00006 \text{ m}$$

   3. The final length is

$$L = L_0 + \Delta L = 1.00006 \text{ m}$$

Practice Questions

   1. You're heating a 1.0-meter aluminum bar, coefficient of linear expansion $2.3 \times 10^{-5}°C^{-1}$, raising its temperature by 100°C. What is the final length of the bar?

   2. You're heating a 2.0-meter gold bar, coefficient of linear expansion $1.4 \times 10^{-5}°C^{-1}$, raising its temperature by 200°C. What is the final length of the bar?

   3. You're heating a 1.5-meter copper bar, coefficient of linear expansion $1.7 \times 10^{-5}°C^{-1}$, raising its temperature by 300°C. What is the final length of the bar?

Practice Answers

1. **1.0023 m.** Use this equation:

$$\Delta L = \alpha L_0 \Delta T$$

Plug in the numbers:

$$\Delta L = \alpha L_0 \Delta T = \left(2.3 \times 10^{-5}{}^\circ C^{-5}\right)(1.0 \text{ m})(100^\circ C) = 2.3 \times 10^{-3} \text{ m}$$

The final length is

$$L = L_0 + \Delta L = 1.0023 \text{ m}$$

2. **2.0056 m.** Use this equation:

$$\Delta L = \alpha L_0 \Delta T$$

Plug in the numbers:

$$\Delta L = \alpha L_i \Delta T = \left(1.4 \times 10^{-5}{}^\circ C^{-5}\right)(2.0 \text{ m})(200^\circ C) = 5.6 \times 10^{-3} \text{ m}$$

The final length is

$$L = L_i + \Delta L = 2.0056 \text{ m}$$

3. **1.5077 m.** Use this equation:

$$\Delta L = \alpha L_0 \Delta T$$

Plug in the numbers:

$$\Delta L = \alpha L_0 \Delta T = \left(1.7 \times 10^{-5}{}^\circ C^{-5}\right)(1.5 \text{ m})(300^\circ C) = 7.7 \times 10^{-3} \text{ m}$$

The final length is

$$L = L_0 + \Delta L = 1.5077 \text{ m}$$

Volume expansion: Taking up more space

Linear expansion, as the name indicates, takes place in one dimension, but the world comes with three dimensions. If an object undergoes a small temperature change of just a few degrees, you can say that the volume of the solid or liquid will change in a way proportionate to the temperature change. As long as the temperature differences involved are small, the fraction by which the solid expands, $\Delta V/V_0$, is proportional to the change in temperature, ΔT (where ΔV represents the change in volume and V_0 represents the original volume).

With volume expansion, the constant involved is called the *coefficient of volume expansion.* This constant is given by the symbol β, and, like α, it's often measured in $°C^{-1}$. Using β, here's how you can express the equation for volume expansion:

$$\frac{\Delta V}{V_0} = \beta \Delta T$$

Remember: When you solve for ΔV, you get the volume-expansion equation in standard form:

$$\Delta V = \beta V_0 \Delta T$$

You've created the analog (or equivalent) of the equation $\Delta L = \alpha L_0 \Delta T$ for linear expansion (see the earlier section "Linear expansion: Getting longer").

Tip: If the lengths and temperature changes are small, you find that $\beta = 3\alpha$ for most solids. This makes sense, because you go from one dimension to three. For example, for steel, α is $1.2 \times 10^{-5}°C^{-1}$ and β is $3.6 \times 10^{-5}°C^{-1}$. Liquids also undergo linear volume expansion, but the preceding relation between β and α doesn't apply generally.

Example

Q. The foreman of a gasoline refinery notices that his workers are filling the radiators of the trucks full up to the brim. Each radiator holds 15 quarts of coolant, which is 1.4×10^{-2} cubic meters, and has a 1-quart coolant reservoir, which is 9.5×10^{-4} cubic meters. The β for the coolant is $\beta = 4.1 \times 10^{-4}°C^{-1}$. The radiator is made of copper so $\beta = 5.1 \times 10^{-5}°C^{-1}$. The reservoir is made of plastic and has a negligible β. The radiator starts at $20°C$ and heats up to its working temperature of $92°C$. Will the radiator overflow? Will a 1-quart $\left(9.5 \times 10^{-4} \text{ m}^3\right)$ coolant reservoir be enough to catch the overflow?

A. The correct answer is yes, it will overflow and, yes, the reservoir is large enough.

1. Use this formula for the expansion of the coolant:

$$\Delta V_c = \beta_c V_{0c} \Delta T$$

where the change in temperature is $\Delta T = 92°C - 20°C = 72°C$.

2. Plug in the numbers:
$$\Delta V_c = \beta_c V_{0c} \Delta T = 4.2 \times 10^{-4} \text{ m}^3$$

3. The radiator will also expand, meaning it can hold more coolant.

4. Because the radiator is made of copper, it'll expand as though it were made of solid copper. Here's the change in volume for the radiator:

$$\Delta V_r = \beta_r V_{0r} \Delta T$$

5. Plug in your numbers:

$$\Delta V_r = \beta_r V_{0r} \Delta T$$
$$= \left(5.1 \times 10^{-5}{}^\circ C^{-1}\right)\left(1.4 \times 10^{-2} \text{ m}^3\right)\left(72^\circ C\right)$$
$$\approx 5.2 \times 10^{-5} \text{ m}^3$$

6. The total overflow is equal to the coolant expansion minus the amount the radiator expands, so plug in your numbers:

$$\Delta V = \Delta V_c - \Delta V_r$$
$$= 4.2 \times 10^{-4} \text{ m}^3 - 5.2 \times 10^{-5} \text{ m}^3$$
$$\approx 3.7 \times 10^{-4} \text{ m}^3$$

So each radiator will overflow a little more than a third of a quart, and the overflow reservoir is 1 quart in volume.

Practice Questions

1. You're heating a 2.0-cubed-meter aluminum block, coefficient of volume expansion $6.9 \times 10^{-5}{}^\circ C^{-1}$, raising its temperature by $30^\circ C$. What is the final volume of the block?

2. You're heating a 2.0-cubed-meter copper block, coefficient of volume expansion $5.1 \times 10^{-5}{}^\circ C^{-1}$, raising its temperature by $20^\circ C$. What is the final volume of the block?

3. You're heating a 1.0-cubed-meter glass block, coefficient of volume expansion $1.0 \times 10^{-5}{}^\circ C^{-1}$, raising its temperature by $27^\circ C$. What is the final volume of the block?

Practice Answers

1. **2.0041 m³.** Use this equation:

$$\Delta V = \beta V_0 \Delta T$$

Plug in the numbers:

$$\Delta V = \beta V_0 \Delta T = \left(6.9 \times 10^{-5}{}^\circ C^{-1}\right)\left(2.0 \text{ m}^3\right)\left(30^\circ C\right) = 0.0041 \text{ m}^3$$

The final volume is

$$V = V_0 + \Delta V = 2.0041 \text{ m}^3$$

2. **2.0020 m³.** Use this equation:

$$\Delta V = \beta V_0 \Delta T$$

Plug in the numbers:

$$\Delta V = \beta V_0 \,\Delta T = \left(5.1 \times 10^{-5}{}^{\circ}C^{-1}\right)\left(2.0 \text{ m}^3\right)\left(20^{\circ}C\right) = 0.0020 \text{ m}^3$$

The final volume is

$$V = V_0 + \Delta V = 2.0020 \text{ m}^3$$

3.   **1.00027 m³.** Use this equation:

$$\Delta V = \beta V_0 \,\Delta T$$

Plug in the numbers:

$$\Delta V = \beta V_0 \,\Delta T = \left(1.0 \times 10^{-5}{}^{\circ}C^{-1}\right)\left(1.0 \text{ m}^3\right)\left(27^{\circ}C\right) = 0.00027 \text{ m}^3$$

The final volume is

$$V = V_0 + \Delta V = 1.00027 \text{ m}^3$$

HEAT: GOING WITH THE FLOW (OF THERMAL ENERGY)

What, really, is heat? When you touch a hot object, heat flows from the object to you, and your nerves record that fact. When you touch a cold object, heat flows from you to that object, and again, your nerves keep track of what's happening. Your nerves record why objects feel hot or cold — because heat flows from them to you or from you to them.

To understand heat, you need to understand thermal energy. *Thermal energy* is the energy that a body has in the vibrations of its molecules — the energy stored in the internal molecular motion of an object. The temperature of a body usually increases with its thermal energy.

When two bodies are brought into thermal contact, thermal energy is free to be exchanged between them. If no thermal energy flows between them, they are in *thermal equilibrium*. In other words, they are in a kind of balance. Two objects in thermal equilibrium are said to have the same *temperature*. If thermal energy *does* flow between them — an object at a higher temperature is in thermal contact with an object at a lower temperature and the thermal energy flows from the hotter body to the cooler one — they're not in thermal equilibrium.

Remember: In physics terms, *heat* is thermal energy that flows from objects of higher temperatures to objects of lower temperatures. The unit of this energy in the MKS system is the *joule* (J) — the same unit you use for other forms of energy and work (see Chapter 9).

This section covers heat and how the change in energy affects temperature. We also discuss phase changes, special cases in which a substance can absorb heat without changing temperature.

Getting specific with temperature changes

At a given temperature, different materials can hold different amounts of thermal energy. For instance, if you warm up a potato, it can hold its heat longer (as your tongue can testify) than a lighter material such as cotton candy. Why? Because the potato stores more thermal energy for a given change in temperature; therefore, more heat has to flow to cool the potato than is needed to cool the cotton candy. The measure of how much heat an object of a given mass can hold at a given temperature is called its *specific heat capacity.*

Suppose that you see someone making a pot of coffee. You measure exactly 1.0 kilogram of brewed coffee in the pot, and then you get down to the real measurements. You find out that you need 4,186 joules of heat energy to raise the temperature of the coffee by 1°C, but you need only 840 joules to raise 1.0 kilogram of glass by 1°C; the coffee and glass have different specific heat capacities. The energy goes into the substance being heated, which stores the energy as internal energy until it leaks out again. (***Note:*** If you need 4,186 joules to raise 1.0 kilogram of coffee by 1°C, you need double that, 8,372 joules, to raise 2.0 kilograms of coffee by 1°C or to raise 1.0 kilogram of coffee by 2°C.)

Remember: The following equation relates the amount of heat needed to raise an object's temperature to the change in temperature and the amount of mass involved:

$$Q = cm\Delta T$$

Here, Q is the amount of heat energy involved (measured in joules if you're using the MKS system), m is the mass, ΔT is the change in temperature, and c is a constant called the *specific heat capacity,* which is measured in joules per kilogram-degree Celsius, or J/(kg·°C). In Chapter 16, you can find a calculation of the specific heat capacity for the special case of an ideal gas, but usually physicists calculate specific heat capacity through experiment, so most problems give you c or refer you to a table of specific-heat values for various materials.

E

Example

Q. You're heating a 1.0-kilogram copper block, specific heat capacity of 387 joules per kilogram-degree Celsius, raising its temperature by 45°C. What amount of heat do you have to apply?

A. The correct answer is 17,400 joules.

   1. Use this equation:

$$Q = cm\Delta T$$

   2. Plug in the numbers:

$$Q = cm\Delta T = \left(387 \text{ J/}\left[\text{kg}\cdot°\text{C}\right]\right)\left(1.0 \text{ kg}\right)\left(45°\text{C}\right) = 17,400 \text{ J}$$

Practice Questions

1. You're heating a 15.0-kilogram copper block, specific heat capacity of 387 joules per kilogram-degree Celsius, raising its temperature by 100°C. What heat do you have to apply?

2. You're heating a 10.0-kilogram steel block, specific heat capacity of 562 joules per kilogram-degree Celsius, raising its temperature by 170°C. What heat do you have to apply?

3. You're heating a 3.0-kilogram glass block, specific heat capacity of 840 joules per kilogram-degree Celsius, raising its temperature by 60°C. What heat do you have to apply?

4. You're cooling a 80.0-kilogram glass block, specific heat capacity of 840 joules per kilogram-degree Celsius, lowering its temperature by 16°C. What heat do you have to extract?

5. You put 7,600 joules into a 14-kilogram block of silver, specific heat capacity of 235 joules per kilogram-degree Celsius. How much have you raised its temperature?

6. You add 10,000 joules into a 8.0-kilogram block of copper, specific heat capacity of 387 joules per kilogram-degree Celsius. How much have you raised its temperature?

Practice Answers

1. **5.8 × 10⁵ J.** Use this equation:

$$Q = cm\Delta T$$

Plug in the numbers:

$$Q = cm\Delta T = \left(387 \text{ J}/\left[\text{kg}\cdot{}^\circ\text{C}\right]\right)(15.0 \text{ kg})(100^\circ\text{C}) = 5.8\times10^5 \text{ J}$$

2. **9.6 × 10⁵ J.** Use this equation:

$$Q = cm\Delta T$$

Plug in the numbers:

$$Q = cm\Delta T = \left(562 \text{ J}/\left[\text{kg}\cdot{}^\circ\text{C}\right]\right)(10.0 \text{ kg})(170^\circ\text{C}) = 9.6\times10^5 \text{ J}$$

3. **1.5 × 10⁵ J.** Use this equation:

$$Q = cm\Delta T$$

Plug in the numbers:

$$Q = cm\Delta T = \left(840 \text{ J}/\left[\text{kg}\cdot°\text{C}\right]\right)\left(3.0 \text{ kg}\right)\left(60°\text{C}\right) = 1.5\times10^5 \text{ J}$$

4. **1.1 × 10⁶ J.** Use this equation:

$$Q = cm\Delta T$$

Plug in the numbers:

$$Q = cm\Delta T = \left(840 \text{ J}/\left[\text{kg}\cdot°\text{C}\right]\right)\left(80.0 \text{ kg}\right)\left(-16°\text{C}\right) = -1.1\times10^6 \text{ J}$$

So the heat extracted is 1.1 × 10⁶ joules.

5. **2.3°C.** Use this equation:

$$Q = cm\Delta T$$

Solve for ΔT:

$$\frac{Q}{cm} = \Delta T$$

Plug in the numbers:

$$\frac{Q}{cm} = \Delta T = \frac{7,600 \text{ J}}{\left(235 \text{ J}/(\text{kg}\cdot°\text{C})\right)\left(14 \text{ kg}\right)} = 2.3°\text{C}$$

6. **3.2°C.** Use this equation:

$$Q = cm\Delta T$$

Solve for ΔT:

$$\frac{Q}{cm} = \Delta T$$

Plug in the numbers:

$$\frac{Q}{cm} = \Delta T = \frac{10,000 \text{ J}}{\left(387 \text{ J}/(\text{kg}\cdot°\text{C})\right)\left(8.0 \text{ kg}\right)} = 3.2°\text{C}$$

Just a new phase: Adding heat without changing temperature

Remember: *Phase changes* occur when materials change state, going from liquid to solid (as when water freezes), solid to liquid (as when rocks melt into lava), liquid to gas (as when you boil water for tea), and so on. When the material in question changes to a new state — liquid, solid, or gas (you can also factor in a fourth state: plasma, a superheated gas-like state) — some heat goes into or comes out of the process without changing the temperature.

You can even have solids that turn directly into gas. As dry ice (frozen dioxide gas) gets warmer, it turns into carbon dioxide gas. This process is called *sublimation*.

Imagine that you're calmly drinking your lemonade at an outdoor garden party. You grab some ice to cool your lemonade, and the mixture in your glass is now half ice, half lemonade (which you can assume has the same specific heat as water), with a temperature of exactly 0°C.

As you hold the glass and watch the action, the ice begins to melt — but the contents of the glass don't change temperature. Why? The heat (thermal energy) going into the glass from the outside air is melting the ice, not warming the mixture up. So does this make the equation for heat energy $\left(Q = cm\Delta T\right)$ useless? Not at all — it just means that the equation doesn't apply for a phase change.

In this section, you see how heat affects temperature before, during, and after phase changes.

Breaking the ice with phase-change graphs

If you graph the heat added to a system versus the system's temperature, the graph usually slopes upward; adding heat increases temperature. However, the graph levels out during phase changes, because on a molecular level, making a substance change state requires energy. After all the material has changed state, the temperature can rise again.

Imagine that someone has taken a bag of ice and thoughtlessly put it on the stove. Before it hit the stove, the ice was at a temperature below freezing (–5°C), but being on the stove is about to change that. You can see the change taking place in graph form in Figure 14-2.

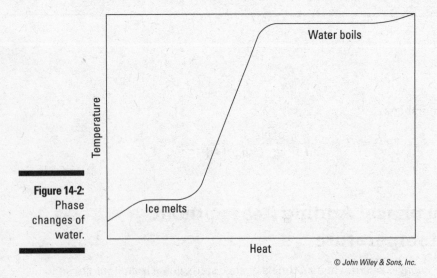

Figure 14-2:
Phase changes of water.

Water boils

Ice melts

Temperature

Heat

As long as no phase change takes place, the equation $Q = cm\Delta T$ holds (the specific heat capacity of ice is around 2.0×10^{-3} J/kg·°C), which means that the temperature of the ice will increase linearly as you add more heat to it, as you see in the graph.

However, when the ice reaches 0°C, the ice is getting too warm to hold its solid state, and it begins to melt, undergoing a phase change. When you melt ice, breaking up the crystalline ice structure requires energy, and the energy needed to melt the ice is supplied as heat. That's why the graph in Figure 14-2 levels off in the middle — the ice is melting. You need heat to make the ice change phase to water, so even though the stove adds heat, the temperature of the ice doesn't change as it melts.

As you watch the bag of ice on the stove, however, you note that all the ice eventually melts into water. Because the stove is still adding heat, the temperature begins to rise, which you see in Figure 14-2. The stove adds more and more heat to the water, and in time, the water starts to bubble. "Aha," you think. "Another phase change." And you're right: The water is boiling and becoming steam. The bag holding the ice seems pretty resilient, and it expands while the water turns to steam.

You measure the temperature of the water. Fascinating — although the water boils, turning into steam, the temperature doesn't change. Once again, you need to add heat to incite a phase change — this time from water to steam. You can see in Figure 14-2 that as you add heat, the water boils, but the temperature of that water doesn't change.

What's going to happen next, as the bag swells to an enormous volume? You never get to find out, because the bag finally explodes. You pick up a few shreds of the bag and examine them closely. How can you account for the heat that's needed to change the state of an object? How can you add something to the equation for heat energy to take into account phase changes? That's where the idea of latent heat comes in.

Understanding latent heat

Latent heat is the heat per kilogram that you have to add or remove to make an object change its state; in other words, latent heat is the heat needed to make a phase change happen. Its units are joules per kilogram (J/kg) in the MKS system.

Remember: Physicists recognize three types of latent heat, corresponding to the changes of phase between solid, liquid, and gas:

- **The latent heat of fusion, L_f:** The heat per kilogram needed to make the change between the solid and liquid phases, as when water turns to ice or ice turns to water

- **The latent heat of vaporization, L_v:** The heat per kilogram needed to make the change between the liquid and gas phases, as when water boils or when steam condenses into water

- **The latent heat of sublimation, L_s:** The heat per kilogram needed to make the change between the solid and gas phases, as when dry ice evaporates

Water's latent heat of fusion of water, L_f, is 3.35×10^5 joules per kilogram, and its latent heat of vaporization, L_v, is 2.26×10^6 joules per kilogram. In other words, you need 3.35×10^5 joules to melt 1 kilogram of ice at 0°C (just to melt it, not to change its temperature). And you need 2.26×10^6 joules to boil 1 kilogram of water into steam.

Remember: Here's the formula for heat transfer during phase changes, where ΔQ is the change in heat, m is the mass, and L is the latent heat:

$$\Delta Q = mL$$

Here, L takes the place of the ΔT (change in temperature) and c (specific heat capacity) terms in the temperature-change formula.

Example

Q. You have a glass of 50.0 grams of water at room temperature, 25°C, but you'd prefer ice water at 0°C. How much ice at 0.0°C do you need to add?

A. The correct answer is 15.6 grams.

1. The heat absorbed by the melting ice must equal the heat lost by the water you want to cool. Here's the heat lost by the water you're cooling:

$$\Delta Q_{water} = mc\Delta T = mc\left(T_f - T_0\right)$$

2. Plug in the numbers:

$$\begin{aligned}
\Delta Q_{water} &= mc\Delta T \\
&= mc\left(T_f - T_0\right) \\
&= (0.050 \text{ kg})(4{,}186 \text{ J/kg°C})(0°C - 25°C) \\
&= -5.23 \times 10^3 \text{ J}
\end{aligned}$$

3. So the water needs to lose 5.23×10^3 joules. How much ice would that melt? That looks like this, where L_m is the latent heat of melting:

$$\Delta Q_{ice} = m_{ice}L_m$$

4. You know that for water, L_m is 3.35×10^5 joules per kilogram, so you get this:

$$\Delta Q_{ice} = m_{ice}L_m = m_{ice}\left(3.35 \times 10^5 \text{ J/kg}\right)$$

5. You know that equation has to be equal to the heat lost by the water, so you can set it to

$$\Delta Q_{ice} = \Delta Q_{water}$$

In other words:

$$m_{ice} = \frac{\Delta Q_{water}}{L_m} = \frac{5.23 \times 10^3 \text{ J}}{3.35 \times 10^5 \text{ J/kg}}$$

6. You can now evaluate how much ice needs to be melted:

$$m_{ice} = \frac{5.23 \times 10^3 \text{ J}}{3.35 \times 10^5 \text{ J/kg}} = 1.56 \times 10^{-2} \text{ kg}$$

So you need 1.56×10^{-2} kilograms, or 15.6 grams of ice.

Practice Questions

1. You have 100.0 grams of coffee in your mug at 80°C. How much ice at 0.0°C would it take to cool 100.0 grams of coffee at 80°C to 65°C?

2. You have 200.0 grams of cocoa at 90°C. How much ice at 0.0°C do you have to add to the cocoa (assuming that it has the specific heat capacity of water) to cool it down to 60°C?

Practice Answers

1. **10.3 g.** Calculate how much heat has to be lost by the coffee. Assuming it has the same specific heat capacity as water, that's

$$\Delta Q_{coffee} = mc\Delta T = mc\left(T_f - T_0\right)$$

Plug in the numbers:

$$\begin{aligned} \Delta Q_{coffee} &= mc\Delta T \\ &= mc\left(T_f - T_0\right) \\ &= \left(0.10 \text{ kg}\right)\left(4{,}186 \text{ J/kg°C}\right)\left(65°C - 80°C\right) \\ &= -6.28 \times 10^3 \text{ J} \end{aligned}$$

How much ice do you need to remove 6.28×10^3 joules? In this case, the heat supplied to the ice not only must melt the ice, $\Delta Q_{ice} = m_{ice}L_m$, but also needs to raise the temperature of the water that comes from melting the ice from 0°C to 65°C, so you have to add this:

$$\Delta Q_{ice} = m_{ice}L_m + m_{ice}c\Delta T = m_{ice}L_m + m_{ice}c\left(T_f - T_0\right)$$

This has to be equal to the heat lost by the coffee, so you get

$$6.28 \times 10^3 \text{ J} = m_{ice}[L_m + c\left(T_f - T_0\right)]$$

Solve for m_{ice}:

$$m_{ice} = \frac{6.28 \times 10^3 \text{ J}}{\left[L_m + c\left(T_f - T_0\right)\right]}$$

Plug in the numbers:

$$m_{ice} = \frac{6.28 \times 10^3 \text{ J}}{\left[L_m + c(T_f - T_0) \right]}$$

$$= \frac{6.28 \times 10^3 \text{ J}}{\left[\left(3.35 \times 10^5 \text{ J/kg} \right) + \left(4{,}186 \text{ J/kg°C} \right) \left(65°C - 0°C \right) \right]}$$

$$= 0.0103 \text{ kg}$$

So you need 1.03×10^{-2} kilograms, or 10.3 grams of ice.

2.   **43 g.** Find how much heat has to be lost by the cocoa. Assuming it has the same specific heat capacity as water, that's

$$\Delta Q_{cocoa} = mc\Delta T = mc(T_f - T_0)$$

Plug in the numbers:

$$\Delta Q_{cocoa} = mc\Delta T$$

$$= mc(T_f - T_0)$$

$$= (0.20 \text{ kg})(4{,}186 \text{ J/kg°C})(60°C - 90°C)$$

$$= -2.5 \times 10^4 \text{ J}$$

How much ice do you need to remove 2.5×10^4 joules from the cocoa? In this case, the heat supplied to the ice not only must melt the ice, $\Delta Q_{ice} = m_{ice}L_m$, but also needs to raise the temperature of the water that comes from melting the ice from 0°C to 60°C, so you have to add this:

$$\Delta Q_{ice} = m_{ice}L_m + m_{ice}c\Delta T = m_{ice}L_m + m_{ice}c(T_f - T_0)$$

This has to be equal to the heat lost by the cocoa, so you get

$$2.5 \times 10^4 \text{ J} = m_{ice}[L_m + c(T_f - T_0)]$$

Solve for m_{ice}:

$$m_{ice} = \frac{2.5 \times 10^4 \text{ J}}{\left[L_m + c(T_f - T_0) \right]}$$

Plug in the numbers:

$$m_{ice} = \frac{2.5 \times 10^3 \text{ J}}{\left[L_m + c(T_f - T_0) \right]}$$

$$= \frac{2.5 \times 10^3 \text{ J}}{\left[\left(3.35 \times 10^5 \text{ J/kg} \right) + \left(4{,}186 \text{ J/kg°C} \right) \left(60°C - 0°C \right) \right]}$$

$$= 0.043 \text{ kg}$$

So you need 4.3×10^{-2} kilograms, or 43 grams of ice.

15

HERE, TAKE MY COAT: HOW HEAT IS TRANSFERRED

Heat is the transfer of thermal energy from one point to another (see Chapter 14). You witness the transfer of heat every day. You cook some pasta, and you see currents of water cycling the noodles in the pan. You pick up the pan without a hand towel, and you burn your hand. You look to the sky on a summer day, and you feel your face warming up. You give your coat to your date, and you watch his or her feelings for you warm up (through radiation, of course!).

In this chapter, we discuss the three primary ways in which heat can be transferred. You find out how to predict how quickly pot handles get hot, see why hot fluids rise, and discover how the sun warms the Earth.

CONVECTION: LETTING THE HEAT FLOW

Convection is a means of transferring thermal energy (heat) in a fluid. In *convection,* the flowing fluid carries energy along, mixing with the rest of the fluid and thereby transferring the thermal energy. Through this mixing, thermal energy moves from a higher-temperature region to a lower-temperature region.

Remember: Convection occurs in both liquids and gases, because both liquids and gases are fluids. *Buoyancy,* which is the upward force on the part of the fluid that's less dense than the surrounding fluid, often drives the fluid's motion. Fluids expand when you heat them, changing the fluid's density (refer to Chapter 14 for info on thermal expansion). Cooler, denser regions of fluid tend to sink as warmer, less dense regions rise, causing the fluid to flow.

Figure 15-1 shows a cross-section of a pan of water coming to a boil. The water at the bottom heats up, expands slightly, and then rises in the pan by buoyant forces. The warm fluid carries the thermal energy from the bottom of the pan to the top.

Convection may be natural or forced, and the following subsections give you the story on both.

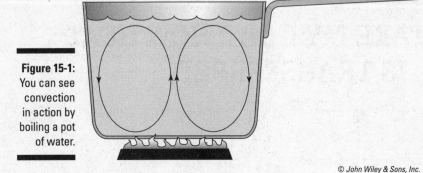

Figure 15-1: You can see convection in action by boiling a pot of water.

Hot fluid rises: Putting fluid in motion with natural convection

You may have heard the maxim "heat rises," which is all about convection. However, a more accurate statement is that "hot fluid rises." In substances where convection is free to take place — that is, in gases and liquids — hotter material naturally ends up on top and cooler material ends up on the bottom because of buoyancy.

If your house has two floors, you often end up with the bottom floor being cooler than the top floor. The warmer air rises by buoyancy, which drives the convection. Physicists refer to this type of convection as *natural convection* because it isn't externally driven.

Tip: To understand how natural convection works, look at the microscopic picture. Any substance is made up of molecules, tiny particles that zip around at varying speeds. When a gas or liquid becomes hot, its molecules move faster. If you have a heating element that contacts the bottom of the substance — such as a wood stove at the bottom of a room or a stove element that's heating a kettle of water — the molecules near the heating element become hot. Hotter molecules have more kinetic energy and so can zip around faster and hit other molecules harder.

Because they move faster and hit harder, hotter molecules make the substance in their immediate area less dense. That is, they have more energy with which to push other molecules out of the way. The molecules that have been hit also have more energy to push other molecules out of the way, so the substance in the immediate vicinity of the heating element becomes less dense.

A unit volume of material that's less dense weighs less than a unit volume of the surrounding material, and if that material is a gas or liquid, the less dense stuff rises. Because the denser material has more mass per volume, it sinks under the influence of gravity.

Controlling the flow with forced convection

With natural convection, you rely on the fact that hot fluid rises to transfer heat. But sometimes natural convection is the opposite of what you want. With *forced convection,* you control the movement of the warm or cool fluid, often using a fan or pump.

For example, take a room on a cold winter's day. Because hot fluids rise, the hotter air in the room drifts up to the ceiling, while the cooler air in the room settles near the bottom of the room, where you are. So in time, all the hot air in the room collects near the ceiling, and all the cold air collects near the floor. Although you were originally quite cozy, you may now be getting pretty cold — all as a result of natural convection.

What can you do? You can turn on your room's ceiling fan in reverse! Ceiling fans force the air to circulate, so the hot air near the top of the room moves downward. The warmer air at the top of the room now ends up at the bottom of the room again, where you are. Just make sure you choose a low speed so you don't create a breeze.

You find forced convection all around you. The fans in a desktop computer, for example, cause forced convection (and the lack of room for a fan in laptop computers has caused plenty of overheating problems). Refrigerators use fans to blow away heat, again relying on forced convection.

Here's another example, this time of forced convection happening twice in the same system. Cars generate a lot of heat when they're running. To keep the engine cool, a pump circulates coolant throughout the engine. The liquid coolant transfers heat away from the engine to the radiator to keep the car from overheating. And the radiator itself is another example of forced convection, moving the air not with a cooling fan but with the motion of the car itself: the car drives air through the radiator, cooling it as the car moves. When the car isn't moving, the engine produces less heat, so there's less need to dissipate heat from the radiator.

TOO HOT TO HANDLE: GETTING IN TOUCH WITH CONDUCTION

Conduction transfers heat through material directly, through contact. Take a look at the metal pot in Figure 15-2 and its metal handle; the pot has been boiling for 15 minutes. Would you want to lift it off the fire by grabbing the handle without an oven mitt? Probably not. The handle is hot because of conduction of heat through the metal handle.

On the molecular level, the molecules near the heat source are heated and begin vibrating faster. They bounce off nearby molecules and cause them to vibrate faster. That increased bouncing is what heats a substance.

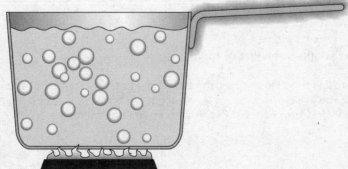

Figure 15-2:
Conduction heats the pot that holds the boiling water.

© *John Wiley & Sons, Inc.*

How the elephant got its ears: A physics lesson in body design

As bodies become larger, their volume grows faster than their surface area. Cooling a larger body becomes more difficult because for every unit of volume of the body, there's less surface area through which the heat can escape. This idea also applies to animals, and it partly explains why the elephant needs such large ears. Because an elephant has such a large body, it has lots of heat to conduct through its body and then from its skin to the air; but relative to its large volume, the elephant doesn't have much surface area through which to conduct the heat. So the elephant sports two great big ears, with a large surface area, through which to conduct away its heat.

Some materials, such as most metals, conduct heat better than others, such as porcelain, wood, or glass. The way substances conduct heat depends a great degree on their molecular structures, so different substances react differently.

Finding the conduction equation

You have to take different properties of objects into account when you want to examine the conduction that takes place. If you have a bar of steel, for example, you have to consider the bar's area and length, along with the temperature at different parts of the bar.

Take a look at Figure 15-3, where a bar of steel is being heated on one end and the heat is traveling by conduction toward the other side. Can you find out the thermal energy transferred? No problem.

Figure 15-3:
Conducting
heat in a bar
of steel.

© John Wiley & Sons, Inc.

Here are the factors that affect the rate of conduction:

- **Temperature difference:** The greater the difference in temperature between the two ends of the bar, the greater the rate of thermal energy transfer, so more heat is transferred. The heat, Q, is proportional to the difference in temperature, ΔT:

$$Q \propto \Delta T$$

- **Cross-sectional area:** A bar twice as wide conducts twice the amount of heat. In general, the amount of heat conducted, *Q,* is proportional to the cross-sectional area, *A,* like this:

$$Q \propto A$$

- **Length (distance heat must travel):** The longer the bar, the less heat that will make it all the way through. Therefore, the conducted heat is inversely proportional to the length of the bar, *l:*

$$Q \propto \frac{1}{l}$$

- **Time:** The amount of heat transferred, *Q,* depends on the amount of time that passes, *t* — twice the time, twice the heat. Here's how you express this idea mathematically:

$$Q \propto t$$

Now you can put the variables together, using *k* as a constant of proportionality that's yet to be determined.

Remember: Here's the equation for heat transfer by conduction through a material:

$$Q = \frac{kA\Delta Tt}{l}$$

This equation represents the amount of heat transferred by conduction in a given amount of time, *t,* down a length *l,* where the cross-sectional area is *A.* Here, *k* is the material's *thermal conductivity,* measured in joules per second-meters-degrees Celsius, or J/(s·m·°C).

Remember: Different materials (such as glass, steel, copper, and bubble gum) conduct heat at different rates, so the thermal conductivity constant depends on the material in question. Lucky for you, physicists have measured the constants for various materials already. Check out some of the values in Table 15-1.

Table 15-1   Thermal Conductivities for Various Materials

Material	*Thermal Conductivity (J/[s·m·°C])*
Diamond	1,600
Silver	420
Copper	390
Brass	110
Lead	35
Steel	14.0
Glass	0.80
Water	0.60
Body fat	0.20

continued

Table 15-1 *Continued*

Material	Thermal Conductivity (J/[s·m·°C])
Wood	0.15
Wool	0.04
Air	0.0256
Styrofoam	0.033

Example

Q. The vacationing Johnson family wants to know whether they have enough ice in their ice chest to last for 16 hours while they're out camping. The outside temperature is 35°C. The walls of the Styrofoam ice chest are 4.0 centimeters thick, and the total surface area of the ice chest is 0.66 square meters. If the Johnsons load the ice chest with 1.5 kilograms of ice at 0°C, will the ice last for 16 hours? (Water's latent heat of fusion is 3.34×10^5 joules per kilogram.)

A. The correct answer is no, the ice will last 7.3 hours.

1. Start with the conduction equation:

$$Q = \frac{kA\Delta Tt}{l}$$

2. Solve the equation for the time, t:

$$t = \frac{Ql}{kA\Delta T}$$

3. You know everything on the right-hand side of the equation except for the heat, Q. Use water's latent heat of fusion to figure out how much heat you need to change the ice from a solid to liquid state (see Chapter 14 for details on heat and phase changes). In general, you need the following amount of heat to melt ice at 0°C:

$$Q = mL$$

where Q is the amount of heat needed, m is the mass of ice, and L is water's latent heat of fusion, 3.34×10^5 joules per kilogram.

4. Plug in the numbers to find the heat:

$$Q = mL = (1.5 \text{ kg})(3.34 \times 10^5 \text{ J/kg}) \approx 5.0 \times 10^5 \text{ J}$$

5. Now you know Q, so you have enough info to use the conduction equation. Plug the numbers in the conduction equation (solved for time) to find the answer:

$$t = \frac{(5.0 \times 10^5 \text{ J})(0.040 \text{ m})}{(0.033 \text{ J/[s·m·°C]})(0.66 \text{ m}^2)(35°C)} \approx 26,200 \text{ s}$$

6. Convert your answer to hours:

$$t = \frac{26{,}200 \ \cancel{s}}{1} \times \frac{1 \ \cancel{min}}{60 \ \cancel{s}} \times \frac{1 \ hr}{60 \ \cancel{min}} \approx 7.3 \ hr$$

The ice will last for only 7.3 hours.

Practice Questions

1. The thermal conductivity of silver is 420 joules per second-meters-degrees Celsius. How much heat is conducted per second between two points on a silver rod, cross section 0.40 square meters, 1.5 meters apart, with a temperature difference of 91°C?

2. If you have a steel rod, thermal conductivity 79 joules per second-meters-degrees Celsius, with a cross section 0.30 square meters and a length of 50 centimeters, how long do you have to wait for 1,000.0 joules to be transferred from one end to the other if the temperature difference is 90°C?

3. If you have a brass rod, thermal conductivity 110 joules per second-meters-degrees Celsius, with a cross section 0.50 square meters and a length of 73 centimeters, how long do you have to wait for 3,000.0 joules to be transferred from one end to the other if the temperature difference is 180°C?

4. You have a mystery substance, length 2.0 meters, cross section 0.40 square meters, and apply a temperature difference of 69°C across its length. If the substance conducts 1,000.0 joules of heat in 0.50 seconds, what is its thermal conductivity?

Practice Answers

1. **10,000 J/s.** Use this equation:

$$Q = \frac{k \cdot A \cdot \Delta T \cdot t}{l}$$

Solve for Q/t:

$$\frac{Q}{t} = \frac{k \cdot A \cdot \Delta T}{l}$$

Plug in the numbers:

$$\frac{Q}{t} = \frac{k \cdot A \cdot \Delta T}{l} = \frac{\left(420 \ J/[s \cdot m \cdot °C]\right)\left(0.40 \ m^2\right)\left(91°C\right)}{1.5 \ m} = 10{,}000 \ J/s$$

2. **0.23 s.** Use this equation:

$$Q = \frac{k \cdot A \cdot \Delta T \cdot t}{l}$$

Solve for t:

$$t = \frac{l \cdot Q}{k \cdot A \cdot \Delta T}$$

Plug in the numbers:

$$t = \frac{l \cdot Q}{k \cdot A \cdot \Delta T} = \frac{(0.5 \text{ m})(1,000.0 \text{ J})}{(79 \text{ J/[s} \cdot \text{m} \cdot {}^\circ\text{C]})(0.30 \text{ m}^2)(90^\circ\text{C})} = 0.23 \text{ s}$$

3. **0.22 s.** Use this equation:

$$Q = \frac{k \cdot A \cdot \Delta T \cdot t}{l}$$

Solve for t:

$$t = \frac{l \cdot Q}{k \cdot A \cdot \Delta T}$$

Plug in the numbers:

$$t = \frac{l \cdot Q}{k \cdot A \cdot \Delta T} = \frac{(0.73 \text{ m})(3,000.0 \text{ J})}{(110 \text{ J/[s} \cdot \text{m} \cdot {}^\circ\text{C]})(0.50 \text{ m}^2)(180^\circ\text{C})} = 0.22 \text{ s}$$

4. **145 J/(s · m · °C).** Use this equation:

$$Q = \frac{k \cdot A \cdot \Delta T \cdot t}{l}$$

Solve for k:

$$k = \frac{l \cdot Q}{t \cdot A \cdot \Delta T}$$

Plug in the numbers:

$$k = \frac{l \cdot Q}{t \cdot A \cdot \Delta T} = \frac{(2.0 \text{ m})(1,000.0 \text{ J})}{(0.5 \text{ s})(0.4 \text{ m}^2)(69^\circ\text{C})} = 145 \text{ J/(s} \cdot \text{m} \cdot {}^\circ\text{C)}$$

Considering conductors and insulators

Materials with high *thermal conductivity,* such as copper, conduct heat well. For example, you may have seen copper wire in indoor/outdoor thermometers. The wires conduct heat in from outside so that the thermometer can measure the outside temperature. Diamond is a far better conductor of heat than copper, as you can see in Table 15-1 (but building indoor/outdoor thermometers with diamonds would be a little pricey).

On the other end of the scale, some materials act as heat *insulators* because their thermal conductivity is so low. For example, body fat has a low thermal conductivity, as you can see in Table 15-1, and so it's a natural insulator. Thus, body fat can help keep you warm on cold days.

Tip: Of course, in terms of conduction, the champion thermal insulator of all time is a vacuum. Conduction relies on the movement of heat through material, and if there's no material, there can't be any conduction. Nothing can conduct the heat in that case.

For that reason, people use vacuum flasks to keep foods hot or cold. These flasks have a double wall with a vacuum between the walls, so no heat can be conducted from inside to outside or outside to inside. Therefore, your soup stays hot, or your iced tea stays cold.

Some thermal conductivity does exist between inside and outside in a vacuum flask. There's always some path for heat to take, such as the stopper of the flask itself. Because some heat is conducted, vacuum flasks keep hot foot hot or cold food cold only for a little while. Theoretically, if you had a capsule of food floating in a vacuum, there'd be no heat loss or gain through conduction at all — but there'd still be heat loss or gain through radiation.

Cool to the touch

Why do metals feel cold to the touch when they're at room temperature? If you know something about thermal conduction, the phenomenon makes sense. Metals are such good thermal conductors that they carry away the heat from your fingers very quickly, which leads to a drop in temperature in your skin. The nerves in your skin detect this drop in temperature and send the message to your brain that you're touching something cold. Wood, on the other hand, is not a good conductor (it's an insulator), so little heat is conducted away from your fingertips. Your brain interprets this to mean that the wood is warmer than the metal, when they're actually both in thermal equilibrium with the room — that is, the same temperature!

RADIATION: RIDING THE (ELECTROMAGNETIC) WAVE

Radiation is another way to transfer heat. You experience radiation personally whenever you get out of the shower soaking wet in the dead of winter and bask in the warmth of the heat lamp in your bathroom. Why? Because of a little physics, of course. The heat lamp, which you see in Figure 15-4, beams out heat to you and keeps you warm through radiation.

With *radiation,* electromagnetic waves carry the energy (you can find plenty of info on electromagnetic waves in *Physics II For Dummies*). Electromagnetic radiation comes from accelerating electric charges. On a molecular level, that's what happens as objects warm up — their molecules move around faster and faster and bounce off other molecules hard.

Heat energy transferred through radiation is as familiar as the light of day; in fact, it *is* the light of day. The sun is a huge thermal reactor about 93 million miles away in space, and neither conduction nor convection can produce any of the energy that arrives to Earth through the vacuum of space. The sun's energy gets to the Earth through radiation, which you can confirm on a sunny day just by standing outside and letting the sun's rays warm your face.

Mutual radiation: Giving and receiving heat

Every object around you is continually radiating, unless its temperature is at absolute zero (which is a little unlikely because you can't physically get to a temperature of absolute zero, with no molecular movement). A scoop of ice cream, for example, radiates. Even you radiate all the time, but that radiation isn't visible as light because it's in the infrared part of the spectrum. However, that light is visible to infrared scopes, as you've probably seen in the movies or on television.

Figure 15-4:
An incandescent light bulb radiates heat into its environment.

© John Wiley & Sons, Inc.

You radiate heat in all directions all the time, and everything in your environment radiates heat back to you. When you have the same temperature as your surroundings, you radiate as fast and as much to your environment as it does to you. When two things are in thermal contact but no thermal energy is exchanged between them, they're in *thermal equilibrium*. If two things are in thermal equilibrium, they have the same temperature.

Remember: If your environment didn't radiate heat back to you, you'd freeze, which is why space is considered so "cold." There's nothing cold to touch in space, and you don't lose heat through conduction or convection. Rather, the environment doesn't radiate back at you, which means that the heat you radiate away is lost. You can freeze very fast from the lost heat.

When an object heats up to about 1,000 kelvins, it starts to glow red (which may explain why, even though you're radiating, you don't glow red in the visible light spectrum). As the object gets hotter, its radiation moves up in the spectrum through orange, yellow, and so on up to white hot at somewhere around 1,700°K (about 2,600°F).

Blackbodies: Absorbing and reflecting radiation

Humans understand heat radiation and absorption in the environment intuitively. For example, on a hot day, you may avoid wearing a black t-shirt, because you know it would make you hotter. A black t-shirt absorbs light from the environment while reflecting less of it back than a white t-shirt. The white t-shirt keeps you cooler because it reflects more radiant heat back to the environment.

Some objects absorb more of the light that hits them than others. Objects that absorb all the radiant energy that strikes them are called *blackbodies*. A blackbody absorbs 100 percent of the radiant energy striking it, and if it's in equilibrium with its surroundings, it emits all the radiant energy as well.

In terms of reflection and absorption of radiation, most objects fall somewhere between mirrors, which reflect almost all light, and blackbodies, which absorb all light. The middle-of-the-road objects absorb some of the light striking them and emit it back into their surroundings. Shiny objects are shiny because they reflect most of the light, which means they don't have to emit as much heat radiantly into the room as other objects. Dark objects appear dark because they don't reflect much light, which means they have to emit more as radiant heat (usually lower down in the spectrum, where the radiation is infrared and can't be seen).

The Stefan-Boltzmann constant

How much heat does a blackbody emit when it's at a certain temperature? The amount of heat radiated is proportional to the time you allow — twice as long, twice as much heat radiated, for example. So you can write the heat relation, where *t* is time, as follows:

$$Q \propto t$$

And as you may expect, the amount of heat radiated is proportional to the total area doing the radiating. So you can also write the relation as follows, where *A* is the area doing the radiating:

$$Q \propto At$$

Temperature, T, has to be in the equation somewhere — the hotter an object, the more heat radiated. Experimentally, physicists found that the amount of heat radiated is proportional to T to the fourth power, T^4. So now you have the following relation:

$$Q \propto AtT^4$$

To show the exact relationship between heat and the other variables, you need to include a constant, which physicists measured experimentally. To find the heat emitted by a blackbody, you use the Stefan-Boltzmann constant, σ, which goes in the equation like this:

$$Q = \sigma AtT^4$$

The value of σ is 5.67×10^{-8} J/(s·m²·K⁴). Note, however, that this constant works only for blackbodies that are perfect emitters.

The Stefan-Boltzmann law of radiation

Most objects aren't perfect emitters, so you have to add another constant most of the time — one that depends on the substance you're working with. The constant is called *emissivity, e*.

The *Stefan-Boltzmann law of radiation* says the following:

$$Q = e\sigma AtT^4$$

where e is an object's emissivity, σ is the Stefan-Boltzmann constant 5.67×10^{-8} J/(s·m²·K⁴), A is the radiating area, t is time, and T is the temperature in kelvins.

Example

Q.   A person has a surface area of 1.8 square meters, an emissivity of 0.65, and a temperature of 37°C. How much power does this person radiate?

A.   The correct answer is 610 watts.

   1.   Use this equation:

$$Q = e\sigma AtT^4$$

   2.   Solve for power, Q/t:

$$\frac{Q}{t} = e \cdot \sigma \cdot A \cdot T^4$$

   3.   Plug in the numbers, remembering to convert from Celsius to Kelvin:

$$\frac{Q}{t} = e \cdot \sigma \cdot A \cdot T^4$$
$$= (0.65)(5.67 \times 10^{-8} \, \text{J/[s·m}^2 \cdot \text{K}^4]\text{)}(1.8 \, \text{m}^2)(37 \, \text{K} + 273 \, \text{K})^4$$
$$= 610 \, \text{W}$$

Practice Questions

1. A coffeepot has a temperature of 90°C and a surface area of 0.20 square meters. If its emissivity is 0.95, what power is it radiating?

2. A toaster has a temperature of 120°C and a surface area of 0.15 square meters. If its emissivity is 0.90, what power is it radiating?

3. If a 200-gram coffeepot holds 300 grams of coffee, its temperature is 80°C, and its surface area is 0.18 square meters, how long will it take for the coffee to cool to 75°C, given that the pot's emissivity is 0.90 (omitting the fact that the pot is heated by radiation coming from its surroundings)?

4. If a 100-gram teacup holds 200 grams of tea, its temperature is 82°C, and its surface area is 0.07 square meters, how long will it take for the tea to cool to 76°C, given that the cup's emissivity is 0.60 and the specific heat capacity of the cup is 840 J/(kg · °C) (omitting the fact that the cup is heated by radiation coming from its surroundings)?

Practice Answers

1. **190 W.** Use this equation:

$$Q = e \cdot \sigma \cdot A \cdot t \cdot T^4$$

Solve for power, which is Q/t:

$$\frac{Q}{t} = e \cdot \sigma \cdot A \cdot T^4$$

Plug in the numbers:

$$\frac{Q}{t} = e \cdot \sigma \cdot A \cdot T^4$$
$$= (0.95)(5.67 \times 10^{-8} \text{ J/[s} \cdot \text{m}^2 \cdot \text{K}^4])(0.2 \text{ m}^2)(90 \text{ K} + 273 \text{ K})^4$$
$$= 190 \text{ J/s} = 190 \text{ W}$$

2. **180 W.** Use this equation:

$$Q = e \cdot \sigma \cdot A \cdot t \cdot T^4$$

Solve for power, which is Q/t:

$$\frac{Q}{t} = e \cdot \sigma \cdot A \cdot T^4$$

Plug in the numbers:

$$\frac{Q}{t} = e \cdot \sigma \cdot A \cdot T^4$$

$$= (0.90)\left(5.67 \times 10^{-8} \text{ J/[s} \cdot \text{m}^2 \cdot \text{K}^4]\right)\left(0.15 \text{ m}^2\right)\left(120 \text{ K} + 273 \text{ K}\right)^4$$

$$= 180 \text{ J/s} = 180 \text{ W}$$

3. **29 s.** Use this equation (which works because the temperature difference involved, 80°C to 75°C, is small):

$$Q = e \cdot \sigma \cdot A \cdot t \cdot T^4$$

Solve for time:

$$t = \frac{Q}{e\sigma A T^4}$$

Solve for the amount of heat you need to lose:

$$\Delta Q = cm\Delta T$$

Plug in the numbers, assuming that coffee has the specific heat of water and ignoring the specific heat of the pot:

$$\Delta Q = cm\Delta T$$

$$= (4{,}186 \text{ J/[kg} \cdot \text{°C])}(0.2 \text{ kg})(80° - 75°)$$

$$= 4{,}200 \text{ J}$$

Put in the numbers to find the time:

$$t = \frac{Q}{e\sigma A T^4}$$

$$= \frac{4{,}200 \text{ J}}{(0.90)\left(5.67 \times 10^{-8} \text{J/[s} \cdot \text{m}^2 \cdot \text{K}^4]\right)(0.18 \text{ m}^2)\left(80 \text{ K} + 273 \text{ K}\right)^4}$$

$$= 29 \text{ s}$$

4. **146 s.** Use this equation (which works because the temperature difference involved, 82°C to 76°C, is small):

$$Q = e \cdot \sigma \cdot A \cdot t \cdot T^4$$

Solve for time:

$$t = \frac{Q}{e\sigma A T^4}$$

Solve for the amount of heat you need to lose:

$$\Delta Q = \text{heat lost by tea} + \text{heat lost by teapot}$$

Plug in the numbers, assuming that tea has the specific heat of water and using the given value for the specific heat capacity of the cup:

$$\Delta Q = c_1 m_1 \Delta T + c_2 m_2 \Delta T$$

$$= (4{,}186 \text{ J/[kg} \cdot {}^{\circ}\text{C])}(0.200 \text{ kg})(82^{\circ} - 76^{\circ}) + (840 \text{ J/[kg} \cdot {}^{\circ}\text{C])}(0.100 \text{ kg})(82^{\circ} - 76^{\circ})$$

$$= 5{,}530 \text{ J}$$

Put in the numbers to find the time:

$$t = \frac{Q}{e\sigma A T^4}$$

$$= \frac{5{,}530}{(0.60)\left(5.67 \times 10^{-8} \text{ J/[s} \cdot \text{m}^2 \cdot \text{K}^4]\right)\left(0.07 \text{ m}^2\right)\left(82 \text{ K} + 273 \text{ K}\right)^4}$$

$$= 146 \text{ s}$$

16

IN THE BEST OF ALL POSSIBLE WORLDS: THE IDEAL GAS LAW

Gas gets just about everywhere — in balloons, in the wind, in your stove, even in your lungs. In physics, an atom-by-atom (or molecule-by-molecule) knowledge of these gases is essential when you start working with heat, pressure, volume, and more.

Get out your stomach pills, because in this chapter, you get gassy! This chapter focuses on the ideal gas law, which explains the relationship among pressure, heat, volume, and the amount of a gas. But first, we introduce you to the mole, a measurement that helps you work with gases on the molecular level.

DIGGING INTO MOLECULES AND MOLES WITH AVOGADRO'S NUMBER

To look at gases on the molecular level, you need to know how many molecules you have in a certain sample. Counting the molecules is impractical, so instead, physicists use a measurement called a *mole* to relate the mass of a sample to the number of molecules it contains.

A *mole* (abbreviated *mol*) is the number of atoms in 12.0 grams of carbon isotope 12. *Carbon isotope 12* — also called carbon-12, or just carbon 12 — is the most common version of carbon. *Isotopes* of an element have the same number of protons but different numbers of neutrons. Carbon-12 has six protons and six neutrons (a total of 12 particles); however, some carbon atoms (isotopes) have a few more neutrons in them — carbon-13, for example, has seven neutrons. The average mass of a mole of a mixture of the carbon isotopes works out to be 12.011 grams.

Remember: The number of atoms in one mole (in 12.0 grams of carbon-12) has been measured as 6.022×10^{23}, which is called *Avogadro's number, N_A*.

Do you find the same number of atoms in, say, 12.0 grams of sulfur? Nope. Each sulfur atom has a different mass from each carbon atom, so even if you have the same number of grams, you have a different number of atoms.

How much more mass does an atom of sulfur have than an atom of carbon-12? If you check the periodic table of elements hanging on the wall in a science lab, you find that the *atomic mass* of sulfur is 32.06. (*Note:* The atomic mass usually appears under the element's symbol.) But 32.06 what? It's 32.06 *atomic mass units*, u, where each atomic mass unit is 1/12 of the mass of a carbon-12 atom.

A mole of carbon-12 (6.022×10^{23} atoms of carbon-12) has a mass of 12.0 grams, and the mass of your average sulfur atom is bigger than the mass of a carbon-12 atom:

Sulfur mass = 32.06 u

Carbon-12 mass = 12 u

Therefore, a mole of sulfur atoms must have this mass:

$$\frac{32.06 \text{ u}}{12 \text{ u}}(12.0 \text{ g}) = 32.06 \text{ g}$$

Tip: How convenient! A mole of an element has the same mass in grams as its atomic mass in atomic units. You can read the atomic mass of any element in atomic units off any periodic table. For instance, you can find that a mole of silicon (atomic mass: 28.09 units) has a mass of 28.09 grams, a mole of sodium (atomic mass: 22.99 units) has a mass of 22.99 grams, and so on. Each of those moles contains 6.022×10^{23} atoms.

Now you can determine the number of atoms in a diamond, which is solid carbon (atomic mass: 12.01 units). A mole is 12.01 grams of diamond, so when you find out how many moles you have, you multiply that times 6.022×10^{23} atoms. Then if you like, you can work out how many atoms of carbon are in a 1-carat diamond: 1 carat is equal to 0.200 grams, so here's how many atoms you have:

$$\frac{0.200 \text{ g}}{12.01 \text{ g}}\left(6.022 \times 10^{23}\right) \approx 1.00 \times 10^{22}$$

Remember: Not every object is made up of a single kind of atom. When atoms combine, you have molecules. For example, water is made up of two hydrogen atoms for every one oxygen atom (H_2O). Instead of the atomic mass, you look for the *molecular mass,* which is also measured in atomic mass units. For example, the molecular mass of water is 18.0153 atomic mass units, so 1 mole of water molecules has a mass of 18.0153 grams.

Some physics problems provide the molecular mass; others require you to calculate molecular mass by using the atomic mass and the compound's molecular formula — that is, you add the atomic masses of the individual atoms in the molecule.

Example

Q. How many molecules are in 10.0 grams of water?

A. The correct answer is 3.3×10^{23} molecules.

1. You know that 1 mole of water has the following mass:

1 mole of water = 18.0 g

2.   A mole has 6.022×10^{23} molecules, so 10.0 grams has this many molecules:

$$6.022 \times 10^{23} \frac{(10.0)}{(18.0)} = 3.3 \times 10^{23}$$

Practice Questions

1.   You have 10.0 grams of calcium, atomic mass 40.08 units. How many atoms do you have?

2.   You have 29.0 grams of zinc, atomic mass 65.41 units. How many atoms do you have?

Practice Answers

1.   **1.5×10^{23} atoms.** You know that 1 mole of calcium has a mass of 40.08 grams, so you have this many moles:

$$n = \frac{10.0}{40.08} = 0.25 \text{ moles}$$

With this many moles, you have this many molecules:

$$N = nN_A = 0.25(6.022 \times 10^{23}) = 1.5 \times 10^{23} \text{ atoms}$$

2.   **2.6×10^{23} atoms.** You know that 1 mole of zinc has a mass of 65.41 grams, so you have this many moles:

$$n = \frac{29.0}{65.41} = 0.44 \text{ moles}$$

With this many moles, you have this many molecules:

$$N = nN_A = 0.44(6.022 \times 10^{23}) = 2.6 \times 10^{23} \text{ atoms}$$

RELATING PRESSURE, VOLUME, AND TEMPERATURE WITH THE IDEAL GAS LAW

When you start working atom-by-atom and molecule-by-molecule, you begin working with gases from a physics point of view. For example, you can relate the temperature, pressure, volume, and number of moles together for a gas. The relation we introduce in this section doesn't always hold true, but it always works for ideal gases.

Remember: *Ideal gases* are those gases for which the ideal gas law holds. This law is an idealized model in which the gas's particles are small compared to the average distance between them and only interact by colliding elastically. It so happens that there is no such "ideal" gas, but real gases best approximate this scenario when the pressure is low and the temperature is high. Ideal gases are very light, like helium.

Forging the ideal gas law

By using the *ideal gas law,* you can predict the pressure of an ideal gas if given how much gas you have, its temperature, and the volume you've enclosed it in. Here's how the various factors affect pressure:

- **Temperature:** Experiments show that if you keep the volume constant and heat a gas, the pressure goes up linearly, as you see in Figure 16-1. In other words, at a constant volume, where T is the temperature measured in kelvins and P is the pressure, the pressure is proportional to temperature:

$$P \propto T$$

- **Volume:** If you let the volume vary, you also find that the pressure is inversely proportional to the volume:

$$P \propto \frac{T}{V}$$

  For instance, if the volume of a gas doubles, its pressure is cut in half.

- **Moles:** When the volume and temperature of an ideal gas are constant, the pressure is proportional to the number of moles of gas you have — twice the amount of gas, twice the pressure (see the earlier section "Digging into Molecules and Moles with Avogadro's Number" for info on moles). If the number of moles is $n,$ then you can say the following:

$$P \propto \frac{nT}{V}$$

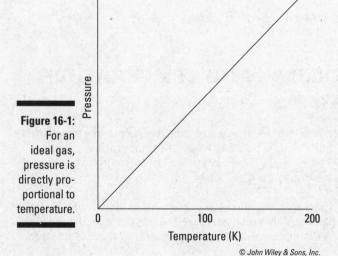

Figure 16-1:
For an ideal gas, pressure is directly proportional to temperature.

© John Wiley & Sons, Inc.

Remember: Adding a constant, R — *the universal gas constant,* which has a value of 8.31 joules/mole-kelvin (J/mol·K) — gives you the *ideal gas law,* which relates pressure, volume, number of moles, and temperature:

$$PV = nRT$$

Here, P is pressure, n is the number of moles of gas you have, R is the universal gas constant, and T is the temperature, measured in kelvins. The volume, V, is measured in cubic meters, m^3. Sometimes, you'll see volume given in liters, where 1.0 liters = 10^{-3} cubic meters. Using this law, you can predict the pressure of a gas, given how much you have of it, its temperature, and the volume you've enclosed it in.

You may come across a special set of conditions when talking about gases — *standard temperature and pressure,* or STP. The standard pressure is 1 atmosphere or 1.013×10^5 newtons per square meter, N/m^2, which is given its own units, pascals, abbreviated Pa.

The standard temperature is 0°C (or 273.15°K).

You can use the ideal gas law to calculate that, at STP, 1.0 mole of an ideal gas occupies 22.4 liters of volume (1.0 liter is 1×10^{-3} cubic meters). How do you get 22.4 liters? You know that $PV = nRT$, and solving for V gives you

$$V = \frac{nRT}{P}$$

Plug in STP conditions and do the math:

$$V = \frac{(1.0)(8.31 \text{ J/[mol·K]})(273.15 \text{ K})}{1.013 \times 10^5 \text{ Pa}} \approx 22.4 \times 10^{-3} \text{ m}^3$$

And that's 22.4 liters, the volume 1 mole of ideal gas occupies at STP conditions.

You can also express the ideal gas law a little differently by using the total number of molecules, N, and Avogadro's number, N_A (see the earlier section "Digging into Molecules and Moles with Avogadro's Number"):

$$PV = nRT = \left(\frac{N}{N_A}\right)RT$$

Tip: The constant R/N_A is also called *Boltzmann's constant, k,* and it has a value of 1.38×10^{-23} joules per kelvin. Using this constant, the ideal gas law becomes

$$PV = NkT$$

The pressure you use in the ideal gas law is the *absolute pressure*. You'll have a hard time measuring the absolute pressure using an ordinary pressure gauge. Usually air is pressing on the outside of the gauge, so what you'll end up measuring is the difference in pressure between the inside and the outside of your pressure gauge. The *gauge pressure* is the difference between the pressure inside a system and the pressure outside (which is usually atmospheric pressure) — precisely what you measure. For example, an ordinary tire gauge measures the difference between the air pressure inside the tire and the air pressure outside the tire.

Example

Q. You have 1.0 mole of air in your tire at 0°C, volume 10.0 liters. What is the gauge pressure of the tire?

A. The correct answer is 1.3×10^5 pascals.

1. Use this equation:

$$PV = nRT$$

2. Solve for P:

$$P = nRT/V$$

3. Plug in the numbers to get the pressure pushing out:

$$P = nRT/V = (1.0)(8.31 \text{ J/[mol·K]})(273 \text{ K})/(10 \times 10^{-3} \text{ m}^3) = 2.3 \times 10^5 \text{ Pa}$$

4. Subtract the pressure from the surrounding air, which pushes in, assuming that the air is at 0°C, too:

$$P = 2.3 \times 10^5 \text{ Pa} - 1.013 \times 10^5 \text{ Pa} = 1.3 \times 10^5 \text{ Pa}$$

Practice Questions

1. You have 2.3 moles of air in your tire at 0°C, volume 12.0 liters. What pressure is the tire inflated to?

2. You have a bottle of 2.0 moles of gas, volume 1.0 liters, temperature 100°C. What is the pressure inside the bottle?

Practice Answers

1. **3.3×10^5 Pa.** Use this equation:

$$PV = nRT$$

Solve for P:

$$P = nRT/V$$

Plug in the numbers:

$$P = nRT/V = (2.3)(8.31 \text{ J/[mol·K]})(273 \text{ K})/(12.0 \times 10^{-3} \text{ m}^3) = 4.3 \times 10^5 \text{ Pa}$$

Subtract the pressure from the surrounding air, which pushes in, assuming that the air is at 0°C, too:

$$P = 4.3 \times 10^5 \text{ Pa} - 1.013 \times 10^5 \text{ Pa} = 3.3 \times 10^5 \text{ Pa}$$

2. **6.2 × 10⁶ Pa.** Use this equation:

$$PV = nRT$$

Solve for P:

$$P = nRT/V$$

Plug in the numbers:

$$P = nRT/V = (2.0)(8.31 \text{ J/[mol·K]})(373 \text{ K})/(1.0 \times 10^{-3} \text{ m}^3) = 6.2 \times 10^6 \text{ Pa}$$

Boyle's and Charles's laws: Alternative expressions of the ideal gas law

You can often express the ideal gas law in different ways. For example, you can express the relationship between the pressure and volume of an ideal gas before and after one of those quantities changes at a constant temperature like this:

$$P_f V_f = P_i V_i$$

This equation, called *Boyle's law*, says that all other factors being the same, the product of pressure and volume (PV) will be conserved.

Remember: At constant pressure, you can say that the following relationship is true for an ideal gas:

$$\frac{V_f}{T_f} = \frac{V_i}{T_i}$$

This equation, called *Charles's law,* says that the ratio of volume to temperature (V/T) will be conserved for an ideal gas, all other factors being the same.

Example

Q. A helium balloon has a volume of 3.2 liters when filled at sea level, where the outside pressure is 1.01×10^5 pascals. What is the volume of the balloon if it rises to an altitude where the pressure outside is only 2.9×10^4 pascals if the temperature doesn't change?

A. The correct answer is 11.1 liters.

1. Use Boyle's law because the temperature of the sample is constant:

$$P_f V_f = P_i V_i$$

2. Solve the formula for the final volume:

$$V_f = \frac{P_i V_i}{P_f}$$

3. Plug in the numbers:

$$V_f = \frac{(1.01 \times 10^5 \, \text{Pa})(3.2 \, \text{L})}{2.9 \times 10^4 \, \text{Pa}}$$

$$V_f = 11.1 \, \text{L}$$

The balloon will expand to 11.1 liters unless it pops first.

Practice Questions

1. An ideal gas occupies 22.4 liters at 1.0 atmosphere of pressure. If temperature is constant, how much volume would it occupy at 2.5 atmospheres of pressure?

2. A sample of an ideal gas is compressed from 3.0 liters to 1.25 liters. If the final pressure of the gas is 3.0×10^5 pascals, what was the initial pressure?

3. A 6.04-liter sample of helium at −56°C is heated until it has a volume of 8.23 liters. Assuming constant pressure, what is the final temperature of the sample?

4. A sample of helium occupies a volume of 2.00 liters at 20.0°C. Assuming constant pressure, at what temperature would it occupy twice as much space?

Practice Answers

1. **9.0 L.** Use Boyle's law: $P_i V_i = P_f V_f$. Solve the equation for V_f:

$$V_f = \frac{P_i V_i}{P_f}$$

Plug in the values from the question:

$$V_f = \frac{P_i V_i}{P_f}$$

$$V_f = \frac{(1.0 \, \text{atm})(22.4 \, \text{L})}{2.5 \, \text{atm}}$$

$$V_f = 9.0 \, \text{L}$$

If the gas occupies 22.4 liters at 1 atmosphere of pressure, it will occupy only 9 liters at 2.5 atmospheres.

2. **1.3×10^5 Pa.** Solve Boyle's law for P_i:

$$P_i V_i = P_f V_f$$

$$P_i = \frac{P_f V_f}{V_i}$$

Substitute the values from the question:

$$P_i = \frac{(3.0 \times 10^5 \text{ Pa})(1.25 \text{ L})}{3.0 \text{ L}}$$
$$P_i = 1.3 \times 10^5 \text{ Pa}$$

The initial pressure was $P_i = 1.3 \times 10^5$ pascals.

3. **23°C.** Use Charles's law because pressure is constant:

$$\frac{V_f}{T_f} = \frac{V_i}{T_i}$$

Solve Charles's law for final temperature, T_f:

$$T_f = \frac{V_f T_i}{V_i}$$

Plug in the numbers, remembering that all temperatures should be in Kelvin:

$$T_f = \frac{(8.23 \text{ L})([-56 + 273.15] \text{ K})}{6.04 \text{ L}}$$
$$T_f = 295.88 \text{ K}$$
$$T_f = (295.88 - 273.15)°C$$
$$T_f = 23°C$$

The final temperature is 23°C.

4. **313°C.** Use Charles's law because pressure is constant:

$$\frac{V_f}{T_f} = \frac{V_i}{T_i}$$

Solve Charles's law for final temperature, T_f:

$$T_f = \frac{V_f T_i}{V_i}$$

Convert degrees Celsius to kelvins to calculate, and then convert back to degrees Celsius to get the answer:

$$T_f = \frac{(4.00 \text{ L})([20.0 + 273.15] \text{ K})}{2.00 \text{ L}}$$
$$T_f = 586.3 \text{ K}$$
$$T_f = (586.3 - 273.15)°C$$
$$T_f = 313°C$$

The trick to this problem is not being fooled into using Charles's law with degrees Celsius. It almost looks like the temperature should be 40°C because double the volume should relate to half the temperature. Unfortunately, that doesn't work, because Celsius isn't an absolute temperature scale like Kelvin. The final temperature is 313°C.

TRACKING IDEAL GAS MOLECULES WITH THE KINETIC ENERGY FORMULA

You can examine certain properties of molecules of an ideal gas as they zip around. For instance, you can calculate the average kinetic energy of each molecule with a very simple equation:

$$KE_{avg} = \frac{3}{2}kT$$

where k is Boltzmann's constant, 1.38×10^{-23} joules per kelvin (J/K), and T is the temperature in kelvins. And because you can determine the mass of each molecule if you know which gas you're dealing with (see the section "Digging into Molecules and Moles with Avogadro's Number," earlier in this chapter), you can figure out the molecules' speeds at various temperatures.

Example

Q. What is the average speed of nitrogen molecules at 28°C?

A. The correct answer is 518 meters per second.

1. Use this equation:

$$KE_{avg} = \frac{3}{2}kT$$

2. Plug in the numbers:

$$KE_{avg} = \frac{3}{2}kT = \frac{3}{2}\left(1.38 \times 10^{-23} \text{ J/K}\right)\left(301 \text{ K}\right) = 6.23 \times 10^{-21} \text{ J}$$

3. You know

$$KE = \frac{1}{2}mv^2$$

4. Solve for v:

$$v = \sqrt{\frac{2KE}{m}}$$

5. Plug in the numbers. Each nitrogen atom has a mass of about 4.65×10^{-26} kilograms:

$$v = \sqrt{\frac{2KE}{m}} = \sqrt{\frac{2(6.23 \times 10^{-21} \text{ J})}{4.65 \times 10^{-26} \text{ kg}}} = 518 \text{ m/s} = 1,160 \text{ mph}$$

Note: Not all nitrogen molecules will have the same speed. Some molecules will have greater kinetic energy, and thus higher speed, and others will have less. The speed given here, 518 meters per second, corresponds to the average kinetic energy.

Practice Questions

1. What is the speed of air molecules at 43°C?

2. What is the speed of air molecules at 900°C?

Practice Answers

1. **530 m/s.** Use this equation:

$$KE_{avg} = \frac{3}{2} kT$$

Plug in the numbers:

$$KE_{avg} = \frac{3}{2} kT = \frac{3}{2} \left(1.38 \times 10^{-23} \text{ J/K}\right)\left(273 \text{ K} + 43 \text{ K}\right) = 6.5 \times 10^{-21} \text{ J}$$

Here's the equation for kinetic energy:

$$KE = \frac{1}{2} mv^2$$

Solve for v:

$$v = \sqrt{\frac{2KE}{m}}$$

Plug in the numbers, assuming the molecular mass of nitrogen, $m = 4.65 \times 10^{-26}$ kilograms (which you can figure out by using Avogadro's number and the mass of 1 mole of nitrogen):

$$v = \sqrt{\frac{2KE}{m}} = \sqrt{\frac{2(6.5 \times 10^{-21} \text{ J})}{4.65 \times 10^{-26} \text{ kg}}} = 530 \text{ m/s}$$

2. **1,020 m/s.** Use this equation:

$$KE_{avg} = \frac{3}{2}kT$$

Plug in the numbers:

$$KE_{avg} = \frac{3}{2}kT = \frac{3}{2}\left(1.38 \times 10^{-23} \text{ J/K}\right)\left(273 \text{ K} + 900 \text{ K}\right) = 2.4 \times 10^{-20} \text{ J}$$

Here's the equation for kinetic energy:

$$KE = \frac{1}{2}mv^2$$

Solve for v:

$$v = \sqrt{\frac{2KE}{m}}$$

Plug in the numbers, assuming the molecular mass of nitrogen, $m = 4.65 \times 10^{-26}$ kilograms (which you can figure out by using Avogadro's number and the mass of 1 mole of nitrogen):

$$v = \sqrt{\frac{2KE}{m}} = \sqrt{\frac{2(2.4 \times 10^{-20} \text{ J})}{4.65 \times 10^{-26} \text{ kg}}} = 1,020 \text{ m/s}$$

17

HEAT AND WORK: THE LAWS
OF THERMODYNAMICS

IN THIS CHAPTER

* Achieving thermal equilibrium
* Storing heat and energy under different conditions
* Revving up heat engines for efficiency
* Dropping close to absolute zero

If you've ever had an outdoor summer job, you know all about heat and work, a relationship encompassed by the term *thermodynamics*. This chapter brings together those two cherished topics, which we cover in detail in Chapter 9 (work) and Chapter 14 (heat).

Thermodynamics has laws one through three, much like Newton, but it does Newton one better: Thermodynamics also has a zeroth law. You may think that odd, because few other sets of everyday objects start off that way ("Watch out for that zeroth step — it's a doozy . . ."), but you know how physicists love their traditions.

In this chapter, we cover thermal equilibrium (the zeroth law), heat and energy conservation (the first law), heat flow (the second law), and absolute zero (the third law). Time to throw the book at thermodynamics.

THERMAL EQUILIBRIUM: GETTING TEMPERATURE WITH THE ZEROTH LAW

Two objects are in *thermal equilibrium* if heat can pass between them but no heat is actually doing so. For example, if you and the swimming pool you're in are at the same temperature, no heat is flowing from you to it or from it to you (although the possibility is there). You're in thermal equilibrium. On the other hand, if you jump into the pool in winter, cracking through the ice covering, you won't be in thermal equilibrium with the water. And you don't want to be. (Don't try this physics experiment at home!)

To check for thermal equilibrium (especially in cases of frozen swimming pools that you're about to jump into), you should use a thermometer. You can check the temperature of the pool with the thermometer and then check your temperature. If the two temperatures agree — in other words, if you're in thermal equilibrium with the thermometer, and the thermometer is in thermal equilibrium with the pool — you're in thermal equilibrium with the pool.

Remember: The *zeroth law of thermodynamics* says that if two objects are in thermal equilibrium with a third, then they're in thermal equilibrium with each other. Then you can say that each of these objects has a thermal property that they all share — this property is called *temperature.*

Among other jobs, the zeroth law sets up the idea of temperature as an indicator of thermal equilibrium. The two objects mentioned in the zeroth law are in equilibrium with a third, giving you what you need to set up a scale such as the Kelvin scale.

CONSERVING ENERGY: THE FIRST LAW OF THERMODYNAMICS

The *first law of thermodynamics* deals with energy conservation. One of the forms of energy involved is the *internal energy* that resides in the motion of the atoms and molecules (vibrations and random jostling). Another of the terms in this law is *heat,* which is a transfer of thermal energy. And finally, there is *work,* which is a transfer of mechanical energy; for example, work is done on a gas when it is compressed. The first law of thermodynamics states that these energies, together, are conserved. The initial internal energy in a system, U_i, changes to a final internal energy, U_f, when heat, Q, is absorbed or released by the system and the system does work, $-W$, on its surroundings (or the surroundings do work on the system), such that

$$U_f - U_i = \Delta U = Q - W$$

For mechanical energy to be conserved (see Chapter 9), you have to work with systems where no energy is lost to heat — there could be no friction, for example. All of that changes now. Now you can break down the total energy of a system, which includes heat, work, and the internal energy of the system.

These three quantities — heat, work, and internal energy — make up all the energy you need to consider. When you add heat, Q, to a system, and that system doesn't do work, the amount of internal energy in the system, which is given by the symbol U, changes by Q. A system can also lose energy by doing work on its surroundings, such as when an engine lifts weight at the end of a cable. When a system does work on its surroundings and gives off no waste heat, its internal energy, U, changes by W. In other words, you're in a position to think in terms of heat as energy, so when you take into account all three quantities — heat, work, and the internal energy — energy is conserved.

The first law of thermodynamics is a powerful one because it ties all the quantities together. If you know two of them, you can find the third.

Calculating with conservation of energy

The most confusing part about using $\Delta U = Q - W$ is figuring out which signs to use. The quantity Q (heat transfer) is positive when the system absorbs heat and negative when the system releases heat. The quantity W (work) is positive when the system does work on its surroundings and negative when the surroundings do work on the system.

Tip: To avoid confusion, don't try to figure out the positive or negative values of every mathematical quantity in the first law of thermodynamics; work from the idea of energy conservation instead. Think of values of work and heat flowing out of the system as negative:

- **The system absorbs heat:** $Q > 0$.
- **The system releases heat:** $Q < 0$.
- **The system does work on the surroundings:** $W > 0$.
- **The surroundings do work on the system:** $W < 0$.
- **The system's energy increases if ΔU is positive.**
- **The system's energy decreases if ΔU is negative.**

Example

Q. Say that a motor does 1,000 joules of work on its surroundings while releasing 3,000 joules of heat. By how much does its internal energy change?

A. The correct answer is –4,000 joules.

1. You know the motor does 1,000 joules of work on its surroundings, so you know its internal energy decreases by 1,000 joules.

2. The motor also releases 3,000 joules of heat while doing its work, so the internal energy of the system decreases by an additional 3,000 joules. Think of negative values of work and heat as energy flowing out of the system, making the total change of internal energy this:

$$\Delta U = -1{,}000 - 3{,}000 = -4{,}000 \text{ J}$$

Practice Questions

1. You have a motor that absorbs 3,000 joules of heat while doing 2,000 joules of work. What is the change in the motor's internal energy?

2. You have a motor that absorbs 2,500 joules of heat while doing 1,700 joules of work. What is the change in the motor's internal energy?

Practice Answers

1. **1,000 J.** Use this equation:

$$U_f - U_i = \Delta U = Q - W$$

Plug in the numbers:

$$\Delta U = 3{,}000 \text{ J [heat coming in]} - 2{,}000 \text{ J [work going out]} = 1{,}000 \text{ J}$$

2. **800 J.** Use this equation:

$$U_f - U_i = \Delta U = Q - W$$

Plug in the numbers:

$$\Delta U = 2{,}500 \text{ J [heat coming in]} - 1{,}700 \text{ J [work going out]} = 800 \text{ J}$$

Working with gases

You come across a number of quantities in this chapter — volume, pressure, temperature, and so on. The ways in which these quantities vary as work is done determine the final state of the system. For example, if a gas is doing work while you keep its temperature constant, the amount of work performed and the intermediate and final states of the system will be different from when you keep the gas's pressure constant instead.

Remember: A gas performs work only if the gas expands. You can show this idea mathematically. First, note that work, W, equals force, F, times distance, s (see Chapter 9):

$$W = Fs$$

In turn, force equals pressure, P, times area, A (see Chapter 8). This means that you can write work as pressure times area times distance:

$$W = PAs$$

Finally, the area times the distance (As) equals the change in volume, ΔV, so here's the new work equation:

$$W = P\Delta V$$

Tip: The formula $W = P\Delta V$ makes graphs of pressure versus volume very useful in thermodynamics. The curve you draw shows how pressure and volume change in relation to each other, and the area under the curve shows how much work is done.

In the next four sections, we cover four standard conditions under which work is performed in thermodynamics: constant pressure, constant volume, constant temperature, and constant heat. We also graph pressure and volume for each of these processes and show you what work looks like. *Note:* When anything changes in these processes, the change is assumed to be *quasi-static,* which means the change comes slowly enough that the pressure and temperature are the same throughout the system's volume.

At constant pressure: Isobaric processes

When you have a process where the pressure stays constant, it's called *isobaric* (*baric* means "pressure"). In Figure 17-1, you see a cylinder with a piston being lifted by a quantity of gas as the gas gets hotter. The volume of the gas is changing, but the weighted piston keeps the pressure constant.

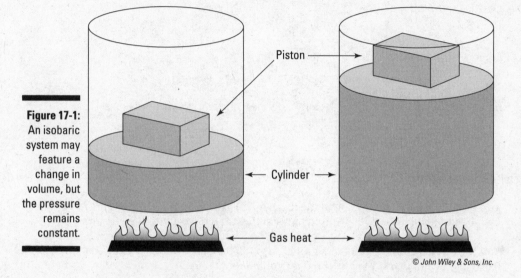

Figure 17-1: An isobaric system may feature a change in volume, but the pressure remains constant.

Piston

Cylinder

Gas heat

© John Wiley & Sons, Inc.

Graphically, you can see what the isobaric process looks like in Figure 17-2, where the volume is changing while the pressure stays constant. Because $W = P\Delta V$, the work is the shaded area beneath the graph.

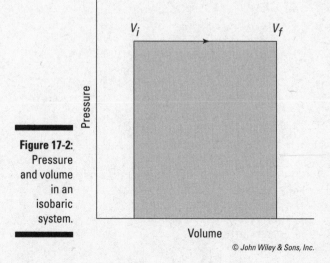

Figure 17-2: Pressure and volume in an isobaric system.

V_i V_f

Pressure

Volume

© John Wiley & Sons, Inc.

Example

Q. You have 60.0 cubic meters of an ideal gas at 200.0 pascals and heat the gas until it expands to a volume of 120 cubic meters. How much work did the gas do?

A. The correct answer is 12,000 joules.

1. Use this equation:

$$W = P\Delta V$$

2. Plug in the numbers:

$$W = P\Delta V = (200.0\ \text{Pa})\ (120\ \text{m}^3 - 60\ \text{m}^3) = 12,000\ \text{J}$$

The gas did 12,000 joules of work as it expanded under constant pressure.

Practice Questions

1. You have 50.0 cubic meters of an ideal gas at 1,000.0 pascals and heat the gas until it expands to a volume of 300.0 cubic meters. How much work did the gas do?

2. You have 100.0 cubic meters of an ideal gas at 300.0 pascals, and it expands, doing 6,000.0 joules of work. What is the final volume of the gas?

Practice Answers

1. **250,000 J.** Use this equation:

$$W = P\Delta V$$

Plug in the numbers:

$$W = P\Delta V = (1,000.0\ \text{Pa})(300.0\ \text{m}^3 - 50.0\ \text{m}^3) = 250,000\ \text{J}$$

2. **120 m³.** Use this equation:

$$W = P\Delta V = P(V_f - V_i)$$

Solve for V_f:

$$V_f = \frac{W}{P} + V_i$$

Plug in the numbers:

$$V_f = \frac{W}{P} + V_i = \frac{6,000.0 \text{ J}}{300.0 \text{ Pa}} + 100.0 \text{ m}^3 = 120 \text{ m}^3$$

At constant volume: Isochoric processes

What if the pressure in a system isn't constant? You may see a simple closed container, which can't change its volume. In this case, the volume is constant, so you have an *isochoric* process.

In Figure 17-3, someone has neglectfully tossed a spray can onto a fire. As the gas inside the spray can heats up, its pressure increases, but its volume stays the same (unless, of course, the can explodes).

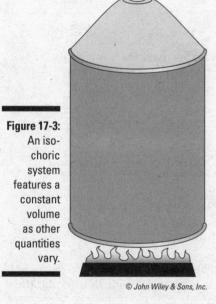

Figure 17-3:
An iso-
choric
system
features a
constant
volume
as other
quantities
vary.

© John Wiley & Sons, Inc.

How much work does the fire do on the spray can? Look at the graph in Figure 17-4. In this case, the volume is constant, so *Fs* (force times distance) equals 0. No work is being done — the area under the graph is 0.

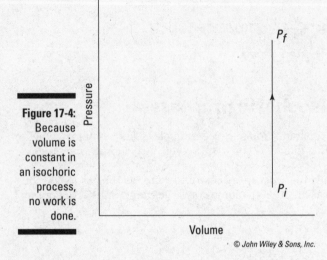

Figure 17-4:
Because
volume is
constant in
an isochoric
process,
no work is
done.

© John Wiley & Sons, Inc.

Example

Q. You have 60.0 cubic meters of an ideal gas at 200.0 pascals and heat the gas until its pressure is 300.0 pascals at the same volume. How much work did the gas do?

A. The correct answer is none.

1. Use this equation:

$$W = P\Delta V$$

2. Because $\Delta V = 0$, no work was done. This process is an isochoric process.

Practice Questions

1. You have 50.0 cubic meters of an ideal gas at 1,000.0 pascals and heat the gas until it has a pressure of 3,000.0 pascals, still at the same volume. How much work did the gas do?

2. You have 300.0 cubic meters of an ideal gas at 1,500.0 pascals and heat the gas until it has a pressure of 6,000.0 pascals. How much work did the gas do if the volume is constant?

Practice Answers

1. **0 J.** Use this equation:

$$W = P\Delta V = P(V_f - V_i)$$

Because no change in volume occurred, this process was isochoric, and no work was done.

2. **0 J.** Use this equation:

$$W = P\Delta V = P(V_f - V_i)$$

Because no change in volume occurred, this process was isochoric, and no work was done.

At constant temperature: Isothermal processes

In an *isothermal system,* the temperature remains constant as other quantities change. Look at the remarkable apparatus in Figure 17-5. It's specially designed to keep the temperature of the enclosed gas constant, even as the piston rises. When you apply heat to this system, the piston rises or lowers slowly in such a way as to keep the product of pressure times volume constant. Because $PV = nRT$ (see Chapter 14), the temperature stays constant as well. (Remember that n is the number of moles of gas that remains constant, and R is the gas constant.)

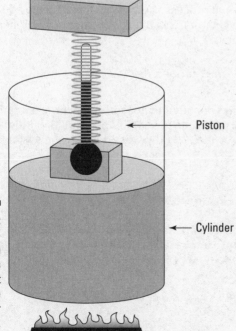

Piston

Cylinder

Figure 17-5: An isothermal system maintains a constant temperature amidst other changes.

© John Wiley & Sons, Inc.

What does the work look like as the volume changes? Because $PV = nRT$, the relation between P and V is

$$P = \frac{nRT}{V}$$

You can see this equation graphed in Figure 17-6, which shows the work done as the shaded area underneath the curve. But what the heck is that area?

Figure 17-6:
The area
under
the curve
shows
the work
done in an
isothermal
process.

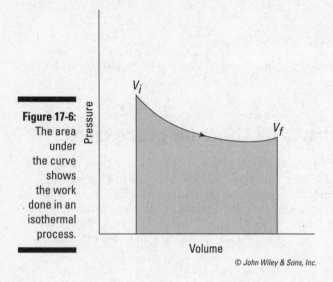

© John Wiley & Sons, Inc.

Figure 17-6: The area under the curve shows the work done in an isothermal process.

Remember: The work done in an isothermal process is given by the following equation, where *ln* is the natural log (*ln* on your calculator), R is the gas constant (8.31 J/mol·K), V_f is the final volume, and V_i is the initial volume:

$$W = nRT \ln\left(\frac{V_f}{V_i}\right)$$

Remember: Because the temperature stays constant in an isothermal process and because the internal energy for an ideal gas equals $(3/2)nRT$ (see Chapter 16), the internal energy doesn't change. Therefore, you find that heat equals the work done by the system:

$$\Delta U = Q - W$$
$$0 = Q - W$$
$$Q = W$$

If you immerse the cylinder you see in Figure 17-5 in a heat bath, what would happen? The heat, Q, would flow into the apparatus, and because the temperature of the gas stays constant, all that heat would become work done by the system.

Example

Q.   You have 1.0 mole of ideal gas at a constant temperature of 20°C, and you expand the gas from $V_i = 0.010$ cubic meters to $V_f = 0.020$ cubic meters. What work did the gas do in expanding?

A.   The correct answer is 1,700 joules.

1.   Use this equation:

$$W = nRT \ln(V_f/V_i)$$

2.   Plug in the numbers:

$$W = nRT \ln(V_f/V_i) = (1.0)(8.31 \text{ J/K})(273.15 \text{ K} + 20 \text{ K}) \ln(0.020/0.010) = 1{,}690 \text{ J}$$

which rounds to 1,700 joules with significant figures.

3.   The work done by the gas was 1,700 joules. The change in the internal energy of the gas is 0 joules, as it must be in isothermal processes. Because $Q = W$, the amount the gas was heated is equal to 1,700 joules.

Practice Questions

1.   You have 1.0 mole of ideal gas at a constant temperature of 30.0°C, and you expand the gas from $V_i = 2.0$ cubic meters to $V_f = 3.0$ cubic meters. What work did the gas do in expanding?

2.   You have 0.60 mole of ideal gas at a constant temperature of 25°C, and you expand the gas from $V_i = 1.0$ cubic meters to $V_f = 3.0$ cubic meters. What heat was supplied to the gas in expanding?

Practice Answers

1.   **1,020 J.** Use this equation:

$$W = nRT \ln(V_f/V_i)$$

Plug in the numbers:

$$W = nRT \ln(V_f/V_i) = (1.0)(8.31 \text{ J/K})(273.15 \text{ K} + 30.0 \text{ K}) \ln(3.0/2.0) = 1{,}020 \text{ J}$$

The work done by the gas was 1,020 joules. The change in the internal energy of the gas is 0 joules, as it must be in isothermal processes. Because $Q = W$, the heat added to the gas is equal to 1,020 joules.

2.   **1,630 J.** Use this equation:

$$W = nRT \ln(V_f/V_i)$$

Plug in the numbers:

$$W = nRT \ln(V_f/V_i) = (0.60)(8.31 \text{ J/K})(273.15 \text{ K} + 25 \text{ K}) \ln(3.0/1.0) = 1,630 \text{ J}$$

The work done by the gas was 1,630 joules. The change in the internal energy of the gas is 0 joules, as it must be in isothermal processes. Because $Q = W$, the heat added to the gas is equal to 1,630 joules.

At constant heat: Adiabatic processes

In an *adiabatic process,* no heat flows from or to the system. Take a look at Figure 17-7, which shows a cylinder surrounded by an insulating material. The insulation prevents heat from flowing into or out of the system, so any change in the system is adiabatic.

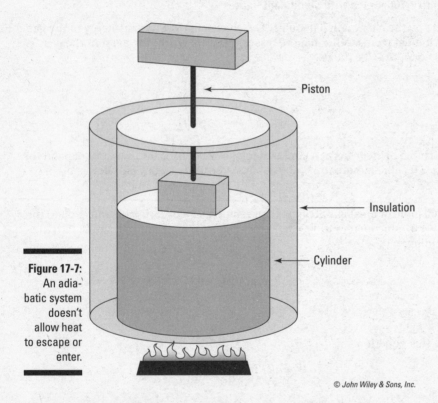

Piston

Insulation

Cylinder

Figure 17-7: An adia-batic system doesn't allow heat to escape or enter.

Examining the work done during an adiabatic process, you can say $Q = 0$, so ΔU (the change in internal energy) equals $-W$. Because the internal energy of an ideal gas is $U = (3/2)nRT$ (see Chapter 14), the work done is the following:

$$W = \frac{3}{2}nR(T_f - T_i)$$

where T_f represents the final temperature and T_i represents the initial temperature. So if the gas does work, that work comes from a change in temperature — if the temperature goes down, the gas does work on its surroundings.

You can see what a graph of pressure versus volume looks like for an adiabatic process in Figure 17-8. The adiabatic curve in this figure, called an *adiabat,* is different from the isothermal curves, called *isotherms.* The work done when no heat flows into or out of the system is the shaded area under the curve.

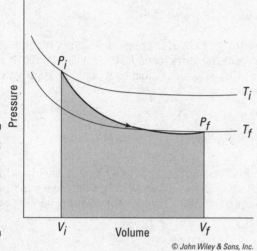

Figure 17-8:
An adiabatic graph of pressure versus volume.

© John Wiley & Sons, Inc.

In adiabatic expansion or compression, you can relate the initial pressure and volume to the final pressure and volume this way:

$$P_i V_i^{\gamma} = P_f V_f^{\gamma}$$

In this equation, γ is the ratio of the specific heat capacity of an ideal gas at constant pressure divided by the specific heat capacity of an ideal gas at constant volume (*specific heat capacity* is the measure of how much heat an object can hold; see Chapter 15):

$$\gamma = \frac{C_P}{C_V}$$

To figure out specific heat capacity, you need to relate heat, Q, and temperature, T. You usually use the formula $Q = cm\Delta T$, where c represents specific heat capacity, m represents the mass, and ΔT represents the change in temperature.

For gases, however, it's easier to talk in terms of *molar specific heat capacity,* which is given by C and whose units are joules per mole-kelvin (J/mol·K). With molar specific heat capacity, you use a number of moles, n, rather than the mass, m:

$$Q = Cn\Delta T$$

To solve for C, you must account for two different quantities, C_P (constant pressure) and C_V (constant volume). Solved for Q, the first law of thermodynamics states that

$$Q = W + \Delta U$$

So if you can get ΔU and W in terms of T, you're set.

First, consider heat at constant volume (Q_V). The work done (W) is $P\Delta V$, so at constant volume, no work is done; $W = 0$, so $Q_V = \Delta U$. And ΔU, the change in internal energy of an ideal gas, is $(3/2)nR\Delta T$ (see Chapter 16). Therefore, Q at constant volume is as follows:

$$Q_V = \Delta U = \frac{3}{2}nR\Delta T$$

Now look at heat at constant pressure (Q_P). At constant pressure, work (W) equals $P\Delta V$. And because $PV = nRT$, you can represent the work as nRT: $W = P\Delta V = nR\Delta T$. At constant pressure, the change in energy, ΔU, is still $(3/2)nR\Delta T$, just as it is at constant volume. Therefore, here's Q at constant pressure:

$$Q_P = W + \Delta U$$
$$= \frac{3}{2}nR\Delta T + nR\Delta T$$

So how do you get the molar specific heat capacities from this? You've decided that $Q = Cn\Delta T$, which relates the heat exchange, Q, to the temperature difference, ΔT, via the molar specific heat capacity, C. This equation holds true for the heat exchange at constant volume, Q_V, so you write

$$Q_V = C_V n\Delta T$$

where C_V is the specific heat capacity at constant volume. You already have an expression for Q_V, so you can substitute into the earlier equation:

$$\frac{3}{2}nR\Delta T = C_V n\Delta T$$

Then you can divide both sides by $n\Delta T$ to get the specific heat capacity at constant volume:

$$C_V = \frac{3}{2}R$$

If you repeat this for the specific heat capacity at constant pressure, you get

$$C_P = \frac{3}{2}R + R = \frac{5}{2}R$$

Now you have the molar specific heat capacities of an ideal gas. The ratio you want, γ, is the ratio of these two equations:

$$\gamma = \frac{C_P}{C_V} = \frac{5}{3}$$

Remember: For an ideal gas, you can connect pressure and volume at any two points along an adiabatic curve this way:

$$P_i V_i^{5/3} = P_f V_f^{5/3}$$

Example

Q. You start with 3.0 liters of gas at 1.0 atmosphere and end up, after an adiabatic change, with 6.0 liters of gas. What is the new pressure?

A. The correct answer is 0.31 atmosphere.

1. Use this equation:

$$P_f = \frac{P_i V_i^{5/3}}{V_f^{5/3}}$$

2. Plug in the numbers:

$$P_f = \frac{P_i V_i^{5/3}}{V_f^{5/3}} = \frac{(1.0 \text{ atm})(3.0 \text{ L})^{5/3}}{(6.0 \text{ L})^{5/3}} = 0.31 \text{ atmosphere}$$

Practice Questions

1. You start with 1.0 liters of gas at 1.0 atmosphere and end up after an adiabatic change with 3.0 liters of gas. What is the new pressure?

2. You have 3.0 moles of ideal gas that undergo an adiabatic change, going from 23°C to 69°C. What work was done on the gas?

Practice Answers

1. **0.16 atmosphere.** Use this equation:

$$P_f = \frac{P_i V_i^{5/3}}{V_f^{5/3}}$$

Plug in the numbers:

$$P_f = \frac{P_i V_i^{5/3}}{V_f^{5/3}} = \frac{(1.0 \text{ atm})(1.0 \text{ L})^{5/3}}{(3.0 \text{ L})^{5/3}} = 0.16 \text{ atmosphere}$$

2. **1,720 J.** Use this equation:

$$W = \frac{3}{2} nR (T_i - T_f)$$

Plug in the numbers:

$$W = \frac{3}{2} nR (T_i - T_f) = (1.5)(3.0)(8.31 \text{ J/K})(23\,°C - 69\,°C) = -1,720 \text{ J}$$

The work done by the gas is –1,720 joules, so the work done on the gas is 1,720 joules.

FLOWING FROM HOT TO COLD: THE SECOND LAW OF THERMODYNAMICS

The *second law of thermodynamics* says that heat flows naturally from an object at a higher temperature to an object at a lower temperature, and heat doesn't flow in the opposite direction of its own accord.

The law is certainly borne out in everyday observation — when was the last time you noticed an object getting colder than its surroundings unless another object was doing some kind of work? You can force heat to flow away from an object when it would naturally flow into it if you do some work — as with refrigerators or air conditioners — but heat doesn't go in that direction by itself.

Heat engines: Putting heat to work

You have many ways to turn heat into work. You may have a steam engine, for example, that has a boiler and a set of pistons, or you may have an atomic reactor that generates superheated steam that can turn a turbine.

Engines that rely on a heat source to do work are called *heat engines;* you can see the principle behind a heat engine in Figure 17-9. A heat source provides heat to the engine, which does work. The waste heat left over goes to a *heat sink,* which effectively has an infinite heat capacity, because it can take such a large amount of heat energy without changing temperature. The heat sink could be the surrounding air, or it could be a water-filled radiator, for example. As long as the heat sink is at a lower temperature than the heat source, the heat engine can do work — at least theoretically.

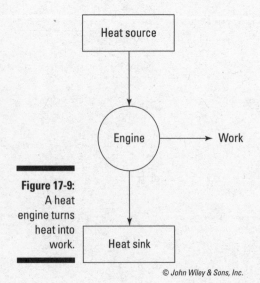

Figure 17-9:
A heat engine turns heat into work.

© *John Wiley & Sons, Inc.*

Heat supplied by a heat source is given the symbol Q_h (for the hot source), and heat sent to a heat sink is given the symbol Q_c (for the cold heat sink). With some calculations, you can find the efficiency of a heat engine. The efficiency is the ratio of the work the engine does, W, to the input amount of heat — the fraction of the input heat that the engine converts to work:

$$\text{Efficiency} = \frac{\text{Work}}{\text{Heat Input}} = \frac{W}{Q_h}$$

If the engine converts all the input heat to work, the efficiency is 1.0. If no input heat is converted to work, the efficiency is 0.0. Often, the efficiency is given as a percentage, so you express these values as 100 percent and 0 percent.

Remember: Because total energy is conserved, the heat into the engine must equal the work done plus the heat sent to the heat sink, which means that $Q_h = W + Q_c$. Therefore, you can rewrite the efficiency in terms of just Q_h and Q_c:

$$\text{Efficiency} = \frac{W}{Q_h} = \frac{Q_h - Q_c}{Q_h} = 1 - \frac{Q_c}{Q_h}$$

Example

Q. Your car is powered by a heat engine and does 3.0×10^7 joules of work getting you up a small hill. If the heat engine is 25 percent efficient, how much heat did it use, and how much did it exhaust?

A. The correct answer is 1.2×10^8 joules, 9×10^7 joules.

1. Use this equation:

$$\text{Efficiency} = \frac{W}{Q_h} = \frac{3.0 \times 10^7 \text{ J}}{Q_h} = 0.25$$

2. Solve for Q_h:

$$Q_h = \frac{3.0 \times 10^7 \text{ J}}{0.25} = 1.2 \times 10^8 \text{ J}$$

So the input heat was 1.2×10^8 joules.

3. Use this equation:

$$Q_h = W + Q_c$$

4. Solve for Q_c:

$$Q_h - W = Q_c$$

5. Plug in the numbers:

$$Q_h - W = 1.2 \times 10^8 \text{ J} - 3.0 \times 10^7 \text{ J} = 9 \times 10^7 \text{ J} = Q_c$$

So the output heat was 9×10^7 joules.

Practice Questions

1. A car running on a heat engine does 7.0×10^7 joules of work. If the heat engine is 26 percent efficient, how much heat did it use, and how much did it exhaust?

2. An 87 percent efficient heat engine does 4.5×10^{10} joules of work. How much heat did it use, and how much did it exhaust?

3. A heat engine does 4.6×10^7 joules of work when supplied 8.9×10^7 joules. What is its efficiency?

4. A heat engine does 8.1×10^7 joules of work when supplied 10.9×10^7 joules. What is its efficiency?

Practice Answers

1. **2.7×10^8J, 2.0×10^8J.** Use this equation:

$$\text{Efficiency} = \frac{W}{Q_h} = \frac{7.0 \times 10^7 \text{ J}}{Q_h} = 0.26$$

Solve for Q_h:

$$Q_h = \frac{7.0 \times 10^7 \text{ J}}{0.26} = 2.7 \times 10^8 \text{ J}$$

So the input heat was 2.7×10^8 joules.

Use this equation:

$$Q_h = W + Q_c$$

Solve for Q_c:

$$Q_h - W = Q_c$$

Plug in the numbers:

$$Q_h - W = 2.7 \times 10^8 \text{ J} - 7.0 \times 10^7 \text{ J} = 2.0 \times 10^8 \text{ J} = Q_c$$

So the output heat was 2.0×10^8 joules.

2. **5.2×10^{10}J, 7×10^9J.** Use this equation:

$$\text{Efficiency} = \frac{W}{Q_h} = \frac{4.5 \times 10^{10} \text{ J}}{Q_h} = 0.87$$

Solve for Q_h:

$$Q_h = \frac{4.5 \times 10^{10} \text{ J}}{0.87} = 5.2 \times 10^{10} \text{ J}$$

So the input heat was 5.2×10^{10} joules.

Use this equation:

$$Q_h = W + Q_c$$

Solve for Q_c:

$$Q_h - W = Q_c$$

Plug in the numbers:

$$Q_h - W = 5.2 \times 10^{10} \text{ J} - 4.5 \times 10^{10} \text{ J} = 7 \times 10^9 \text{ J} = Q_c$$

So the output heat was 7×10^9 joules.

3.   **52 percent.** Use this equation:

$$\text{Efficiency} = \frac{W}{Q_h}$$

Plug in the numbers:

$$\text{Efficiency} = \frac{W}{Q_h} = \frac{4.6 \times 10^7 \text{ J}}{8.9 \times 10^7 \text{ J}} = 52\%$$

4.   **74 percent.** Use this equation:

$$\text{Efficiency} = \frac{W}{Q_h}$$

Plug in the numbers:

$$\text{Efficiency} = \frac{W}{Q_h} = \frac{8.1 \times 10^7 \text{ J}}{10.9 \times 10^7 \text{ J}} = 74\%$$

Limiting efficiency: Carnot says you can't have it all

Given the amount of work a heat engine does and its efficiency, you can calculate how much heat goes in and how much comes out (along with a little help from the law of conservation of energy, which ties work, heat in, and heat out together). But why not create 100 percent efficient heat engines? Converting all the heat that goes into a heat engine into work would be nice, but the real world doesn't work that way. Heat engines inevitably lose energy to their surroundings, such as through friction on the pistons in a steam engine.

Studying this problem, Sadi Carnot (a 19th-century engineer) came to the conclusion that the best you can do, effectively, is to use an engine that has no such losses. If the engine experiences no losses, the system will return to the state it was in before the process took place. This is called a *reversible process.* For example, if a heat engine loses energy overcoming friction, it doesn't have a reversible process, because it doesn't end up in the same state when the process is complete. You have the most efficient heat engine when the engine operates reversibly.

Carnot's principle says that no nonreversible engine can be as efficient as a reversible engine and that all reversible engines that work between the same two temperatures have the same efficiency. Here's the kicker: A perfectly reversible engine doesn't exist, so Carnot came up with an ideal one.

No real engine can operate reversibly, so Carnot imagined a kind of ideal, reversible engine. In the *Carnot engine,* the heat that comes from the heat source is supplied at a constant temperature T_h. Meanwhile, the rejected heat goes into the heat sink, which is at a constant temperature T_c. Because the heat source and the heat sink are always at the same temperature, you can say that the ratio of the heat provided and rejected is the same as the ratio of those temperatures (expressed in kelvins):

$$\frac{Q_c}{Q_h} = \frac{T_c}{T_h}$$

Remember: And because the efficiency of a heat engine is Efficiency $= 1-(Q_c/Q_h)$, the efficiency of a Carnot engine is

$$\text{Efficiency} = 1 - \frac{Q_c}{Q_h} = 1 - \frac{T_c}{T_h}$$

This equation represents the *maximum possible efficiency* of a heat engine. You can't do any better than that. And as the third law of thermodynamics states (see the final section in this chapter), you can't reach absolute zero; therefore, T_c is never 0, so the efficiency is always 1 minus some number. You can never have a 100 percent efficient heat engine.

Example

Q. Assume that you use the surface of the sun (about 5,778 kelvins) as the heat source and interstellar space (about 2.725 kelvins) as the heat sink for a Carnot heat engine. What is the engine's efficiency?

A. The correct answer is 0.9995.

1. Use this equation:

$$\text{Efficiency} = 1 - \frac{T_c}{T_h}$$

2. Plug in the numbers:

$$\text{Efficiency} = 1 - \frac{T_c}{T_h} = 1 - \left(\frac{2.725\ \text{K}}{5,778\ \text{K}}\right) \approx 0.9995$$

Practice Questions

1. The heat source for a Carnot engine is at 87°C, and the heat sink is at 49°C. What is the engine's efficiency?

2. The heat source for a Carnot engine is at 67°C, and the heat sink is at 29°C. What is the engine's efficiency?

Practice Answers

1. **11%.** Use this equation:

$$\text{Efficiency} = 1 - \frac{T_c}{T_h}$$

Plug in the numbers:

$$\text{Efficiency} = 1 - \frac{T_c}{T_h} = 1 - \frac{(273\ \text{K} + 49\ \text{K})}{(273\ \text{K} + 87\ \text{K})} = 11\%$$

2. **11%.** Use this equation:

$$\text{Efficiency} = 1 - \frac{T_c}{T_h}$$

Plug in the numbers:

$$\text{Efficiency} = 1 - \frac{T_c}{T_h} = 1 - \frac{(273\ \text{K} + 29\ \text{K})}{(273\ \text{K} + 67\ \text{K})} = 11\%$$

Going against the flow with heat pumps

Usually, Carnot engines take heat from the hot reservoir (Q_h), do work, and then dump the leftover heat in the cold reservoir (Q_c). But what if you swapped the hot and cold reservoirs and actually did some work on the Carnot engine (instead of having it do work on you)? Then you could "pump heat uphill," from the cold reservoir to the hot reservoir. You can do this if you connect the input of a Carnot engine to the cold reservoir and connect the exhaust to the hot reservoir.

Why on Earth would you want to move heat? Think about a cold room on an even colder day. If you connect a Carnot engine to the outside — which is colder than the inside — doing some work on the Carnot engine can drive heat into the room. This use of a Carnot engine is called a *heat pump,* because you put in work to drive heat uphill, from the cold reservoir to the hot one.

Why are heat pumps a good way to warm your house? Consider heating with electric heat instead. If you use enough electricity to add 1,000 joules to the heat inside your house, you have to pay for 1,000 joules of energy. But if you pump the heat from outside to inside, most of the heat comes from the cold reservoir, and you only have to provide the work needed to pump the heat into the hot reservoir.

A heat pump can be used to move heat in the other direction, too, to bring about cooling. In this case, mechanical work is used to pump heat from a source at a higher temperature to a lower temperature. Your refrigerator uses electrical energy to drive a compressor unit, which forms part of the refrigeration cycle.

Operating heat pumps requires less work than the heat they transfer. So how much work do you need to pump Q_h of heat into the hot reservoir? You can start from this equation:

$$Q_h = W + Q_c$$

Here, Q_c is the heat taken from the cold reservoir. W is the amount of work you need to supply to the heat pump. You're trying to solve for the work, so rearrange the equation:

$$W = Q_h - Q_c$$

For a Carnot engine, $Q_c / Q_h = T_c / T_h$. Therefore, here's the formula for the heat taken from the outside:

$$Q_c = Q_h \left(\frac{T_c}{T_h} \right)$$

Now plug this value of Q_c into the work equation $\left(W = Q_h - Q_c \right)$ and simplify:

$$W = Q_h \left(1 - \frac{T_c}{T_h} \right)$$

The heat delivered to you from a heat pump is more than the work you put into the heat pump. You can measure how much more heat you get out of a heat pump than the work you put in using the *coefficient of performance (COP)*:

$$\text{COP} = \frac{Q_h}{W}$$

The coefficient of performance tells you how much heat you get out of a heat pump per work you have to put in to it.

For something like electric heat, where you have to pay for all the heat you get, the coefficient of performance is 1. But for a heat pump, the coefficient can be a lot higher than 1, indicating that you get more heat out of the pump than the work you put in.

The coefficient of performance depends on the inside and outside temperatures. You can put the coefficient of performance into a form that makes its dependence on temperature explicit.

Because $W = Q_h - Q_c$, the coefficient of performance equation becomes

$$\text{COP} = \frac{Q_h}{W} = \frac{Q_h}{Q_h - Q_c}$$

Or if you multiply both the numerator and denominator by $1/Q_h$, you can express this like so:

$$\text{COP} = \frac{Q_h}{W} = \frac{1}{1 - Q_c/Q_h}$$

For a Carnot engine, $Q_c/Q_h = T_c/T_h$, so you end up with

$$\text{COP} = \frac{1}{1 - Q_c/Q_h} = \frac{1}{1 - T_c/T_h}$$

Example

Q. Suppose that you're vacationing in your woodland cabin, which is at 20°C. You want to pump some heat in from outside, which is at 10°C. You decide you need about 4,000 joules of heat. How much work would you need to pump 4,000 joules into your house?

A. The correct answer is 136 joules.

1. Use this equation:

$$W = Q_h\left(1 - \frac{T_c}{T_h}\right)$$

2. Plug in the numbers (with all temperatures in kelvins)

$$W = 4,000 \text{ J}\left(1 - \frac{283 \text{ K}}{293 \text{ K}}\right) \approx 136 \text{ J}$$

Practice Questions

1. The hot reservoir for a Carnot heat pump is at 27°C, and the cold reservoir is at 0.0°C. What is the heat pump's coefficient of performance?

2. The hot reservoir for a Carnot heat pump is at 37°C, and the cold reservoir is at –23°C. How much heat can you pump using 373 joules of work?

Practice Answers

1. **11.** Use this equation:

$$COP = \frac{1}{1 - T_c / T_h}$$

Plug in the numbers (with all temperatures in kelvins):

$$COP = \frac{1}{1 - T_c / T_h}$$

$$= \frac{1}{1 - 273 \text{ K} / 300 \text{ K}}$$

$$\approx 11$$

2. **1,900 J.** Use this equation:

$$W = Q_h \left(1 - \frac{T_c}{T_h} \right)$$

Solve for the heat Q_h:

$$W = Q_h \left(1 - \frac{T_c}{T_h} \right)$$

$$\frac{W}{\left(1 - T_c / T_h \right)} = Q_h$$

Plug in the numbers (with all temperatures in kelvins):

$$Q_h = \frac{W}{\left(1 - T_c / T_h \right)}$$

$$= \frac{373 \text{ J}}{\left(1 - 250 \text{ K} / 310 \text{ K} \right)}$$

$$\approx 1,900 \text{ J}$$

GOING COLD: THE THIRD (AND ABSOLUTE LAST) LAW OF THERMODYNAMICS

Absolute zero is the lower limit for the temperature of any system, and the third law of thermodynamics can be formulated in terms of this temperature. The *third law of thermodynamics* is pretty straightforward — it just says that you can't reach absolute zero (0 kelvins, or about −273.15°C) through any process that uses a finite number of steps. In other words, you can't get down to absolute zero at all. Each step in the process of lowering an object's temperature to absolute zero can get the temperature a little closer, but you can't get all the way there.

PART V

the part of

TENS

WEB EXTRA *You might be surprised at the people who have made notable contributions to the field of physics. Go to www.dummies.com/extras/ucanphysics1 for an article that lists ten folks who made physics what it is today.*

IN THIS PART . . .

- Find out what the ten most-common physics problem-solving mistakes are.

- Discover ten astonishing physics theories.

18

TEN COMMON MISTAKES PEOPLE MAKE WHEN SOLVING PROBLEMS

IN THIS CHAPTER

- Handling common mistakes
- Knowing how to spot potential problems

This chapter discusses the most common errors people make when working out physics problems. In our many years of teaching physics, certain types of problems stand out, and you see them here.

MIXING UNITS

The most common error made in solving physics problems involves mixing the units from one system with another system. If the problem is given to you in inches, kilograms, and seconds, convert it into a consistent system of units before proceeding to work out the answer. For example, if you want to use the MKS system, convert *everything* into MKS before working out the problem.

EXPRESSING THE ANSWER IN THE WRONG UNITS

If the problem asks for the answer in the MKS system, don't give it in CGS units. You'd be surprised at how common these errors are; people are so relieved that they've solved the problem that they goof it up in the last step. Also, don't forget your units on your final answer.

SWAPPING RADIANS AND DEGREES

Degrees are commonly used in physics problems — except when it comes to angular velocity and acceleration. That's when you have to make sure you're working with radians. Use the $180°/\pi$ conversion factor to convert from radians to degrees when needed.

Tip: If you are using a graphing calculator, make sure that it is set to the correct mode (degrees or radians) for the type of problem that you are solving.

GETTING SINES AND COSINES MIXED UP

Physics students often make the mistake of interchanging sines and cosines. Take a look at Figure 18-1 and keep the following relationships in mind:

$$\sin \theta = y/h = \text{opposite/hypotenuse}$$

$$\cos \theta = x/h = \text{adjacent/hypotenuse}$$

$$\tan \theta = y/x = \text{opposite/adjacent}$$

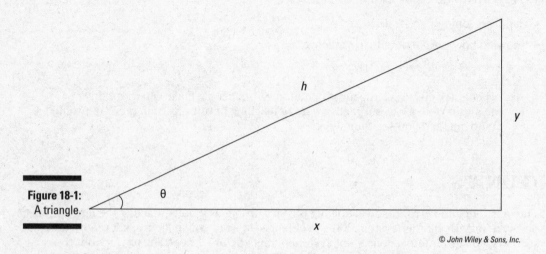

Figure 18-1:
A triangle.

© John Wiley & Sons, Inc.

NOT TREATING VECTORS AS VECTORS

When you add vectors, use vector addition. That means resolving vectors into components. Too many people just add the magnitudes of the vectors without realizing that they should be adding components instead.

MIXING UP THE SIGNS OF THE COMPONENTS OF VECTORS

Before you write down equations involving components of vectors, you should choose a coordinate system by deciding which directions are the positive x and positive y directions. Most of the time, you choose up or north to be the positive y direction and right or east to be the positive x direction. If you're not sure about the direction, you should guess — if you guess wrong, you'll end up with a minus sign.

GETTING THE DIRECTION OF FORCES WRONG

When you are solving a force problem, you can keep track of the directions of the forces by drawing a free-body diagram. The gravitational force exerted by the Earth always points toward the center of the Earth. The normal force is always perpendicular to the surface (not necessarily up!). The frictional force opposes the relative motion of surfaces in contact.

NEGLECTING LATENT HEAT

When you're faced with a problem that involves a phase change, such as from ice to water, don't forget to take the latent heat into account. When ice changes into water, it absorbs latent heat that you have to account for in your solution.

USING THE WRONG TEMPERATURE IN THE IDEAL GAS LAW

When you use the ideal gas law, always convert the temperature to kelvins. Often you are given the temperature in Celsius. To convert from degrees Celsius to kelvins, just add 273.15.

GETTING THE SIGNS WRONG IN THE FIRST LAW OF THERMODYNAMICS

You can figure out the signs in the first law of thermodynamics by thinking about the conservation of energy. The internal energy of the system increases if the system absorbs heat or if the surroundings do work on the system.

19

TEN WILD PHYSICS THEORIES

This chapter gives you ten outside-the-box physics facts that you may not hear or read about in a classroom. As with anything in physics, however, you shouldn't really consider these "facts" as final truth — they're just the current state of many theories. Theories in physics often turn out to be approximations to better theories developed later. And in this chapter, some of the theories get pretty wild, so don't be surprised to see them superseded in the coming years.

HEISENBERG SAYS YOU CAN'T BE CERTAIN

You may have heard of the uncertainty principle, but you may not have known that a physicist named Heisenberg first suggested it. Of course, that explains why it's called the *Heisenberg uncertainty principle,* we suppose. The principle had its beginnings in the wave nature of matter, as Louis de Broglie suggested. Matter is made up of particles, such as electrons. But particles also act as waves, much like light waves — you just don't normally notice it because particles have such small wavelengths.

Particles have wave-like properties, and the more localized the wave is, the more certain you can be about the position of the particle. However, the wavelength of the wave is directly related to the momentum of the particle. The more definite the wavelength of a wave, the more certain you can be about the momentum of the particle. But because of the nature of waves, the more definite the wavelength, the more spread out in space it becomes. That's why the more sure you are of the momentum, the less sure you can be of the position, and vice versa. You can also say that the more precisely you measure their locations, the less precisely you know their momenta.

BLACK HOLES DON'T LET LIGHT OUT

Black holes are created when particularly massive stars use up all their fuel and collapse inwardly to form super-dense objects, much smaller than the original stars. Only very large stars end up as black holes. Stars that aren't quite massive enough to collapse that far often end up as neutron stars instead. A neutron star occurs when gravity has smashed together all the electrons, protons, and neutrons, effectively creating a single mass of neutrons with the density of an atomic nucleus.

Black holes go even further than that. They collapse so far that not even light can escape their intense gravitational pulls. How's that? The photons that make up light aren't supposed to have any mass. How can they possibly be trapped in a black hole?

Photons are indeed affected by gravity, a fact predicted by Einstein's theory of general relativity. Tests have experimentally confirmed that light passing next to massive objects in the universe is bent by their gravitational fields. Gravity affects photons, and the gravitational pull of a black hole is so strong that photons can't escape it.

GRAVITY CURVES SPACE

Isaac Newton gave physicists a great theory of gravitation, and from him came the following famous equation:

$$F = \frac{Gm_1m_2}{r^2}$$

where F represents force, G represents the universal gravitational constant, m_1 represents one mass, m_2 represents another mass, and r represents the distance between the masses (see Chapter 7). Newton was able to show that what made an apple fall also kept the planets in orbit. But Newton had one problem he could never figure out: how gravity could operate instantaneously at a distance.

Enter Einstein, who created the modern take on this problem. Instead of thinking of gravity as a simple force, Einstein suggested in his theory of general relativity that space and time are actually different aspects of a single entity called *spacetime*. Mass and energy curve the spacetime, and this curvature *is* gravity!

What's really happening when a planet orbits the sun is that it's simply following the shortest path through the curved spacetime through which it travels. The mass of the sun curves the spacetime around it, and the planets follow that curvature.

YOU CAN MEASURE A SMALLEST DISTANCE

Physicists now have a theory that a "smallest distance" exists. It's the *Planck length,* named after the physicist Max Planck. The length is the smallest division that, theoretically, you can divide space into. However, the Planck length — about 1.6×10^{-35} meters, or about 10^{-20} times the approximate size of a proton — is really just the smallest amount of length with any physical significance, given the current understanding of the universe. Smaller than this, and the whole notion of distance breaks down.

MATTER AND ANTIMATTER DESTROY EACH OTHER

One of the coolest things about high-energy physics, also called *particle physics,* is the discovery of antimatter. Antimatter is sort of the reverse of matter. The counterparts of electrons are called *positrons* (which are positively charged), and the counterparts of protons are *antiprotons* (which are negatively charged). Even neutrons have an antiparticle: *antineutrons.* A neutron is made up of smaller particles called *quarks* and *gluons,* which have antiparticle versions, too. So the antineutron has no charge just like the neutron, but each of the quarks and gluons it's made of is the anti-version of the quarks and gluons in the neutron.

In physics terms, matter is sort of on the plus side, and antimatter sort of on the negative side. When the two come together, they destroy each other, leaving pure energy — light waves of great energy, called *gamma waves.* And like any other radiant energy, gamma waves can be considered heat energy, so if you have a pound of matter and a pound of antimatter coming together, you'll have quite a bang.

That bang, pound for pound, is much stronger than a standard atomic bomb, where only 0.7 percent of the fissile material is turned into energy. When matter hits antimatter, 100 percent is turned into energy.

If antimatter is the opposite of matter, shouldn't the universe have as much antimatter as it does matter? Observations show that the universe contains very little antimatter. That's a puzzler. High-energy experiments show that the universe treats matter and antimatter a little different. These small differences could explain why the universe is mostly matter today, even if it had the same amount of matter and antimatter in the past.

SUPERNOVAS ARE THE MOST POWERFUL EXPLOSIONS

What's the most energetic action that can happen anywhere, throughout the entire universe? What event releases the most energy? What's the all-time champ when it comes to explosions? Your not-so-friendly neighborhood supernova. A *supernova* occurs when a very massive star explodes. The star's fuel is used up, and its structure is no longer supported by an internal release of energy. At that point, the star collapses in on itself, and if the star is massive enough, the gravitational potential energy that the star had is suddenly released upon the collapse.

Among the 100 billion stars in the Milky Way, the last known supernova occurred nearly 400 years ago. (We say *known* because light takes quite a while to reach Earth; a star could've gone supernova 100 years ago, but if it's far enough from Earth, no one would know it yet.)

Most of the star that becomes a supernova explodes at speeds of about 10,000,000 meters per second, or about 22,300,000 miles per hour. By comparison, even the highest of explosives on Earth detonate at speeds of 1,000 to 10,000 meters per second.

Because the physics of how a star explodes is quite well understood, physicists can observe the apparent brightness of a supernova in a distant galaxy and work out how far away that galaxy must be. This development has led to the most-accurate measurements of the rate of expansion of the universe!

THE UNIVERSE STARTS WITH THE BIG BANG AND ENDS WITH THE GNAB GIB

The first ideas about the large-scale nature of the universe tended to hypothesize that the universe was steady and unchanging and that it had existed for all time and would continue to do so.

Astronomer Edwin Hubble measured the velocities of galaxies and found that they were all moving away from us and that the more distant the galaxy, the faster it was moving away. This could only mean one thing: The universe is expanding. (The best way to imagine this is to think of the galaxies as dots drawn on a balloon that's being inflated. Each of the dots moves away from all the others as the balloon expands, and the wider the separation between the dots, the faster they move away from each other.) This means the universe yesterday was slightly smaller than it is today, and so on and so on backward in time until the universe was all concentrated into a single point! This is the point at which space and time were concentrated into what's called a *singularity.* It's from this singularity, in a single violent event called the *Big Bang,* that space, time, and the universe expanded to what it is today.

Given that the universe was "born" in the Big Bang, this raised the question of whether it might "die." Or if not, what might the ultimate fate of the universe be? Well, Einstein's theory of general relativity is useful here, because it tells how space and time curve with a given distribution of matter and energy. The theory predicts that the ultimate fate of the universe depends on the density of mass and energy in the universe. If enough mass and energy is in the universe, then that mass and energy may cause enough attraction to halt the expansion of the universe and reverse it — bringing the entire universe back to a single point in an event called the Big Crunch. Otherwise, the universe will continue to expand forever — getting colder and darker. Neither option seems very appealing!

MOST MATTER IS INVISIBLE

Only about 5 percent of the energy in the universe today is in ordinary matter like the protons, neutrons, and electrons that make up atoms. The other 95 percent or so consists of *dark matter* and *dark energy.*

About a quarter of the energy in the universe is in dark matter. Dark matter interacts very weakly with ordinary matter like light, so you can't see it directly. Why would you think it exists if you can't see it? Well, dark matter exerts gravitational forces on the ordinary matter around it. Observations of the motion of stars and galaxies show that there's a lot of invisible matter pulling on the matter you can see in your telescope.

That leaves more than 70 percent of the energy of the universe in dark energy. Dark energy is an unknown form of energy that causes the expansion of the universe to accelerate.

The simplest explanation for dark energy is that the *vacuum,* or empty space, has energy. Roughly, you can think of this energy as coming from particle-antiparticle pairs that pop out of nothing for a very brief period of time and then disappear. The only problem with this explanation is that calculations of the vacuum energy give a number that's about 10^{120} times bigger than the observed value. How embarrassing!

MICROWAVE OVENS ARE HOT PHYSICS

You can find plenty of physics going on in microwaves — everyday items you may have taken for granted in your pre-physics life. What really happens in a microwave oven? A device called a *magnetron* generates waves of radiation similar to those involved in the transport of thermal energy (see Chapter 15). These waves are called *electromagnetic waves,* and they have a similar form to sine waves.

Electromagnetic waves with different wavelengths have very different properties. If they have a wavelength in a particular range, then they're visible as light; in another range, at longer wavelengths, they cause water to heat up. The waves exert forces on the molecules as they pass through water, causing the molecules to oscillate in a way similar to simple harmonic motion (Chapter 13).

You may remember from chemistry that water molecules are polar because of the arrangement of the hydrogen and oxygen atoms and the distribution of the electrons. The hydrogen and oxygen atoms share electrons, but the electrons spend more time by the oxygen nucleus, which has a stronger pull. This means that one end of the molecule has a partial positive charge and the other has a partial negative charge.

A microwave is composed of an oscillating electrical field, and the water molecules, with their partial charges, rotate to align with that changing field. The oscillating water molecules bump and jostle the surrounding molecules that constitute the food. This increased oscillating and jostling motion of the molecules is exactly what's meant by an increased temperature — and your dinner is ready! The frequency of the microwave determines the frequency of the oscillating molecules (the frequency of their simple harmonic motion), and this transfers energy to the molecules at a rate that increases with the frequency (and intensity) of the wave. The frequency of microwaves is just right for increasing the temperature at the rate required for cooking food.

Microwave ovens were invented by accident, during the early days of radar. A man named Percy Spencer put his chocolate bar in the wrong place — near a magnetron used to create radar waves — and it melted. "Aha," thought Percy. "This could be useful." And before he knew it, he had invented not only microwave ovens but also microwave popcorn (no kidding).

The universe is full of microwaves, which are a sort of leftover heat glow from the Big Bang. The discovery of this so-called *cosmic background microwave radiation* in the 1960s was a powerful confirmation of the Big Bang theory. You can read more about microwaves and other forms of electromagnetic radiation in *Physics II For Dummies* (Wiley).

IS THE UNIVERSE MADE TO MEASURE?

Fundamental constants are fixed and written into the laws of physics, which describe the whole universe. These constants describe things such as the strength of gravity and the relative masses of fundamental particles. Physicists hope to develop a theory that can explain why the fundamental physical constants have the values that they do. Physicists would like their final theory of everything to be completely self-contained, leaving nothing unexplained — even the values of the fundamental constants.

Physicists have worked out what the world would be like if the constants were slightly different. What would happen if gravity were slightly weaker? What would happen if the forces holding atoms of matter together were slightly stronger? And the answer they find is that if any of the constants were only slightly different from the values that they have, then people wouldn't be able to live in this universe. For example, if gravity were slightly weaker, then stars couldn't form and we'd have no sun. And if gravity were slightly stronger, then stars would burn their fuel so quickly that life wouldn't have time to evolve! How can people explain why the constants seem to be so finely tuned?

The *anthropic principle* says that the constants have to have the values that they do because if they didn't, then we wouldn't be here to measure them. This is a very curious argument that many people are not happy with!

Another puzzle related to the constants is the question of why gravity is so weak. Gravity is exceedingly weak compared to other forces, such as electrical forces (the same kind of force that makes your hair stand on end when you rub a balloon on your shirt and bring the balloon near your head). This question has led some physicists to contemplate extra dimensions to space and time.

INDEX

U

About the Authors

Award-winning author, **Steven Holzner**, PhD, was a graduate of MIT and Cornell. He later served on faculty at both, teaching thousands of students over the years and earning an average student evaluation over 4.9 out of 5.0. Holzner was a contributing editor at *PC Magazine* and also wrote many books, including *Physics For Dummies, Physics II For Dummies,* and *Quantum Physics For Dummies*. His books have sold more than 1.5 million copies.

Daniel Funch Wohns has been teaching physics for nine years. After studying physics and math at Stanford, Dan earned his PhD in physics from Cornell. When he's not teaching, Dan spends his time skiing or canoeing.

Publisher's Acknowledgments

EXECUTIVE EDITOR:
Lindsay Sandman Lefevere

DEVELOPMENT EDITOR:
Victoria M. Adang

COPY EDITOR: Jennette ElNaggar

TECHNICAL EDITOR: Jeremy Smith

ART COORDINATOR: Alicia B. South

PRODUCTION EDITOR:
Suresh Srinivasan

COVER IMAGE: © Beardwood/Wiley

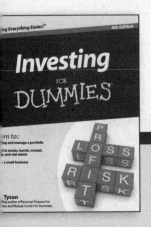

Math & Science

Algebra I For Dummies,
2nd Edition
978-0-470-55964-2

Anatomy and Physiology
For Dummies, 2nd Edition
978-0-470-92326-9

Astronomy For Dummies,
3rd Edition
978-1-118-37697-3

Biology For Dummies,
2nd Edition
978-0-470-59875-7

Chemistry For Dummies,
2nd Edition
978-1-118-00730-3

1001 Algebra II Practice
Problems For Dummies
978-1-118-44662-1

Microsoft Office

Excel 2013 For Dummies
978-1-118-51012-4

Office 2013 All-in-One
For Dummies
978-1-118-51636-2

PowerPoint 2013
For Dummies
978-1-118-50253-2

Word 2013 For Dummies
978-1-118-49123-2

Music

Blues Harmonica
For Dummies
978-1-118-25269-7

Guitar For Dummies,
3rd Edition
978-1-118-11554-1

iPod & iTunes For Dummies,
10th Edition
978-1-118-50864-0

Programming

Beginning Programming
with C For Dummies
978-1-118-73763-7

Excel VBA Programming
For Dummies, 3rd Edition
978-1-118-49037-2

Java For Dummies,
6th Edition
978-1-118-40780-6

Religion & Inspiration

The Bible For Dummies
978-0-7645-5296-0

Buddhism For Dummies,
2nd Edition
978-1-118-02379-2

Catholicism For Dummies,
2nd Edition
978-1-118-07778-8

Self-Help & Relationships

Beating Sugar Addiction
For Dummies
978-1-118-54645-1

Meditation For Dummies,
3rd Edition
978-1-118-29144-3

Seniors

Laptops For Seniors
For Dummies, 3rd Edition
978-1-118-71105-7

Computers For Seniors
For Dummies, 3rd Edition
978-1-118-11553-4

iPad For Seniors
For Dummies, 6th Edition
978-1-118-72826-0

Social Security For Dummies
978-1-118-20573-0

Smartphones & Tablets

Android Phones
For Dummies, 2nd Edition
978-1-118-72030-1

Nexus Tablets For Dummies
978-1-118-77243-0

Samsung Galaxy S 4
For Dummies
978-1-118-64222-1

Samsung Galaxy Tabs
For Dummies
978-1-118-77294-2

Test Prep

ACT For Dummies,
5th Edition
978-1-118-01259-8

ASVAB For Dummies,
3rd Edition
978-0-470-63760-9

GRE For Dummies,
7th Edition
978-0-470-88921-3

Officer Candidate Tests
For Dummies
978-0-470-59876-4

Physician's Assistant Exam
For Dummies
978-1-118-11556-5

Series 7 Exam For Dummies
978-0-470-09932-2

Windows 8

Windows 8.1 All-in-One
For Dummies
978-1-118-82087-2

Windows 8.1 For Dummies
978-1-118-82121-3

Windows 8.1 For Dummies
Book + DVD Bundle
978-1-118-82107-7

Available in print and e-book formats.

Available wherever books are sold. **For more information or to order direct visit www.dummies.com**

Take Dummies with you everywhere you go!

Whether you are excited about e-books, want more from the web, must have your mobile apps, or are swept up in social media, Dummies makes everything easier.

Leverage the Power

For Dummies is the global leader in the reference category and one of the most trusted and highly regarded brands in the world. No longer just focused on books, customers now have access to the For Dummies content they need in the format they want. Let us help you develop a solution that will fit your brand and help you connect with your customers.

Advertising & Sponsorships

Connect with an engaged audience on a powerful multimedia site, and position your message alongside expert how-to content.

Targeted ads • Video • Email marketing • Microsites • Sweepstakes sponsorship

21 Million Monthly Page Views & 13 Million Unique Visitors

gistered trademark of John Wiley & Sons, Inc.